INTRODUCTION TO FOOD AND NUTRITION

Irene Paragas

TAP

Toronto Academic Press

INTRODUCTION TO FOOD AND NUTRITION

Irene Paragas

Toronto Academic Press
4164 Lakeshore Road
Burlington ON L7L 1A4
Canada
www.tap-books.com

Email: orders@arclereducation.com

ISBN: 978-1-77956-321-7

ABOUT THE AUTHOR

Irene Paragas is a dedicated medical doctor and licensed nutritionist-dietitian with a diverse background. Her experience encompasses general practice, nutrition coaching, and medical writing. Presently, Irene provides consultations as a General Physician and offers nutrition and health writing services, highlighting her versatility and proficiency in delivering comprehensive healthcare services. Irene attained her Bachelor of Science in Nutrition from the University of the Philippines and holds licenses in nutrition and medicine. Her wide-ranging work experience and dedication to continuous learning propel her to provide high-quality healthcare and nutrition services.

Table of Contents

List of Figures

List of Tables

PREFACE

Food and nutrition are essential for providing the energy our bodies need to function. We must replenish our nutrients daily to maintain our health, with water being a crucial component. Our diet should include fats, proteins, and carbohydrates, along with key vitamins and minerals. This is especially important for pregnant women and adults over 50, who should pay particular attention to vitamins like vitamin D and minerals such as calcium and iron, both in their food choices and possibly through dietary supplements.

A healthy diet primarily consists of natural foods. A significant portion should include fruits and vegetables, particularly those that are red, orange, or dark green. Whole grains, like whole wheat and brown rice, should also be part of your diet. For adults, dairy products should be non-fat or low-fat. Protein sources can include lean meat and poultry, seafood, eggs, beans, legumes, soy products like tofu, and unsalted seeds and nuts.

Good nutrition also means avoiding certain types of foods. Processed foods often contain high levels of sodium, which can be harmful to individuals with high blood pressure. The USDA recommends that adults consume less than 300 milligrams of cholesterol per day, found in foods like meat and full-fat dairy products. Fried foods, solid fats, and trans fats (present in margarine and many processed foods) are detrimental to heart health. Refined grains (such as white flour and white rice) and refined sugars (like table sugar and high fructose corn syrup) should be limited, especially for those with diabetes. Alcohol should be consumed in moderation—no more than one serving per day for women and two per day for men.

Many high-quality, free guidelines are available that provide detailed information on portion sizes, total calorie intake, and recommendations on what to eat more or less to achieve and maintain good health. Here we address the issues of nutrition, food and diet by analyzing the biological importance of macro- and micro-nutrients including hormetins, discussing the health claims for various types of food, and reviewing the general principles of healthy dietary patterns, including meal timing, caloric restriction, and intermittent fasting.

Contents/Structure of the Book

"Introduction to Food and Nutrition" is a comprehensive guide designed to offer an in-depth understanding of the vital role that food and nutrition play in maintaining health and well-being. The book is structured into 10 chapters, each addressing a critical aspect of food science and nutrition.

Chapter 1 explores the fundamental relationship between foods, nutrition, and overall health. It emphasizes the importance of a balanced diet and how different types of foods contribute to our physical and mental well-being. The chapter introduces key concepts such as macronutrients and micronutrients, and their respective roles in the body.

Chapter 2 is focused on the biological processes involved in digestion, absorption, and nutrient utilization. It provides a detailed overview of the digestive system, explaining how food is broken down, nutrients absorbed, and wastes excreted. The chapter also highlights the roles of various enzymes and digestive organs, offering insights into how nutrients are converted into energy and utilized by the body for various functions.

Chapter 3 covers the six essential nutrients: carbohydrates, proteins, fats, vitamins, minerals, and water. Each section discusses the sources, functions, and recommended daily intake of these nutrients. The chapter also addresses the health consequences of deficiencies and excesses, helping readers understand how to balance their diets to meet nutritional needs effectively.

Chapter 4 focuses on the specialized nutritional requirements of athletes and physically active individuals. It discusses the importance of proper nutrition in enhancing athletic performance, recovery, and overall fitness.

Nutritional disorders such as obesity, malnutrition, and eating disorders are examined in chapter 5. It provides an overview of the causes, symptoms, and treatments of these conditions, emphasizing the role of balanced nutrition in preventing and managing health issues. The chapter also touches on the psychological and social aspects of nutritional disorders.

Effective meal planning is crucial for maintaining a healthy diet. Chapter 6 offers practical advice on creating balanced meal plans that cater to individual dietary needs and preferences. It includes tips on portion control, food group distribution, and how to incorporate a variety of foods to ensure nutritional adequacy.

Food additives are common in many processed foods. Chapter 7 investigates the health implications of various food additives, including preservatives, colorants, and flavor enhancers. It discusses regulatory standards and guides on making informed choices about food consumption to minimize potential health risks.

Diet therapy involves using specific dietary approaches to manage and treat health conditions. Chapter 8 covers therapeutic diets for diseases such as diabetes, hypertension, and cardiovascular disease. It offers practical advice on tailoring diets to meet the needs of individuals with these conditions, aiming to improve health outcomes through targeted nutritional interventions.

Public health initiatives play a vital role in promoting healthy eating habits. Chapter 9 explores various programs and campaigns designed to improve nutritional awareness and behavior. It highlights the importance of education, policy, and community involvement in fostering healthier eating practices at the population level.

Chapter 10 addresses the regulations governing food packaging to ensure safety and quality. It discusses labeling requirements, the role of packaging in protecting food from contamination, and the impact of sustainable packaging practices on the environment. This chapter underscores the importance of regulatory compliance in maintaining public health standards.

This book serves as an essential resource for students, health professionals, and anyone interested in understanding the profound impact of nutrition on health. Through these 10 chapters, readers will gain valuable insights into making informed dietary choices and promoting overall well-being.

Audience

This book is broad and diverse, encompassing students, health professionals, educators, and anyone interested in improving their understanding of nutrition and its impact on health. This book is particularly valuable for students pursuing degrees in nutrition, dietetics, and health sciences, providing them with foundational knowledge essential for their studies and future careers. Health professionals, including dietitians, nutritionists, and fitness trainers, will find this book a practical resource for updating their knowledge and applying nutritional principles in clinical and community settings. Additionally, educators can use this comprehensive guide to enhance their teaching materials and promote nutritional literacy among their students. Finally, individuals passionate about personal health and wellness will benefit from the clear, evidence-based information presented, empowering them to make informed dietary choices and lead healthier lives

Learning Features

- Start-of-chapter learning objectives
- Key terms in each chapter
- Summary Points at the end of each chapter
- Review Questions at the end of each chapter
- Case Studies and Activities in each chapter for solving real problems
- References at the end of each chapter

CHAPTER 1

Foods, Nutrition, and Health

LEARNING OBJECTIVES

After studying this chapter, you will be able to:

- Understand the functions of food
- Know the functions of nutrients
- Identify food sources

KEY TERMS FROM THIS CHAPTER

Antioxidants	Calorie	Carbohydrates
Cells	Energy	Fats
Glycogen	Growth	Illness
Lactation	Lipids	Macronutrients
Micronutrients	Minerals	Natural toxins
Nutrients	Nutrition	Phospholipids
Protection	Proteins	Rate of reaction
Scarcity	Sterols	Temperature Triglyceride

INTRODUCTION

The need for food to survive has always been a fundamental aspect of human existence. Throughout history, a vast amount of knowledge has been amassed on how food can support healthy growth in children and youth, maintain overall health throughout life, meet the unique requirements of pregnancy and lactation, and aid in the healing process after illness.

Understanding the composition of food is crucial in comprehending its nutritional value. It is common knowledge that certain foods can be detrimental to health, while others are essential for maintaining good health. Therefore, individuals must carefully evaluate their beliefs about food and decide whether to embrace or discard them based on the new information they acquire through the study of food science and nutrition.

The insights gained in this field should be directly applied to one's own life. While a significant portion of the dietary history holds scientific advantages and should be preserved, it is also necessary to adapt and modify certain aspects due to the changes in lifestyle.

1.1. FUNCTIONS OF FOOD

Since food is a necessary condition for life, the main objective of eating and drinking is to allow the body to operate normally. Food is made up of a variety of nutrients. The food we eat provides us with over 40 essential nutrients. Based on their shared characteristics, these nutrients can be divided into five main categories: proteins, carbs, fats, vitamins, and minerals. Both as a food and a nutrient, water is vital. Every nutrient category has a distinct physiological role, which describes how food contributes to the maintenance of particular bodily functions. Food serves social, psychological, and physiological purposes in addition to being a source of nutrients. An old proverb states that we are what we eat. Individual's physical, mental, and nutritional well-being. The food we eat and how we eat affects our mental health.

1.1.1. Physiological Functions

Food fulfills several physiological roles, including providing energy, promoting physical growth, offering protection, and controlling bodily functions. Energy is needed to carry out a variety of tasks, including walking, running, sitting, standing, and doing all of household chores. It is also necessary for the body to perform several functions, such as heartbeat, intestinal contraction, and lung expansion and contraction. Energy must be used for bodily functions to take place. Two essential nutrients, known as body fuel—fats and carbohydrates—perform the role of providing energy in food. When these burn, the energy is released as heat that is held in high-energy biological system compounds like ATP. The body uses these substances to release energy when it needs it the method of producing energy that is comparable to burning coal in the atmosphere. For growth and repair, food is also necessary. Millions of structural and functional units known as cells make up the human body. Growth occurs when new cells are added to the existing ones, which enlarge, grow old, and eventually die, to be replaced by new cells. We refer to this procedure as repair. Proteins are required for both growth and repair. From infancy to adulthood, weight and height gain are only achievable through the process of growth.

Food also serves two other important physiological purposes: it protects the body and regulates bodily functions. Protection is the ability of food to prevent infection by maintaining the healthy operation of the body systems that fight infection; in other words, if a person becomes ill, whether from an infection or something else entirely, food and the nutrients it contains help them recover quickly. An individual with a poor diet requires more time to heal. The role that food plays in regulating bodily processes is referred to as the regulatory function previously discussed. The body goes through several processes, including heartbeat, temperature regulation, and muscle contraction. The temperature is kept at 37°C, or 98.4°F. Certain nutrients are responsible for this. In a similar vein, water plays a significant regulatory role. In the body, several chemical reactions occur. These chemical reactions enable the construction of more complex substances from simpler ones. In a similar way, substances that are complex are reduced to their simpler component. Reaction rates are adjusted based on the body's needs.

Proteins, minerals, and vitamins all function as regulators and have a significant role in regulating these reactions.

All the ingredients of food are given to a person in a manner akin to administering medication via injection or intravenous drip, etc. They still require food to satisfy themselves even then. It indicates that the bodies have stretch receptors, and that only food items can fill the satiety center. This is an additional physiological benefit of eating. The raw material used in production is food. A starving mother is unable to give birth to a child. In the same way, a sheep cannot make wool if it is not fed properly.

1.1.2. Social Functions

The community's social, cultural, religious, and culinary lives have always revolved around food. It has been used as a way to show love, friendship, and joy at social, religious, and family gatherings.

Food brings people from different classes, communities, and religions closer together and fosters the development of social relationships. Notice that food of some kind is served at every social event and function that a person has. Eating together at a birthday, engagement, or marriage celebration facilitates the sharing of joy and happiness. They have probably invited friends and family over for supper, tea, etc., quite a bit. Here, feeding them is a way to show how much they care for them and to show how hospitable they are. Christmas is one of the occasions to share happiness with others, much like when sweets are exchanged during holidays like Easter. These are a few instances that illustrate the social dimensions of these gatherings (Darby, 2000; Whitney, 2011).

1.1.3. Psychological Functions

Food offers a sense of safety. A baby picks up security skills through the feeding habits of his mother. The degree to which he feels secure about his food supply will determine his behavioral patterns. Similar to this, when a young child knows there is food in the house and he will be fed, he feels confident and like he belongs. When people have enough food on hand to sustain them during times of scarcity, they feel fairly secure. When one has to eat outside of the home, familiar foods provide a sense of comfort. Food has social status. Using polished white rice as opposed to unpolished rice is considered prestigious. True white flour and white bread, which have supplanted black bread, work similarly. Because some foods are associated with the so-called upper class, others prefer them because they are more readily available, less expensive, and provide greater nutrition. When we have guests, we serve different special foods and reserve specific foods for the family.

Food serves as a way to express emotions, whether it be through eating to relieve tension or avoiding food altogether. Some individuals find comfort in constant snacking when feeling lonely or bored, while others may lose their appetite when feeling angry or frustrated. Certain foods may be associated with negative memories, and what one person finds delicious, another may find repulsive. These emotional connections to food go beyond simply providing nutrients for the body; they are intertwined with cultural traditions, serving as a source of pleasure and a way to alleviate stress. The emotional responses to food are complex and not easily explained by logic or nutritional science.

1.2. FUNCTIONS OF NUTRIENTS

There are nutrients in the food we eat. The body needs nutrients to carry out its essential processes. Since the human body cannot produce or synthesize nutrients, we must get them from diet. One or more of the three primary roles of nutrients are to supply energy, support bodily structure, and/or control bodily chemical reactions. These fundamental bodily processes enable us to move, expel waste, breathe, grow, and procreate. They also enable us to detect and react to our surroundings. The body needs six different classes of nutrients to function and stay healthy overall. These include water, proteins, fats, carbohydrates, vitamins, and minerals. Foods also contain non-nutrients that can either be beneficial (like antioxidants) or harmful (like natural toxins found in plant foods and additives like some dyes and preservatives).

1.2.1. Macronutrients

Macronutrients are those that are required in high concentrations. Macronutrients fall into three categories: proteins, lipids, and carbohydrates. These can be converted into cellular energy by metabolic processing. The chemical bonds in macronutrients provide them with energy. a person's body uses this chemical energy to transform into cellular energy, which is then used for work and basic bodily functions. Although a person needs a lot of water, it is not a macronutrient in the same way as other nutrients because it does not produce calories.

The calorie is a unit of measurement for food energy. The amount of calories listed on nutrition food labels is actually equal to 1000 times each calorie (spelled with a lowercase "c"). In Canada, the term "Calorie" (spelled with a capital "C") on nutrition food labels is equivalent to one kilocalorie (kcal) (1000 c = 1 kcal = 1 C). A calorie is the quantity of heat energy needed to raise one kilogram of water one degree Celsius. The majority of other nations employ the kilojoule (kJ) as their standard energy unit. The energy required to generate a given amount of force is measured in Joules, which are based on work completed. Joules and calories are equivalent units of measurement for energy; 1 kcal equals 4.18 kJ. Although a person needs a lot of water, it is not a macronutrient in the same way as other nutrients because it does not produce calories (Darby, 2000; Whitney, 2011).

1.2.1.1. Carbohydrates

The molecules that makeup carbohydrates are carbon, hydrogen, and oxygen. Grain products, dairy products, fruits, and starchy vegetables like potatoes are the main food sources of carbohydrates. Carbohydrates are also present in non-starchy vegetables, albeit in smaller amounts. Based on their chemical makeup, carbohydrates are generally divided into two categories: simple carbohydrates, also known as simple sugars, and complex carbohydrates. Simple carbs are made up of one or two building blocks. Simple sugars include glucose, which is found in the blood, and sucrose, which is the kind of sugar in a bowl at breakfast.

Long chains of simple sugars, which can be branched or unbranched, make up complex carbohydrates. Digestible complex carbohydrates are broken down by the body into simple sugars, primarily glucose, during digestion. After that, glucose is carried to every cell in the body, where it is stored, converted to energy, or utilized to assemble macromolecules. Although fiber is a complex carbohydrate, the human intestine's digestive enzymes are unable to break it down. Therefore, unless the bacteria in the colon or large intestine break it down, it passes through the digestive tract undigested.

Four kilocalories of digestible carbs provide the body's cells with the energy they need to function. Not only do carbohydrates supply energy and act as building blocks for larger macromolecules, but they are also necessary for the nervous system, heart, and kidneys to function properly. As previously noted, the body has the ability to store glucose for later use. Plants refer to the molecule that stores carbohydrates as starch, while humans refer to it as glycogen. Starch and glycogen are examples of complex carbs.

1.2.1.2. Lipids

Similar to carbohydrates, lipids are a family of molecules made up of carbon, hydrogen, and oxygen; however, they are not soluble in water. The main sources of lipids are nuts, seeds, dairy products, butter, oils, meats, and many processed foods. Triglycerides (also known as triacylglycerols), phospholipids, and sterols are the three primary categories of lipids. Lipids' primary function is to either produce or store energy. Nine kilocalories per gram of lipids compared to four kilocalories per gram of carbohydrates means that lipids have more energy per gram than carbohydrates. Lipids store energy, but they also play a significant role in cell membranes, envelop and shield organs (in tissues that store fat), act as insulation to help regulate body temperature and control a host of other bodily processes.

1.2.1.3. Proteins

Proteins are macromolecules made up of chains of amino acid subunits. The basic building blocks of amino acids are nitrogen, hydrogen, oxygen, and carbon. Meats, dairy products, seafood, and a wide range of plant-based foods—soy, in particular—are good sources of protein. The Greek word for "of primary importance" is "protein," which aptly describes these macronutrients, also referred to as the "workhorses" of life in common parlance. Although proteins contain four kilocalories of energy per gram, this is not their primary purpose. In addition to giving skin, muscles, and bones structure, proteins are involved in the majority of chemical reactions that occur within the body. The human body is thought to contain more than 100,000 distinct proteins, according to scientists. The sequence in which 20 distinct amino acids are bound together to form thousands of distinct proteins is determined by the genetic codes found in DNA, which are essentially protein recipes (Figure 1.1).

Figure 1.1. *The macronutrients: carbohydrates, lipids, protein, and water.*

Source: By U.S. Department of Agriculture – 20111110-OC-AMW-0012\, Public Domain, https://commons.wikimedia.org/w/index.php?curid=17794846.t

1.2.1.4. Water

One essential nutrient that is required in large quantities is water. Despite not containing carbon, water is made up of two hydrogen atoms and one oxygen atom in each molecule. More than 60% of the body weight consists of water. Water is vital for transporting substances in and out of the body, facilitating chemical reactions, cushioning organs, and regulating body temperature. On average, an adult consumes slightly more than two liters of water daily from both food and beverages.

1.2.1.4.1. Estimating Caloric Content

Because the nutrients that produce energy are made of various kinds of chemical bonds, the energy they contain varies. Foods high in protein or carbohydrates provide about 4 calories per gram, while foods high in fat, or triglycerides, provide 9 calories per gram. 1000 times more work can be done with one kilocalorie of energy than with one calorie. The calories listed for a specific food are actually kilocalories on the Nutrition Facts table that can be found on packaged foods.

Since the Nutrition Facts panel lists the total number of calories in a serving of a specific food, estimating the number of calories in commercially prepared food is fairly simple. Simply total the calories in each food item to find out how many calories a person had for breakfast this morning. For instance, a person's breakfast would have consisted of 150 calories from one serving of yogurt, 209 calories from half a cup of low-fat granola cereal, and 100 calories from a glass of orange juice. Add all of these together, and their total calorie intake would be 459.

1.2.1.4.2. Food Quality: Nutrient Density

The ratio of a food's nutritional content to its energy content is one way to assess its quality. Nutrient-dense foods are ones that have a high nutritional content concerning their calorie content. Foods high in nutrients are the antithesis of "empty-calorie" foods, like sugary, carbonated soft drinks, which are high in calories but low in other nutrients. In addition, the microbial content, taste, texture, appearance, and consumer preference are all linked to food quality.

1.2.2. Micronutrients

Micronutrients are substances that the body needs in smaller quantities to perform biological processes. All of the necessary vitamins and minerals are considered micronutrients. 13 vitamins and 16 necessary minerals are present. Micronutrients work as cofactors or parts of enzymes, which help in the process, but they are not sources of energy (calories), unlike proteins, lipids, and carbs (i.e., coenzymes). Proteins called enzymes catalyze chemical reactions within the body and are involved in every bodily function, including energy production, nutrition digestion, and the synthesis of macromolecules. The body uses micronutrients for a variety of vital functions (Table 1.1).

Table 1.1. Minerals and Their Major Functions

Minerals	**Major Functions**
Macro	
Sodium	Fluid balance, nerve transmission, muscle contraction
Chloride	Fluid balance, stomach acid production
Potassium	Fluid balance, nerve transmission, muscle contraction
Calcium	Bone and teeth health maintenance, nerve transmission, muscle contraction, blood clotting
Phosphorus	Bone and teeth health maintenance, acid-base balance
Magnesium	Protein production, nerve transmission, muscle contraction
Sulfur	Protein production
Trace	
Iron	Carries oxygen, assists in energy production
Zinc	Protein and DNA production, wound healing, growth, immune system function
Iodine	Thyroid hormone production, growth, metabolism
Selenium	Antioxidant
Copper	Coenzyme, iron metabolism
Manganese	Coenzyme
Fluoride	Bone and teeth health maintenance, tooth decay prevention
Chromium	Assists insulin in glucose metabolism
Molybdenum	Coenzyme

Source: https://pressbooks.bccampus.ca/nutr1100/chapter/introduction-2/.

1.2.2.1. Minerals

Minerals are inorganic solid materials that crystallize and are categorized based on relative needs. Only a few milligrams or less of trace minerals, including zinc, iron, iodine, molybdenum, and selenium, are needed. Macrominerals in the hundreds of milligram range are needed, including calcium, magnesium, potassium, sodium, and phosphorus. Numerous minerals are necessary for the proper operation of enzymes; others are needed to keep the body's fluid balance, form bone, produce hormones, transmit nerve impulses, contract and relax muscles, and shield the body from dangerous free radicals that can lead to illnesses like cancer (Darby, 2000; Whitney, 2011)

1.2.2.2. Vitamins

There are two categories for the 13 vitamins: fat-soluble and water-soluble. The B vitamins—thiamine, riboflavin, niacin, pantothenic acid, pyridoxine, biotin, folate, and cobalamin—as well as vitamin C are water-soluble vitamins. The vitamins A, D, E, and K are fat-soluble. The body needs vitamins for a variety of processes, including the synthesis of bone tissue, the production of red blood cells, and the maintenance of healthy nerve, immune, and visual systems (Table 1.2).

Table 1.2. Vitamins and their Major Functions

Vitamins	**Major Functions**
Water-Soluble	
Thiamin (B1)	Coenzyme, energy metabolism assistance
Riboflavin (B2)	Coenzyme, energy metabolism assistance
Niacin (B3)	Coenzyme, energy metabolism assistance
Pantothenic acid (B5)	Coenzyme, energy metabolism assistance
Pyridoxine (B6)	Coenzyme, amino acid synthesis assistance
Biotin (B7)	Coenzyme, amino acid and fatty acid metabolism
Folate (B9)	Coenzyme, essential for growth
Cobalamin (B12)	Coenzyme, red blood cell synthesis
C (ascorbic acid)	Collagen synthesis, antioxidant
Fat-Soluble	
A	Vision, reproduction, immune system function
D	Bone and teeth health maintenance, immune system function
E	Antioxidant, cell membrane protection
K	Bone and teeth health maintenance, blood clotting

Source: https://pressbooks.bccampus.ca/nutr1100/chapter/introduction-2/.

Deficits in certain vitamins can result in serious health issues and even death. For instance, pellagra, a disease more prevalent in the early 20th century, is caused by a niacin deficiency. The "4D's" (diarrhea, dermatitis, dementia, and death) are the typical pellagra signs and symptoms. Prior to researchers discovering that improved diets could alleviate pellagra's symptoms, a large number of sufferers of the illness were admitted to mental institutions where they were kept until their deaths. It has also been discovered that some vitamins, like vitamin D, can prevent rickets, vitamin A can prevent night blindness, and vitamin C can prevent scurvy (Table 1.3).

Table 1.3. Functions of Nutrients

Protein	Necessary for tissue formation, cell reparation, and hormone and enzyme production. It is essential for building strong muscles and a healthy immune system.
Carbohydrates	Provide a ready source of energy for the body and provide structural constituents for the formation of cells.

Fat	Provides stored energy for the body, functions as structural components of cells and also as signaling molecules for proper cellular communication. It provides insulation to vital organs and works to maintain body temperature.
Vitamins	Regulate body processes and promote normal body-system functions.
Minerals	Regulate body processes, are necessary for proper cellular function, and comprise body tissue.
Water	Transports essential nutrients to all body parts, transports waste products for disposal, and aids with body temperature maintenance.

Source: https://pressbooks.bccampus.ca/nutr1100/chapter/introduction-2/.

1.3. FOOD SOURCES

Food creates a web of interlocking chains in a particular ecosystem, with apex predators at the top and primary producers at the bottom. Other aspects of the web include detrovores (that eat detritis) and decomposers (that break down dead organisms). Primary producers are light-dependent organisms such as bacteria, protists, plants, and algae (Allan et al., 2007). The herbivores that eat the plants are the primary consumers; the carnivores that eat those herbivores are the secondary consumers. Some organisms are omnivores because they consume both plants and animals, such as the majority of mammals and birds. The apex predators at the end of the chain are those animals whose ecosystem has no known predators (Wallach et al., 2015). People are regarded as the top predators.

Being omnivores, humans can survive on a diet of fruits, vegetables, cooked meat, dairy products, eggs, seaweed, and mushrooms. More food energy is produced globally from cereal grains than from any other kind of crop, making them a staple diet. Around the world, 87% of all grain production is made up of corn (maize), wheat, and rice. 55% of the world's crops are grown for human consumption, with 36% being grown for animal feed and 9% being grown for biofuels (Plumer, 2014). Bread, wine, cheese, and yogurt are examples of foods that are fermented using the help of fungi and bacteria.

1.3.1. Photosynthesis

The process of photosynthesis involves the absorption of solar energy to convert carbon dioxide and water from the air or soil into glucose and oxygen. After that, the glucose is stored as a source of energy and the oxygen is released. Since photosynthetic plants, algae, and some bacteria are frequently found at the base of food chains, photosynthesis serves as the main energy and food source for almost all life on Earth.

Important minerals and nutrients are also taken up by plants from the soil, natural waters, and the air. The three main nutrients required for plant survival are carbon, oxygen, and hydrogen, which are taken up from the air or water (Morgan and Connolly, 2013). Nitrogen, phosphorus, and potassium are the three main nutrients that plants need to grow. Other essential nutrients that plants also need to grow include calcium, sulfur, magnesium, iron, boron, chlorine, manganese, zinc, copper, molybdenum, and nickel.

1.3.2. Microorganisms

The lower rugs of the food chain are also made up of bacteria and other microorganisms. They get their energy from photosynthesis as well as from decomposing trash, dead organisms, and chemical substances. To get their nutrients, some organisms develop symbiotic relationships with other organisms. Protozoa, which in turn supply food for other organisms like tiny invertebrates, rely on bacteria as a food source. Other creatures that consume bacteria are shellfish, nematodes, fan worms, and certain types of snails.

Many small and large aquatic organisms depend on plankton, which is made up of bacteria, archaea, algae, protozoa, and microscopic fungi, as a vital source of food in the marine environment. Because bacteria transform atmospheric nitrogen into nutrient ammonia, life hardly could exist without them. Ammonia serves as the building block for most vitamins, proteins, and nucleic acids. The majority of ammonia in the world is produced by humans ever since the Haber-Bosch Process, an industrial method for nitrogen fixation, was developed.

1.3.3. Plants

Plants can be used as food in the form of seeds, fruits, vegetables, grains, legumes, and nuts. Plants can belong to different categories; seeds like peas are commonly considered vegetables, while botanically described fruits like tomato, squash, pepper, and eggplant fall under the vegetable category. Foods like seeds, nuts, and grains are considered fruits if the portion consumed is derived from the reproductive tissue. When it comes to food preparation, fruits are typically thought of as the leftovers of botanically defined fruits after grains, nuts, seeds, and fruits that are used as vegetables have been eliminated. Grains are seeds that humans eat or harvest. Pulses are a member of the Fabaceae (legume) family, while cereal grains (oats, wheat, rice, corn, barley, rye, sorghum, and millet) belong to the Poaceae (grass) family. Whole grains are meals that have the bran, germ, and endosperm of the original seed. Nuts are a type of dry fruit that are identified by their woody shell.

Stone fruits (cherries and peaches), pome fruits (apples, pears), berries (blackberries, strawberries), citrus fruits (oranges, lemons), melons (watermelon, cantaloupe), Mediterranean fruits (grapes, fig), and tropical fruits (banana, pineapple) are further categories for fleshy fruits, which can be distinguished from dry fruits like grain, seeds, and nuts. Any other portion of the plant that can be consumed, such as the roots, stems, leaves, flowers, bark, or the entire plant, is referred to as a vegetable. These consist of the following: flowers (broccoli and cauliflower), leafy vegetables (spinach and lettuce), stem vegetables (celery and asparagus), roots (potatoes and carrots), and bulbs (onion family).

Plants vary greatly in terms of their protein, fat, and carbohydrate content. The primary forms of carbohydrates are glucose, fructose, starch, and other sugars. Except for vitamin D and vitamin B12, most vitamins are found in plants. Mineral abundance can also vary. Fruit has a high vitamin C content, high levels of simple sugars that give it a sweet taste, and a water content of up to 90%. Vegetables are low in fat and calories and high in starch, potassium, dietary fiber, folate, and vitamins when compared to fleshy fruit (apart from bananas). Nuts are high in protein, fiber, vitamin B, and vitamin E, while grains are primarily composed of carbohydrates. Because they are plentiful and contain fiber and good fats like omega-3 fats, seeds are a good food source for animals. Complex chemical reactions can increase or decrease a nutrient's bioavailability. Certain sugars and vitamins may not be released as a result of phytotates.

Herbivores are animals that exclusively consume plants; frugivores are those that primarily consume fruits and leaves; folivores, such as pandas, are those that eat shoots; and xylophages, or termites, are those that eat wood. A wide variety of animals, including annelids, elephants, chimpanzees, and numerous birds, are considered frugivores. Approximately 182 fish eat fruit or seeds. The main source of nutrients for animals, both domestic and wild, is the wide variety of grasses that have adapted to different environments. Thousands of plant species are consumed by humans; angiosperms may have up to 75,000 edible species, of which only 7,000 are commonly consumed. Plants can be used to make raw ingredients like sugar, herbs, spices, and oils, or they can be processed into breads, pasta, cereals, juices, and jams. Rich oils, such as sunflower, flaxseed, rapeseed (which includes canola oil), and sesame, are obtained by pressing oilseeds.

Numerous flora and fauna have converged to such an extent that fruit serves as an excellent food source for the animal, which subsequently excretes the seeds at a considerable distance to facilitate maximum dispersal. Given that some seeds can withstand digestion, even seed predation can be advantageous to both parties. The only true seed dispersers are ants, but other insects also consume a lot of seeds. Despite being important seed dispersers, birds eat seeds very infrequently. They are distinguished by the thick beak they use to break through the seed coat. Because their teeth can crush harder and larger seeds, mammals consume a wider variety of seeds.

1.3.4. Animals

Animals are either directly or indirectly used as food. Meat, eggs, seafood, and dairy products like milk and cheese fall under this category. Since they include every essential amino acid required by the human body, they are regarded as complete proteins for human consumption and are a significant source of protein.

The protein content of a 4-ounce (110 g) steak, chicken breast, or pork chop is around 30 grams. There are 7 grams of protein in one large egg. About 15 grams of protein are included in a 4-ounce (110 g) serving of cheese.

And there are roughly 8 grams of protein in a cup of milk. Animal products also contain calories, fat, minerals (such as zinc, iron, calcium, and magnesium), and essential vitamins (such as B12).

Mammary gland milk, which is consumed or processed into dairy products (cheese, butter, etc.) in many cultures, is one of the food products made by animals).

Bees produce honey, a reduced nectar from flowers that is used as a common sweetener in many cultures, and birds and other animals lay eggs that are consumed. In times of food scarcity, some cultures use blood in stews like jugged hare, while others use it in blood sausage, thickening sauces, or in cured, salted form (Darby, 2000; Whitney, 2011)

CASE STUDY

NUTRITION-PHYSICAL ACTIVITY

James has been gaining weight over the last five years in his group home and his doctor has said that he needs to exercise more or he will become obese. James is 62, a large man weighing around 200 pounds, who really doesn't want to be told that he needs to exercise for his health.When James first came to the group home he was much more active than currently. He would go for walks in Sunrise Park by himself when the weather was good and he seemed to enjoy that. He also bowled every week with the other men in the group home and was very good at it. A year later James climbed Mt. Clemons with one of the staff. Unfortunately, the staff person left his job at the group home about three months later and no one else was interested in taking him hiking. He worked out at the YMCA for a period of time but he didn't find other people to workout with him and he quit. We think that James started eating more food within the next year as he started to gain weight.

- What are the behaviors that need to be changed?
- What health promotion actions would help?

CLASS ACTIVITY

Dancing Food Groups

Project Skills:	
	Acquiring knowledge. Communicating. Working with others.

Introduction

- Explain to your group members that in this activity they will remember what they have learned in previous activities and use this knowledge to create songs and dances about food groups and their functions.

Experience

- Ask your group members to form groups of 3 – 6 people.

- Explain that each group should create a song and dance that would help others to learn about the food groups and their functions.
- Let the groups find a quiet place, where the other groups can't hear them, to create a song and practice it with their dance.

Give the groups about 15 minutes to practice.

- Ask the groups to come back and let each group have a turn to perform their song and dance for the others. How can you use what you have learned from this activity?

Process

- Let everyone come together to form one large group.
- Lead a discussion about the performances by asking the following questions:
  - Was it difficult or easy for you to make a song and dance about food groups and their functions?
  - Did you remember the information you learned about these topics?
  - What information did you choose to use in your song and which information did you leave out? Why?
  - What did you learn from that activity?
  - How did creating a song and dance help you learn about nutrition?

Apply

- End the activity with a discussion about using these songs and dances at a community or public meeting to help others learn about nutrition and the 4 Food Groups. Ask questions to guide this discussion.

What examples do you know of ways in which songs and dances tell people messages? (Radio and TV advertising, community theater for HIV/AIDS, etc.)

- How could our songs and dances help our community?
- Where does our community have public meetings so that we could ask to show what we have learned during the meeting?
- Who would be good to ask to learn about such meetings?
- Would it be a good idea to sing our songs after church on Sunday?
- Should we talk to the pastor or church elders?

SUMMARY

- Food is an essential prerequisite for life and as such the primary goal of eating and drinking is to enable the body to function normally. Food consists in part of various nutrients. There are over 40 essential nutrients, which are supplied by food we eat. These nutrients can be classified into five major categories based on their similar features, proteins, carbohydrates, fats, vitamins and minerals.
- The physiological functions performed by food are energy giving, body building, protection and regulation of body processes. Energy is required for performing various types of activities such as sitting, standing, walking, running and performing all our house hold and other functioning. It is also required for carrying out various activities within the body e.g., beating of the heart, contraction of the muscles, etc.
- Food has always been the central part of our community, social, cultural and religious life. It has been an expression of love, friendship and happiness at religious, social and family get-togethers. Food creates an atmosphere where the social relations can be developed and it helps in bringing people from different classes, communities and religions closer.
- Food is a source of security. An infant learns security from the way his mother feeds him. Its behavioral patterns will be influenced by the extent to which he feels secure as regards its food supply. Similarly, a growing child gains confidence and a feeling of belonging when he knows there is food in the house and he will be fed.
- The foods we eat contain nutrients. Nutrients are substances required by the body to perform its basic functions. Nutrients must be obtained from our diet since the human body does not synthesize or produce them. Nutrients have one or more of three basic functions: they provide energy, contribute to body structure, and/or regulate chemical processes in the body.
- Nutrients that are needed in large amounts are called macronutrients. There are three classes of macronutrients: carbohydrates, lipids, and proteins. These can be metabolically processed into cellular energy.
- Micronutrients are nutrients required by the body in lesser amounts but are still essential for carrying out bodily functions. Micronutrients include all the essential minerals and vitamins.
- In a given ecosystem, food forms a web of interlocking chains with primary producers at the bottom and apex predators at the top. Other aspects of the web include detrivores (that eat detritis) and decomposers (that break down dead organisms).
- During photosynthesis energy from the sun is absorbed and used to transform water and carbon dioxide in the air or soil into oxygen and glucose. The oxygen is then released, and the glucose is stored as an energy reserve. Photosynthetic plants, algae and certain bacteria often represent the lowest point the food chains, making photosynthesis the primary source of energy and food for nearly all life on earth.

- Bacteria and other microorganisms also form the lower rugs of the food chain. They obtain their energy from photosynthesis or by breaking down dead organisms, waste or chemical compounds. Some form symbiotic relationships with other organisms to obtain their nutrients.
- Plants as a food source are divided into seeds, fruits, vegetables, legumes, grains and nuts. Where plants fall within these categories can vary, with botanically described fruits such as the tomato, squash, pepper and eggplant or seeds like peas commonly considered vegetables.
- Animals are used as food either directly or indirectly. This includes meat, eggs, shellfish and dairy products like milk and cheese. They are an important source of protein and are considered complete proteins for human consumption as they contain all the essential amino acids that the human body needs.

REVIEW QUESTIONS

1. Describe briefly the energy giving function of food.
2. Explain briefly how food helps in regulating our body processes.
3. What role does water play in the human body?
4. Which vitamin is necessary to keep one's eyesight healthy?
5. What components make up food?
6. What use do carbs serve in our bodies?
7. Why is iron necessary for a balanced diet?
8. Discuss the three categories of physiological functions of food.
9. What is the role of proteins?
10. What is a balanced diet?
11. Based on your gender and age, how many grams of protein should you eat daily?
12. Compare and contrast macronutrients and micronutrients. Give examples of each.
13. What are minerals? Give two examples.
14. What is a good source of vitamin A?
15. Why does a person need vitamin D?

REFERENCES

1. Allan, J. D., & Castillo, M. M. (2007). Primary producers. In *Stream Ecology* (pp. 105–134). Springer Netherlands. https://doi.org/10.1007/978-1-4020-5583-6_6.
2. Barth, M. M., Bell, R. A., & Grimmer, K. (2020). *Public Health Nutrition*. Springer Publishing Company. https://doi.org/10.1891/9780826146854.

3. Chen, Y., Lin, B. H., Mancino, L., Ver Ploeg, M., & Zhen, C. (2020). Nutritional quality of retail food purchases is not associated with participation in the Supplemental Nutrition Assistance Program for nutrition-oriented households. *PLOS ONE, 15*(12), e0240263. https://doi.org/10.1371/journal.pone.0240263.
4. Craig, D. (2005). *Food and Nutrition*. Independence Educational Publishers.
5. Darby, C. (2000). *Nutrition & Health*. Milliken Publishing Co.
6. Han, Y., Cheng, Q., Wu, W., & Huang, Z. (2023). DPF-Nutrition: Food nutrition estimation via depth prediction and fusion. *Foods, 12*(23), 4293. https://doi.org/10.3390/foods12234293.
7. Mossenson, S., Giglia, R., Pulker, C. E., Dhaliwal, S. S., Chester, M., Bigwood, R., & Pollard, C. M. (2024). The nutritional quality of food donated to a Western Australian food bank. *Nutrients, 16*(4), 509. https://doi.org/10.3390/nu16040509.
8. Nordin, S. M., & Nordin, K. J. (2017). Food, the source of nutrition. *World Nutrition, 8*(1), 87–94. https://doi.org/10.26596/wn.20178187-94.
9. Nurbaya, N., Najdah, N., & Irwan, Z. (2023). Education on balanced nutrition and healthy snacks in the school environment to prevent nutritional problems among adolescents. *Poltekita: Jurnal Pengabdian Masyarakat, 4*(3), 821–827. https://doi.org/10.33860/pjpm.v4i3.2618.
10. Perez-Cueto, F. J. A., & Olsen, A. (2020). The multifaceted dimensions of food choice and nutrition. *Nutrients, 12*(2), 502. https://doi.org/10.3390/nu12020502.
11. Plumer, B. (2014). How much of the world's cropland is actually used to grow food? *Vox*. Retrieved from: https://www.vox.com (accessed on 11 September 2024).
12. Anak Agung Ngurah Dwi Ariesta Wijaya Putra, Ida Ayu Putu Ary Widnyani, & Pande P. Elza Fitriani. (2022). Food counseling with balanced nutrition for teenagers. *Jurnal Abdimas ITEKES Bali, 1*(1), 46–50. https://doi.org/10.37294/jai.v1i1.404.
13. Wallach, A. D., Izhaki, I., Toms, J. D., Ripple, W. J., & Shanas, U. (2015). What is an apex predator? *Oikos, 124*(11), 1453–1461. https://doi.org/10.1111/oik.01977.
14. Whitney, E. N. (2011). *Nutrition for Health and Health Care* (4th ed.). Wadsworth.
15. Wiseman, G. (2004). *Nutrition and Health*. Taylor & Francis Group Plc.

CHAPTER 2

Digestion, Absorption, and Utilization of Nutrients

LEARNING OBJECTIVES

After studying this chapter, you will be able to:

- Understand the cell as the basic structural and functional unit of life
- Discuss digestion and absorption of nutrients
- Know how nutrients are essential for organ function
- Focus on the utilization of nutrients in the body.

KEY TERMS FROM THIS CHAPTER

Bacteria	Calories	Carbon dioxide
Cellular respiration	Chyme	Circulatory system
Digestion	Digestive secretions	Duodenum
Endoplasmic reticulum	Esophagus	Gastrointestinal tract
Golgi apparatus	Immune system	Ingestion
Metabolism	Mouth	Muscles flex
Muscular system	Nucleus	Organ systems
Organism	Organs	Pancreas

INTRODUCTION

Your body needs a mechanism to convert food and liquids into nutrients it can absorb and utilize in order to survive. When you see, smell, feel, or taste food, digestion starts. The nervous and hormonal systems alert the gastrointestinal tract to the impending arrival of food. Digestive secretions flow, and muscles flex. Digestion is coordinated by cooperating organs such as the pancreas, liver, gall bladder, stomach, esophagus, mouth, and small and large intestines.

Nutrients must pass through the gastrointestinal tract (GIT) in order for you to receive the nutrition you require. The GIT is a muscular, lengthy tube that connects the mouth to the anus. Foods contain macronutrients, which are absorbed by the cells lining the small intestine after being broken down into smaller units during digestion. In the end, nutrients pass through cells that can absorb them, are released into the bloodstream or lymphatic system, and are then distributed throughout the body. Issues like stomach ulcers, a blocked bile duct, inadequate enzymes, or regurgitation of stomach contents into the esophagus can occasionally occur. Gaining more knowledge about the digestive system can help you stay healthy and steer clear of these issues.

2.1. THE BASIC STRUCTURAL AND FUNCTIONAL UNIT OF LIFE: THE CELL

A living thing carries out biological processes that are self-sustaining. The smallest and most fundamental unit of life is a cell. The cell was discovered in 1665 by Robert Hooke, one of the first scientists to use a light microscope. All living forms, including humans, animals, plants, and bacteria, are classified as having cells as their most fundamental structural and functional unit. The cell theory was developed by scientists based on observations made over the next 150 years and applies to all living things, regardless of how simple or complex. Three concepts are included in the cell theory:

- Cells are the most basic building units of life.
- All living things are composed of cells.
- New cells are made from preexisting cells, which divide into two.

Two cells that joined together inside your mother's womb have determined who you are. To start creating new life, the two cells that together contain all of your genetic information (DNA) join together. The body's various organs, systems, blood, blood vessels, bone, tissue, and skin were formed as a result of cells dividing and differentiating into other cells with distinct roles. You are made up of trillions of cells as an adult. Every single one of your body's cells is a small, effective unit of life that can function independently of the others while still depending on them for resources. All of the essential functions of life must be carried out by autonomous single-celled organisms: they must breathe, move, grow, reproduce, detect and react to their surroundings, excrete waste, absorb nutrients (energy capture), and so on. These fundamental functions require organization, even in a single-celled organism. From the atomic level to all of its larger forms, every cell is structured. Water is a molecule made up of hydrogen and oxygen atoms (H_2O). Larger macromolecules are created when molecules join together. Because it can easily form long chains and more complex macromolecules by bonding with four other elements, the carbon atom is frequently referred to as the "backbone of life." Proteins, lipids, carbohydrates, and nucleic acids are the four macromolecules that comprise all of the structural and functional components of cells. Despite being the "most basic" unit of life according to our definition, the cell has intricate structural and functional details. Imagine a cell as a miniature organism made up of minuscule organs known as organelles. The structural and functional components of the organelles are made of multiple macromolecules joined together. The nucleus, which contains the genetic material DNA, the mitochondria, which produce energy, the ribosomes, which produce protein, the endoplasmic reticulum, which serves as a facility for packaging and transport, and the Golgi apparatus, which distributes macromolecules, are the organelles found in a typical animal cell. Furthermore, animal cells have tiny digestive pouches called peroxisomes and lysosomes that break down foreign invaders and break down macromolecules. A cytoskeleton holds each organelle firmly in place within the cytoplasm of the cell. The plasma membrane separates the organelles of the cell from its surroundings.

2.1.1. Tissues, Organs, Organ Systems, and Organisms

While the cells of multicellular organisms are arranged into five levels and are dependent upon one another to coordinate their specific functions and carry out all biological processes, unicellular (single-celled) organisms are capable of independent functioning (Table 2.1).

- **Cells:** The fundamental structural and functional component of all life is the cell. Red blood cells and nerve cells are two examples.
- **Tissues:** Tissues are collections of cells that cooperate and have a common structure and function. Human tissues can be divided into four categories: connective, which joins tissues; epithelial, which lines and shields organs; muscle, which contracts to provide support and movement; and nerve, which reacts to signals from the outside world.
- **Organs:** A collection of tissues grouped together in a particular way to support a common physiological function is called an organ. The liver, heart, and brain are a few examples.

Table 2.1. The 11 Organ Systems in the Human Body and Their Major Functions

Organ System	Organ Components	Major Function
Circulatory	heart, blood/lymph vessels, blood, lymph	Transport nutrients and waste products
Digestive	mouth, esophagus, stomach, intestines	Digestion and absorption
Endocrine	all glands (thyroid, ovaries, pancreas)	Produce and release hormones
Immune	white blood cells, lymphatic tissue, marrow	Defend against foreign invaders
Integumentary	skin, nails, hair, sweat glands	Protective, body temperature regulation
Muscular	skeletal, smooth, and cardiac muscle	Body movement
Nervous	brain, spinal cord, nerves	Interprets and responds to stimuli
Reproductive	gonads, genitals	Reproduction and sexual characteristics
Respiratory	lungs, nose, mouth, throat, trachea	Gas exchange
Skeletal	bones, tendons, ligaments, joints	Structure and support
Urinary	kidneys, bladder, ureters	Waste excretion, water balance

Source: https://2012books.lardbucket.org/books/an-introduction-to-nutrition/s07-nutrition-and-the-human-body.html.

- **Organ Systems:** The human body contains 11 organ systems, each consisting of two or more organs working together to support a specific physiological function. Examples of these systems include the digestive system and central nervous system.
- **Organism:** An organism is a whole living system that is able to carry out every biological function of life.

2.1.2. An Organism Requires Energy and Nutrient Input

Molecules need energy to grow into larger macromolecules, which in turn need energy to become organelles and cells, which in turn need energy to become tissues, organs, and organ systems, and lastly an organism. A healthy diet gives the body the nutrients it needs to produce the energy it needs to carry out its functions. Your body uses the nutrients in food to create new macromolecules.

2.1.3. Nutrient and Energy Flow

Chemical bonds in nutrients store energy. Sunlight provides energy, which plants absorb and use to convert atmospheric carbon dioxide into the molecule glucose through a process known as photosynthesis. Energy is released upon the breakage of the glucose bonds. Cellular respiration is a biological process that is used by humans, plants, and bacteria to extract energy from glucose.

This process converts the chemical energy of glucose into cellular energy in the form of the adenosine triphosphate (ATP) molecule. Oxygen, which is needed for aerobic cellular respiration, is produced as a waste product during photosynthesis. Carbon dioxide (CO_2) and water are the waste products of cellular respiration, which plants use to resume photosynthesis. Energy thus flows back and forth between plants and animals continuously. Within it, nutrients are recycled as energy is used up.

2.2. DIGESTION AND ABSORPTION

Food starts to break down even before it reaches your mouth. Your body alerts your brain when it's time to eat when it feels hungry. What your body perceives and smells affects how ready it is for food. Your brain receives a message when you smell food. The mouth is then instructed by the brain to prepare, and you begin to salivate in anticipation of a mouthwatering meal (Figure 2.1).

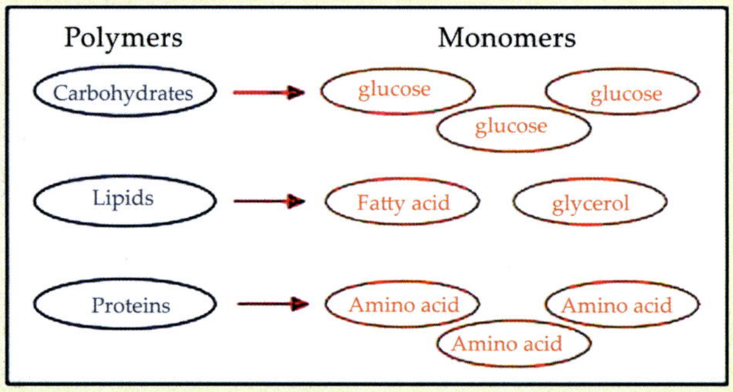

Figure 2.1. *The digestion process. Digestion converts the food we eat into smaller particles, which will be processed into energy or used as building blocks.*

Source: https://2012books.lardbucket.org/books/an-introduction-to-nutrition/s07-nutrition-and-the-human-body.html.

Your digestive system breaks down the food into smaller pieces after you eat. It operates on two levels—mechanical and chemical—to accomplish this. The body's cells will absorb and process the smaller particles to produce energy or to serve as the building blocks for new cells after they have been broken down.

The mouth, pharynx, esophagus, stomach, small intestine, large intestine (or colon), rectum, and anus are among the hollow tube-shaped organs that make up the digestive system, one of the human body's 11 organ systems. Its mucosal tissue lining secretes mucus, which helps food move through the tract, and digestive juices, which aid in the breakdown of food.

The digestive tract is surrounded by smooth muscle tissue, and when this muscle contracts, food is propelled down the tract in waves called peristalsis. Absorption occurs for both nutrients and certain non-nutrients. Fiber and other substances are excreted properly and left behind (Kong and Singh, 2008) (Figure 2.2).

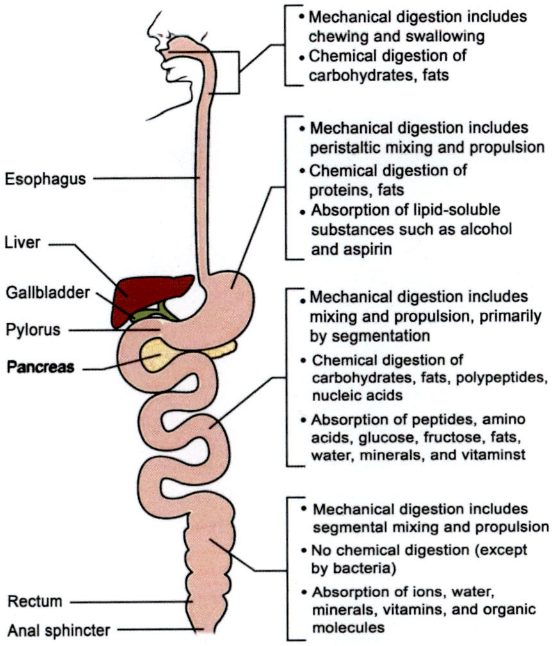

Figure 2.2. *Both, mechanical and chemical digestion take place throughout the gastrointestinal tract as indicated in this diagram, but absorption takes place only in the stomach and small and large intestines.*

Source: https://bio.libretexts.org/Bookshelves/Human_Biology/Human_Biology_(Wakim_and_Grewal)/18%3A_Digestive_System/18.3%3A_Digestion_and_Absorption.

2.2.1. From the Mouth to the Stomach

The process of digestion consists of four steps. Ingestion, or the gathering of food in the digestive tract, is the initial stage. Although it may seem like a straightforward process, ingestion entails thinking about food, smelling food, and uncontrollably releasing saliva into the mouth to prepare it for food entry. The second stage of digestion, which takes place in the mouth, is where food breaks down chemically and mechanically. Enzymes are involved in the chemical breakdown of food, dissolving its constituent ingredients. The stomach, pancreas, small intestine, and salivary glands all secrete these enzymes. The process of mastication, or chewing, in the mouth initiates mechanical breakdown. In contrast, saliva starts the chemical breakdown of food and facilitates its downward movement, while teeth break and grind larger food particles.

The bolus, a slick mixture of partially digested food, travels through the digestive system when you swallow. Because it takes conscious effort to push the food back toward the throat with the tongue, swallowing may initially appear voluntary. However, once it starts, swallowing is automatic and cannot be stopped. The bolus is forced from the mouth into the pharynx and the muscular tube known as the esophagus during swallowing. A tiny flap known as the epiglottis closes as food passes through the pharynx, preventing choking by preventing food from entering the trachea. Food is forced down to the stomach by peristalsis contractions in the esophagus. A sphincter muscle exists at the opening between the stomach and esophagus, and it stays closed until a food bolus is about to pass. Food passes from the esophagus into the stomach as a result of the lower esophageal sphincter opening and relaxing due to the pressure of the food bolus. The muscular contractions of the stomach and small intestine that mash, mix, slosh, and force food down the alimentary canal emphasize the mechanical breakdown of food. Solid food takes four to eight seconds to pass down the esophagus, while liquids pass through in about one second (Kong and Singh, 2008).

2.2.2. From the Stomach to the Small Intestine

Food is mashed, ground, and churned into chyme by strong peristaltic contractions that occur when food enters the stomach, an extremely muscular organ. Chyme is a semiliquid mixture of partially broken-down food and gastric secretions from stomach cells. The hydrochloric acid and pepsin enzymes, which chemically break down food into smaller molecules, are also secreted by stomach cells. The stomach performs three main functions:

1. To store food
2. To mechanically and chemically break down food
3. To empty partially broken-down food into the small intestine

The macronutrient makeup of the meal affects how long food stays in the stomach. Meals high in fat or protein digest more slowly than meals high in carbohydrates. The stomach empties completely after a meal usually takes a few hours.

The duodenum, jejunum, and ileum are the three structural sections that make up the small intestine.

The pancreas and gallbladder are activated and release digestive juices when the chyme reaches the duodenum, the first part of the small intestine. Through a duct into the duodenum, the pancreas secretes up to 1.5 liters of pancreatic juice daily. The majority of this fluid is water, but it also contains enzymes that further break down proteins, carbohydrates, and fats, as well as bicarbonate ions that balance the acidity of the chyme produced in the stomach.

Through a duct that connects to the duodenum, the gallbladder secretes a much smaller amount of bile to aid in the digestion of fats. The liver produces bile, which is then kept in the gall bladder. The components of bile surround fats in a manner akin to how dish soap removes grease from a frying pan, functioning as detergents. This permits fats to flow through the aqueous environment of the small intestine. Food passes through the small intestine in different stages of digestion and is moved and mixed by two distinct types of muscular contractions known as segmentation and peristalsis.

Peristalsis is the movement of food forward caused by circular waves of smooth muscle contraction, much like what happens in the stomach and esophagus. Food is sloshed back and forth in both directions during segmentation, which encourages more chyme mixing. Within the first 25 centimeters of the small intestine, nearly every element of food is completely decomposed into its most basic form. The chyme now contains emulsified fatty acids, monosaccharides, and amino acids instead of proteins, carbohydrates, and lipids.

The remaining portion of the small intestine, known as the ileum (> 5 meters), is where the next stage of digestion, or nutrient absorption, occurs (Figure 2.3).

The small intestine is designed in such a way as to optimize the absorption of nutrients. Its surface area is more than 200 square meters or roughly one tennis court's worth. The small intestine's surface area expands through several folding layers. Villi, which are microscopic projections that resemble fingers and are covered in even smaller projections known as microvilli, cover the inside tissue of the small intestine. Diffusion or specific transport proteins allow the nutrients that have been digested to pass through the intestinal absorptive cells. The intestinal cells release amino acids and monosaccharides (sugars) into capillaries, but first lymphatic vessels—which soon connect with blood vessels—transport the much larger emulsified fatty acids, fat-soluble vitamins, and other lipids (Kong and Singh, 2008).

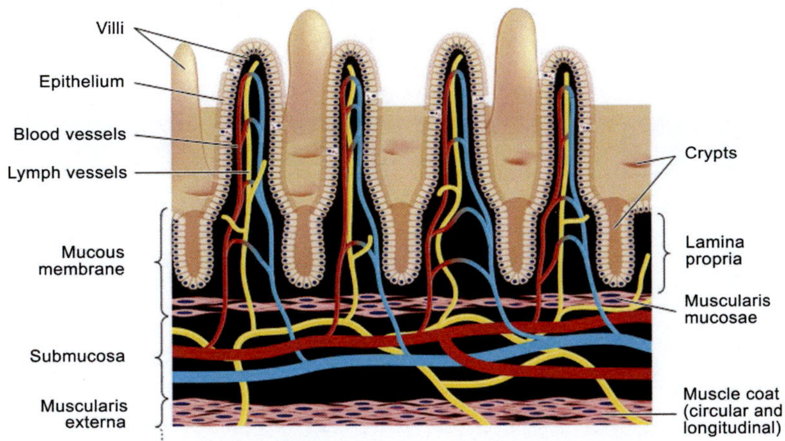

Figure 2.3. *The way the small intestine is structured gives it a huge surface area to maximize nutrient absorption. The surface area is increased by folds, villi, and microvilli. Digested nutrients are absorbed into either capillaries or lymphatic vessels contained within each microvilli.*

Source: https://2012books.lardbucket.org/books/an-introduction-to-nutrition/s07-nutrition-and-the-human-body.html.

2.2.3. From the Small Intestine to the Large Intestine

Digestion is a reasonably efficient process. Through a connecting valve, food that has not yet fully broken down (typically less than 10% of the food consumed) and its indigestible fiber content pass from the small to the large intestine (colon). Reabsorbing water is the large intestine's primary function. Recall that water is found in both solid and liquid foods. During the digestion process, the pancreas adds about 500 milliliters of water and the stomach releases a few 100 milliliters of gastric juice. Reabsorbing the water is essential for the body to conserve water. There is no more chemical or mechanical food breakdown in the large intestine unless it is carried out by the bacteria that live in this part of the digestive system. It is estimated that there are more bacteria in the large intestine than there are cells in the human body overall ($10^{(13)}$), or more than $10^{(14)}$. The vast majority of the bacteria in the large intestine are benign, and some may even be helpful, despite how unpleasant this may sound.

2.2.4. From the Large Intestine to the Anus

The digestion process moves on to step four, the removal of indigestible food as feces, after a few hours in the stomach, three to six hours in the small intestine, and roughly 16 hours in the large intestine.

Indigestible food and gut bacteria make up nearly 50% of the content of feces. It is held in the rectum until defecation causes it to be released through the anus (Kong and Singh, 2008).

2.3. NUTRIENTS ARE ESSENTIAL FOR ORGAN FUNCTION

The body eagerly awaits delivery of nutrients once food has been broken down by the digestive system. The liver is where most nutrients that are absorbed go first. Controlling metabolic homeostasis is one of the liver's main jobs. When the amount of nutrients taken in and absorbed equals the amount of energy needed to carry out the biological functions of life, this state is known as metabolic homeostasis. In other words, energy output from nutrients equals energy intake. From the small intestine to the liver, glucose and amino acids are directly transferred via the body's network of veins and blood vessels. A longer and more complicated path delivers lipids to the liver through the lymphatic system, which is made up of circulatory-like vessels that carry lymph, or white blood cells.

Preserving the body's energy balance is essential because physiological function deteriorates when eating disorders or illnesses disrupt metabolic homeostasis. The only organ in the human body that can export nutrients to other tissues for the production of energy is the liver. Thus, the liver exports nutrients when an individual is fasting and stores nutrients internally when an individual has just finished eating. The input that the liver needs to differentiate between the fed and fasted states and distribute nutrients correctly is provided by blood hormone levels and nutrient levels. The human body's 11 organ systems depend on nutrition to carry out their individual biological tasks. Without energy, no work can be produced. Nutrients power overall health and the body's capacity to perform all essential functions. Organ systems would malfunction without them, preventing human reproduction and the extinction of the species. We will talk about a few of the essential nutrients that support the operations of particular organ systems in this section.

2.3.1. The Circulatory System

One of the human body's 11 organ systems is the circulatory system. Its primary job is to move waste products out of and nutrients into cells. The heart, blood, and blood vessels make up this system. The blood is the fluid used for transportation, and the heart pumps it. All tissues are transported via a complex network of blood vessels made up of capillaries, veins, and arteries. The hepatic portal vein is the primary route via which nutrients absorbed in the small intestine reach the liver. Nutrients leave the liver and go to the heart through the inferior vena cava blood vessel. The nutrient-rich blood is forcefully pumped by the heart to the lungs so that they can absorb oxygen, and then to every other cell in the body. As blood travels toward cells, arteries get smaller and smaller until, at that point, the vessel is known as a capillary because the artery's diameter is so tiny. The blood flow is markedly slowed by the blood vessel's decreased diameter. The significant decrease in blood flow allows cells to absorb nutrients from the blood and exchange waste products from metabolism.

2.3.2. The Central Nervous System

It is estimated that the human brain, weighing only approximately 3 pounds (1,300 kilograms), contains over 100 billion neurons. The central nervous system, which includes the brain, spinal cord, and other nerve bundles throughout the body, is made up primarily of neurons. The central nervous system's primary job is to detect changes in the outside world and initiate a response to them. A sensory neuron on a rose bush, for example, sends a signal up the spinal cord and into the brain when your finger comes into contact with a thorn. A signal from another neuron in the brain causes the muscles in your hand to contract, causing you to jerk your finger away. In less than a tenth of a second, all of this occurs. Charged atoms of sodium, potassium, calcium, and chloride move in response to nerve impulses. These are a few of the vital minerals that we need to consume because the proper operation of the central nervous system depends on them. Neurotransmitters are substances made from amino acids that are used by nerves to communicate with one another. Consuming enough protein from a range of foods will guarantee that the body receives each of the many amino acids that are crucial for the proper operation of the central nervous system.

Over 20% of the energy derived from nutrients is used by the brain on a daily basis. It uses glucose as its primary fuel and won't use anything else unless it is extremely malnourished. Glucose must be routinely supplied to your brain to maintain acute mental alertness and clear thinking. This does not imply that you should down a can of sugar-filled soda right before your next test. Overconsumption of glucose has detrimental effects on the brain in the same way that it does on other organs like the pancreas and kidneys. A prolonged period of high blood glucose can harm brain cells, and excessive blood glucose can impair cognitive function. The cognitive processes of the brain include learning, perceiving, thinking, and processing language. Current research indicates that persistently elevated blood glucose levels significantly increase the likelihood of Alzheimer's disease development, the leading cause of age-related cognitive decline.

2.3.3. The Muscular System

In addition to enabling voluntary movement, the muscular system regulates involuntary organ system movements, such as the heartbeat in the circulatory system and the peristaltic waves in the digestive system. It is made up of more than 600 skeletal muscles, the heart muscle, the smooth muscles encircling your entire gastrointestinal tract, and every arterial blood vessel in your body. Energy transmission to the muscle is necessary for muscle contraction. Every movement depletes cellular energy, and muscles cannot function properly without enough energy. Similar to the liver, muscle can store glucose-derived energy in the form of the big polymeric molecule glycogen.

Muscles, in contrast to the liver, consume all of the energy that is stored within them and do not transfer it to other bodily organs. Muscle contraction becomes weaker when muscle energy stores are reduced. However, because muscle can easily use other fuels, like fatty acids and protein, to produce cellular energy, it is less sensitive to low blood glucose levels than the brain.

2.3.3.1. "Hitting the Wall" or "Bonking"

You may be familiar with the terms "hitting the wall" and "bonking" if you participate in endurance sports. These slang words describe the acute exhaustion that occurs approximately 120 minutes into an endurance activity, like long-distance cycling or marathon running. The physiology behind "hitting the wall" indicates that muscles are now dependent on other nutrients to meet their energy requirements because they have depleted all of their stored glycogen. To make up for the nutritional shortfall, fatty acids are transferred from cells that store fat to the muscle. Fatty acids, on the other hand, require longer to convert into energy than glucose, which lowers performance levels.

A few days before the competition, endurance athletes load up on carbohydrates to prevent "hitting the wall" or "bonking." This will optimize the amount of glycogen that is stored in the muscles and liver of an athlete. It's crucial to remember that not everyone responds well to carb loading. You won't increase your glucose storage if you don't engage in endurance training as well. If you and your friend decide to have a big spaghetti dinner the night before, and you intend to run a five-mile race for fun, the extra carbohydrates you consume will be turned into fat. During an endurance event, athletes can also prevent "hitting the wall" by consuming foods and drinks high in carbohydrates. The average cyclist in the 22-day, 2,400-mile Tour de France actually eats more than 60 grams of carbohydrates an hour.

2.3.4. The Endocrine System

The body's nutritional needs are closely linked to the endocrine system's operations. Along with other duties like reproduction, this organ system is in charge of controlling hunger, nutrient absorption, nutrient storage, and nutrient utilization. The pituitary, thyroid, parathyroid, adrenal, thymus, pineal, pancreas, ovaries, and testes are among the glands that make up the endocrine system. The circulatory system is needed to carry hormones secreted by the glands, which are biological molecules that control cellular processes in other target tissues. Sufficient nourishment is essential for every gland in the endocrine system to operate properly. A lack of protein affects the release of gonadal hormones, which stops reproduction. Teenage girls who are physically fit and have low body fat frequently skip their periods. Malnourished children typically do not produce enough growth hormone, which prevents them from growing to a normal height for their age. The most well-known relationship between nutrition and endocrine system functions is likely the one that links poor dietary habits to obesity and the onset of Type 2 diabetes. As of 2011, the Centers for Disease Control and Prevention (CDC) estimated that 26 million Americans suffered from Type 2 diabetes. This represents 8.3% of the US populace.

What is the underlying mechanism linking overnutrition to Type 2 diabetes? The general consensus is that excessive intake of high-fat and high-sugar foods alters muscle, fat, and liver cells, which in turn reduces the body's ability to respond to the pancreatic hormone insulin. We refer to these cells as "insulin-resistant." Following a meal, the release of insulin directs the liver and other tissues to absorb blood-circulating glucose and fatty acids. Due to insufficient absorption of glucose and fatty acids by insulin-resistant cells, blood levels of these substances stay elevated. The situation is made worse by the blood's persistently high levels of fatty acids and glucose, which hinder the pancreas' ability to release insulin. People with Type 2 diabetes have an increased risk of developing cardiovascular disease, kidney disease, nerve damage, and eye disease due to the long-term elevation of blood glucose and fatty acid levels that damage other tissues.

2.3.5. The Immune System

Numerous varieties of white blood cells circulate in the blood and lymphatic fluid, making up the immune system. Their duties include tracking down, enlisting, attacking, and eliminating alien invaders like viruses and bacteria. Mucus, which entangles and traps microorganisms, the skin, which serves as a barrier, and even the bacteria in the large intestine, which stop harmful bacteria from colonizing the gut, are additional, less well-known aspects of the immune system. Dietary nutrients are absolutely necessary for the immune system to function. Actually, the primary cause of immune system deficiencies globally is malnutrition. The likelihood of contracting an infection increases significantly when immune system functions are compromised. Two distinct syndromes, kwashiorkor and marasmus, are caused by the protein-and/or energy-deficient diets that children in many impoverished, developing nations suffer from. These kids frequently pass away from illnesses that their bodies would have been able to fight off, but because they consume so little protein and/or energy, their immune systems are unable to function properly.

The immune system can benefit from additional nutrients like zinc, selenium, copper, folate, and vitamins A, B_6, C, D, and E. A higher risk of infection and demise can result from deficiencies in these nutrients. By causing damage to skin cells, zinc deficiency suppresses the immune system's barrier functions and is linked to a reduction in the amount of white blood cells in circulation. Zinc supplements given to children under the age of five for longer than three months significantly reduce the incidence and severity of respiratory illnesses and diarrhea, according to a review of multiple studies published in the journal Pediatrics (Aggarwal, Sentz and Miller, 2007). Leprosy, TB, pneumonia, and the common cold have all been found to respond well to zinc supplementation as a therapeutic intervention. It's also critical to keep in mind that numerous studies have demonstrated that getting your vitamins and minerals from a variety of wholesome foods is the best option.

Overnutrition has the same detrimental effects on the immune system as undernutrition. Obese people have a higher chance of acquiring immune system conditions like rheumatoid arthritis, asthma, and certain types of cancer. The amount and quality of fat have an impact on immune system performance. Elevating your intake of omega-3 fatty acids, which are present in salmon and other oily fish, reduces inflammatory responses, while high intakes of saturated and trans fats have a detrimental effect on the immune system. Elevated consumption of omega-3 fatty acids is associated with a decreased likelihood of developing specific autoimmune conditions, like rheumatoid arthritis, and is employed as a component of a comprehensive regimen for the condition.

2.4. UTILIZATION OF NUTRIENTS IN THE BODY

All nutrients are metabolized within our body's cells. In the mitochondria, nutrients are finally converted to carbon dioxide and water through oxidation. The synthesis of ATP uses about 40% of the energy released, with the remaining energy being released as heat. The energy needed for the cell's synthetic reactions comes from the ATP produced. The mitochondria also break down fatty acids. New cells that are created as a result of regular life processes take the place of the dying individual cells.

2.4.1. Calories and Kilocalories: Energy Supply

Imagine the human body as an engine that releases the energy found in the food it breaks down. This energy is used in part for the secretory functions and the mechanical work that the muscles perform, and in part for the tasks required to keep the body's structure and functions intact. Work-related heat production is linked to heat loss, which is managed to maintain a specific body temperature range. But unlike other engines, the human body is always disassembling (catabolizing) and reassembling (anabolizing) its constituent parts. Foods provide the energy required for the chemical reactions involved as well as the nutrients required for the production of the new material.

As energy sources, protein, fat, and carbohydrates are largely interchangeable. Food's energy content is commonly expressed in kilocalories, or calories. A kilocalorie is 1,000 gram-calories, also known as small calories, which are units of measurement for heat energy. Nonetheless, kilocalories are sometimes referred to as "calories" in everyday speech. Put differently, 2,000 calories actually translate to 2,000 kilocalories of potential energy in a diet. The quantity of heat energy needed to raise one kilogram of water from 14.5 to 15.5 degrees Celsius at one atmosphere of pressure is known as one kilocalorie. The joule is another commonly used unit of energy that expresses energy as mechanical work. The energy used to move one kilogram one meter with a force of one Newton is equal to one joule. It is more common for the comparatively higher energy levels in human nutrition to be expressed in megajoules (one megajoule equals 10^6 joules) or kilojoules (1 kilojoule = 10^3 joules). 4.184 kilojoules make up one kilocalorie.

By measuring the amount of heat produced when food is burned (oxidized) in a bomb calorimeter, the amount of energy contained in food can be computed directly. However, the human body is not a calorimeter, and during digestion and metabolism, some potential energy is lost. The following are the corrected physiological values, rounded to whole numbers, for the heats of combustion of the three energy-producing nutrients: 4 kcal (17 kilojoules) for carbohydrates, 4 kcal (17 kilojoules) for proteins, and 9 kcal (38 kilojoules) for fats. Although it is not necessary in the diet, beverage alcohol, or ethyl alcohol, also produces energy—7 kilocalories (29 kilojoules) per gram. Water, vitamins, minerals, and other food ingredients don't contain any energy, but many of them help the body release energy.

If the gram quantities of energy-yielding substances (non-fiber carbohydrate, fat, protein, and alcohol) in a well-digested food are known, the amount of energy it provides can be approximated. For instance, a slice of white bread has 67 kilocalories (280 kilojoules) of energy and contains 12 grams of carbohydrates, 2 grams of protein, and 1 gram of fat. Food labels and food composition tables offer helpful information for assessing a person's diet's nutrient and energy intake. The majority of foods include a variety of nutrients that give you energy in addition to vitamins, minerals, water, and other ingredients. Vegetable oil and table sugar, which are essentially pure fat and carbohydrate (sucrose), respectively, are two noteworthy exceptions (Table 2.2).

Table 2.2. Nutritional Content of Menu Items

Menu Items	Portion Weight g, mL (Min–Max)	Energy kJ/100 g (Min–Max)	Total Fat g/100 g (Min–Max)	Total Sodium g/100 g (Min–Max)	Total Sugar g/100 g (Min–Max)
Ready-to-eat items					
1. Rice cooked in chicken broth, topped with boiled chicken	291 (230–370)	177.2 (135.9–206.2)	5.8 (3.2–8.5)	0.3 (0.2–0.5)	0.0 (0.0–0.0)
2. Rice topped with pork leg stew in gravy	453 (350–600)	153.3 (133.8–218.7)	5.3 (3.1–10.9)	0.3 (0.2–0.5)	2.2 (1.6–3.1)
3. Rice with roasted pork and gravy	355 (150–475)	158.2 (133.9–200.9)	3.8 (4.2–22.8)	0.4 (0.2–0.7)	3.1 (1.6–6.6)
4. Fried rice with pork, vegetable, and egg	344 (260–425)	192.7 (156.3–235.9)	7.6 (4.2–11.7)	0.4 (0.2–0.7)	1.5 (0.0–2.7)
5. Rice with crispy pork and gravy	384 (300–540)	189.1 (152.2–231.6)	7.9 (4.1–14.0)	0.3 (0.2–0.4)	1.9 (0.0–6.6)
6. Rice with stir-fried minced pork, chili, and basil	397 (280–512)	185.0 (139.6–221.0)	7.4 (3.2–11.4)	0.3 (0.1–0.8)	0.9 (0.0–2.3)
7. Rice with spicy chicken salad	383 (290–475)	185.8 (151.2–237.4)	5.3 (3.9–9.0)	0.4 (0.3–0.6)	0.5 (0.0–2.7)
8. Rice topped with pork	377 (270–500)	179.1 (116.5–241.9)	5.9 (1.0–12.5)	0.2 (0.09–0.4)	1.8 (0.0–3.1)

9. Shrimp paste rice with caramelized pork condiment, dried shrimp, green mango, chilies, lime, shallot, cucumber, and long beans	248 (180–355)	198.7 (162.2–240.0)	6.4 (1.9–8.9)	0.6 (0.3–0.9)	0.7 (0.0–3.8)
10. Congee rice with minced pork	578 (490–760)	65.1 (52.5–88.0)	2.3 (1.2–5.6)	0.3 (0.2–0.4)	0.7 (0.0–1.7)
11. Blood tofu soup with gourd leaves	615 (502–730)	44.7 (23.0–104.4)	2.8 (1.0–9.8)	0.4 (0.2–0.5)	0.4 (0.0–2.2)
12. Noodle sheets, soup with meat and tofu	635 (480–800)	77.8 54.3–114.0	3.3 1.7–6.5	0.4 0.2–0.7	1.6 0.0–3.5
13. Rice noodles with pork ball	615 (508–842)	60.9 (36.3–80.4)	1.5 (0.6–2.3)	0.4 (0.3–0.5)	1.4 (0.0–3.3)
14. Rice noodles with fish ball, red sauce, and soup	628 (427–857)	67.5 (46.6–80.4)	1.8 (0.6–3.2)	0.5 (0.3–0.6)	3.0 (1.1–6.0)
15. Egg in five spices soup	281 (190–360)	85.3 (56.9–129.3)	2.9 (2.0–4.0)	0.6 (0.3–0.9)	8.9 (1.7–17.1)
16. Fried streaky pork	185 [1] (90–280)	440.5 (361.2–531.0)	36.3 (23.3–49.9)	0.6 (0.4–0.9)	1.6 (0.0–5.0)
17. Grilled pork	264 (100–420)	307.4 (255.1–347.5)	21.1 (13.5–27.0)	0.8 (0.5–1.3)	10.3 (3.9–14.7)
18. Grilled pork neck	158 (100–260)	374.5 (254.2–461.7)	30.8 (16.0–43.9)	0.5 (0.2–0.8)	2.1 (0.0–7.3)
19. Salmon sushi	177 (100–260)	173.5 (122.7–206.8)	4.4 (0.4–8.4)	0.2 (0.03–0.4)	1.9 (0.0–5.6)
20. Rice topped with salmon	311 (232–379)	177.6 (149.4–203.0)	4.3 (1.5–9.0)	0.1 (0.01–0.5)	2.5 (0.6–0.9)
21. Thai Sukiyaki soup	657 (515–820)	45.0 (30.6–63.5)	1.5 (1.0–2.2)	0.2 0.1–3.5	0.6 (0.0–1.7)
22. Papaya salad, spicy, with dried shrimp and roasted peanuts	302 (230–400)	110.4 (52.6–145.8)	2.6 (1.1–4.9)	0.7 (0.4–1.7)	10.9 (4.3–19.2)
23. Papaya salad, spicy, with fermented crab and fermented fish northeastern style	313 (240–140)	52.4 (38.9–72.9)	0.2 (0.1–0.5)	1.6 (1.3–2.1)	3.6 (0.0–8.5)
24. Pork spicy salad, northeastern style	241 (180–330)	142.2 (91.3–205.9)	7.5 (3.1–13.4)	0.7 (0.4–1.0)	0.7 (0.0–2.3)
25. Grilled pork balls	392 (268–548)	133.9 (86.8–190.1)	5.7 (1.7–11.6)	0.8 (0.4–1.1)	0.7 (0.0–2.6)
Sweets					

1. Chinese pork bun	213 (110–350)	252.6 (213.7–299.8)	9.9 (3.8–18.0)	0.4 (0.2–0.6)	11.5 (7.2–22.8)
2. Deep-fried Chinese dough	105 (40–200)	412.6 (343.8–492.8)	22.6 (12.3–33.3)	0.5 (0.3–0.9)	4.4 (1.1–9.0)
3. Egg tart	187 (90–270)	362.9 (287.5–475.2)	24.0 (16.9–32.3)	0.2 (0.1–0.3)	12.0 (7.1–15.4)
4. Pandan and coconut chiffon cake	307 (70–500)	357.9 (249.9–413.1)	21.2 (15.1–25.0)	0.2 (0.1–0.4)	23.0 (7.1–32.1)
5. Coconut milk ice cream	309 (140–660)	175.4 (120.2–213.5)	10.0 (3.6–14.5)	0.05 (0.09–0.1)	11.3 (6.1–16.0)
Non-alcoholic beverages					
1. Iced espresso	207 (160–260)	106.9 (82.7–128.9)	3.6 (2.3–4.7)	0.03 (0.02–0.04)	13.1 (9.7–16.3)
2. Iced coffee	250 (180–530)	134.6 (89.7–185.1)	4.1 (1.4–6.6)	0.3 (0.1–0.6)	16.9 (8.2–24.3)
3. Iced honey lemon tea	270 (185–360)	88.5 (53.7–113.1)	0.0 (0.0–0.1)	0.03 (0.002–0.02)	19.7 (13.3–24.3)
4. Iced cocoa	236 (200–320)	138.9 (97.9–206.7)	4.7 (3.2–6.6)	0.03 (0.03–0.04)	16.2 (10.3–28.4)
5. Iced Americano	214 (180–269)	24.9 (5.7–47.8)	0.1 (0.0–0.2)	0.004 (0.003–0.01)	6.5 (3.2–10.2)
6. Bubble milk tea	326 (250–460)	128.2 (86.1–186.8)	3.4 (1.6–5.3)	0.03 (0.01–0.04)	12.1 (7.1–18.3)
7. Iced cappuccino	220 (170–270)	110.6 (84.8–138.9)	3.9 (1.5–5.4)	0.04 (0.02–0.05)	12.6 (6.9–15.7)
8. Iced Thai milk tea	231 (174–294)	113.9 (79.3–157.6)	4.1 (2.7–7.3)	0.04 (0.03–0.06)	14.0 (7.2–26.5)
9. Iced green tea Frappuccino	563 (470–680)	67.9 (51.2–96.6)	2.2 (1.4–4.2)	0.02 (0.01–0.05)	8.8 (6.4–12.0)
10. Soy milk	355 (280–460)	47.6 (20.6–71.4)	1.2 (0.1–2.6)	0.009 (0.002–0.03)	4.3 (0.0–7.7)

Note: n = 15 for each menu item. Each menu item was ordered and measured nutritional contents 15 times. Cells highlighted in pink indicate menu items that were extremely high in energy, total fat, total sodium, and total sugar per 100 g.

Source: Jindarattanaporn N, Suya I, Lorenzetti L, Kantachuvesiri S, Thamarangsi T. Nutritional content of popular menu items from online food delivery applications in Bangkok, Thailand: Are they healthy? International Journal of Environmental Research and Public Health. 2023; 20(5):3992. https://doi.org/10.3390/ijerph20053992.

Protein makes up 8–16% of the energy in diets across most of the world, though the proportions of fat and carbohydrates in different populations vary greatly. About 12–15% of energy comes from protein, 30–40% from fat, and 50–60% from carbohydrates in more affluent communities. However, in many impoverished agricultural societies where cereals make up the majority of the diet, carbohydrates account for an even higher proportion of energy, while protein and fat account for a smaller portion. The human body can endure and even flourish on a remarkably diverse range of diets due to its remarkable adaptability. Nonetheless, certain dietary patterns are linked to specific health outcomes.

2.4.2. BMR and REE: Energy Balance

Energy is required when one is not only physically active but also when one's body is at rest. Between 50 and 80% of the energy used each day, depending on physical activity level, goes toward basic metabolic processes (basal metabolism), which allow the body to perform various physiological and biosynthetic functions, such as the synthesis of new tissue in growing children and pregnant and lactating women, as well as breathing and pumping blood. The body uses energy and produces heat during food digestion and subsequent processing. The thermic effect of food, also referred to as diet-induced thermogenesis, is a phenomenon that contributes roughly 10% of daily energy expenditure. The exact percentage depends on the diet's composition and previous dietary habits. Another minor but significant part of energy expenditure is called adaptive thermogenesis, and it represents changes in metabolism brought on by variations in the surrounding temperature, hormone production, emotional stress, or other variables. Physical activity, which includes both voluntary activities like exercise and other activities and involuntary ones like fidgeting, shivering, and maintaining posture, is the last and most variable component of energy expenditure. 20 to 40% of total energy expenditure is attributed to physical activity; this percentage may be even lower in highly sedentary individuals and higher in highly active ones.

The body's metabolically active tissue, lean body mass (mass free of fat and essential fat but excluding storage fat), is largely correlated with basal or resting energy expenditure. Organs with the highest metabolic activity at rest, such as the liver, brain, heart, and kidney, also have the highest energy needs; in contrast, the needs of muscle, bone, and body fat are much lower. In addition to body composition, age, sex, body temperature, and thyroid hormone levels also have an impact on basal metabolism. A comfortable room temperature, at least 12 hours after the last meal, and shortly after awakening in the morning are the controlled and standardized conditions under which the basal metabolic rate (BMR), a precisely defined measure of the energy expenditure required to support life, is determined. Practical reasons make it rare to measure BMR; instead, the resting energy expenditure (REE), which is based on the individual resting comfortably for two to four hours after a meal, is measured under less strict guidelines. The terms BMR and REE are interchangeable and, in practice, differ by no more than 10%, with the REE typically being slightly higher.

Direct calorimetry, or the measurement of heat dissipated from the body, uses devices like water-cooled clothing or insulated rooms big enough to fit a person to determine how much energy is used. However, the less laborious methods of indirect calorimetry—which calculates heat produced by the body from measurements of oxygen inhaled, carbon dioxide exhaled, and urine nitrogen excreted—are typically used to measure energy expenditure. The following formula can be used to approximate the BMR (in kilocalories per day): BMR = 70 × (body weight in kilograms)$^{3/4}$.

Different activities' energy costs have been calculated. Work that requires a lot of energy can require 10 times as many calories per minute as resting. Despite its apparent drain, mental activity has no discernible impact on energy consumption. A man weighing 70 kg (154 pounds), with a potential daily REE of 1,750 kcal, could consume 2,400 kcal on an extremely sedentary day and as much as 4,000 kcal on an extremely active day. Depending on her level of activity, a 55-kg (121-pound) woman whose daily resting energy expenditure may be 1,350 kilocalories could use 1,850 to over 3,000 total kilocalories (Table 2.3).

Table 2.3. Approximate Energy Expenditure for Activity Levels

Activity Category	Energy as Multiple of Resting Energy Expenditure (REE)	Kilocalories Per Minute
Resting (sleeping, reclining)	REE × 1.0	1–1.2
Very light (driving, typing, cooking)	REE × 1.5	up to 2.5
Light (walking on a level surface at 4 to 5 km/hr [2.5 to 3 mph], golf, table tennis)	REE × 2.5	2.5–4.9
Moderate (walking 5.5 to 6.5 km/hr [3.5 to 4 mph], carrying a load, cycling, tennis, skiing, dancing)	REE × 5.0	5.0–7.4
Heavy (walking uphill with a load, basketball, climbing, football, soccer)	REE × 7.0	7.5–12.0

Source: National Academy of Sciences, Recommended Dietary Allowances, 10th ed. (1989).

The principle of conservation of energy states that an individual will gain weight if they consume more energy than they use; conversely, if an individual consumes less energy, they will lose weight because their body will use its stored energy to meet their immediate needs. Small amounts of extra food energy are stored by the body as fat, which is primarily found in adipose tissue, and as glycogen, a form of carbohydrate that is used for short-term storage in muscle and the liver. Although fat makes up the majority of adipose tissue—roughly 87% of it—it also contains some protein and water. About 3,500 kilocalories (14,6 megajoules) of energy deficit are needed to shed 454 grams (one pound) of adipose tissue.

2.4.3. Body Mass, Body Fat, and Body Water

The components of the human body are similar to those in food, but the relative amounts vary depending on an individual's particular life experience and genetic predispositions. A healthy, lean man's body is made up of approximately 62% water, 16% fat, 16% protein, 6% minerals, and less than 1% carbohydrates, with trace amounts of vitamins and other unspecified chemicals. Compared to men of similar weight, women typically carry slightly less of the other components and more fat (about 22% in a healthy, lean woman). To maintain a state of dynamic equilibrium, or homeostasis, the body's various compartments—lean body mass, body fat, and body water—constantly adapt to changes in the internal and external environment. The body constantly breaks down (catabolism) and builds up (anabolism) tissues at different rates. For instance, connective tissue regenerates over several years, red blood cells have a 120-day lifespan, and the epithelial cells lining the digestive tract are replaced at an astounding rate of once every three or four days.

Direct inspection is one method of estimating the percentage of body fat, but it is not very accurate. Indirect measures of body fat include dual-energy X-ray absorptiometry (DXA), total body potassium counting, and underwater weighing. These techniques are fairly accurate but also expensive. However, less precise but more useful techniques are frequently employed. For example, anthropometry measures subcutaneous fat at different locations using skinfold calipers; bioelectrical impedance measures resistance to a low-intensity electrical current to estimate body fat; and near-infrared interactance measures the interaction between fat and protein by shining an infrared light on the biceps. Only on cadavers can the various body compartments be measured directly.

Throughout a lifetime, the body's composition tends to change in somewhat predictable ways, with corresponding changes in nutrient needs during different stages of the life cycle. These changes can be seen during the growing years, during pregnancy and lactation, and as one ages. Frequent exercise can slow down the aging-related loss of lean body mass and rise in body fat.

CASE STUDY

THE CASE OF THE MISSIONARY HEALTH CARE WORKER IN AFRICA

Jenny, a 22-year-old college graduate, has recently decided to spend two years as a missionary in a coastal African village. As part of her duties, she will work in the missionary clinic that has recently been established in the village. This clinic will provide routine medical care and a traveling physician and nurse will visit once each week. Jenny feels fortunate to have the opportunity to use the information that was presented to her in various classes she completed while in college. After completing a 4-month intensive training period (language classes, medical preparation for work in the clinic, and cultural enrichment), she departed for Africa.

Upon her arrival, Jenny found that she would be spending most of her time in the clinic since the village had been without routine medical care for the past 2 years. The previous clinic was closed due to flooding and insufficient funds to rebuild. Initially, Jenny found that she was very dependent upon the physician's weekly visits since her training was limited. Gradually, she found that she was beginning to call upon her training (both clinical and classroom) and she felt comfortable diagnosing and treating some of the more routine cases. One morning, a mother came in with her five-year-old child. The mother reported that the child had been unable to eat or drink anything for the past day because of vomiting. In addition, her child was experiencing severe diarrhea. Jenny asked the mother if any other family members were exhibiting similar symptoms, to which the mother replied that a few other members of the family had similar symptoms, however, not as severe. Jenny's first thought was that the symptoms were very similar to those she had exhibited following a bout of food poisoning while in college. Since the villagers had no refrigeration and poor sanitation, Jenny gave the mother an electrolyte solution containing glucose similar to Gatorade® and told the mother to have the affected children consume this solution and nothing else for the next 24 hours. After that time, if the children were no longer vomiting, she could start feeding them tea and broth. Jenny also cautioned the mother that if the children did not seem to be getting better after 24 hours to bring the children back to the clinic.

The next morning Jenny opened the clinic to find the mother, and not just the child she had seen the previous day, but three more of the woman's children. All of the children were exhibiting similar symptoms that now included muscle cramping and excessive thirst in addition to diarrhea and vomiting. When checking the vital signs of the children, Jenny noticed increases in both the pulse and respiratory rates accompanied by decreased blood pressure. Uncertain as to the appropriate course of action, Jenny contacted the physician by radio. Upon conveying the histories and information to the physician, Jenny received instructions to keep the children at the clinic, start intravenous (IV) infusion with lactated Ringer's solution, and allow the children to drink as much of the electrolyte solution with glucose as they would like. The physician also gave Jenny a list of laboratory tests to run on the blood, urine and stool samples that she should collect. Jenny started the IV infusions and gave each child some of the electrolyte solution. After this, she obtained stool, blood and urine samples from each child and asked the mother to leave the children with her for care and observation. The mother agreed and said she would return later that night to help with the children.

The results of the tests run indicated severe metabolic acidosis, an increased hematocrit, hypokalemia, and the presence of Vibrio cholerae bacteria in the stool samples. Based on the results of these tests, the physician and Jenny diagnosed the children with cholera and obtained a more detailed history in an attempt to determine whether these individuals were the only ones exposed or whether these cases were the first of a possible epidemic. Jenny and the physician found that the family had recently visited relatives in a distant village where similar symptoms had been present in many families that had all celebrated a recent shellfish harvest.

The physician prescribed continued IV infusions with lactated Ringer's solution and electrolyte fluid replacement by mouth. The children were not allowed to consume other foods or drinks, especially coffee or any other beverages containing caffeine. The children were also told that once the vomiting stopped they could start consuming solid foods. The children were all treated with antibiotics and stool samples were taken from other family members to determine whether or not they were infected with Vibrio cholerae. Infected, asymptomatic individuals were treated with antibiotics. The physician later explained to Jenny that caffeine was prohibited because the toxin produced by the bacteria (termed choleragen) binds to the surface of the epithelial cells of the small intestine and activates adenylate cyclase. In addition, the toxin interferes with the active transport of sodium ions in the intestinal lumen.

Answer the following questions about this case

1. Why did Jenny initially suspect that the child and the other family members were experiencing food poisoning?
2. Why did Jenny initially prescribe an electrolyte solution containing glucose?
3. Upon further evaluation on the second day why would the children present with an increased pulse and respiratory rate but a decreased blood pressure?
4. Why did the children present with metabolic acidosis?
5. Why were the children experiencing muscle cramping and complaining of excessive thirst on the second day?
6. Why were the children not allowed to consume any caffeine-containing beverages?
7. Why does the effect of the cholera toxin on the epithelial sodium transport protein contribute to both diarrhea and metabolic acidosis?

CLASS ACTIVITY

HOW FOOD WORKS

Project Skill

To understand the 4 different Food Groups and their functions in the body.

Life Skills

- Recognizing self-worth
- Working with others

Preparation

- Using a marker, write the functions of food across the top of one Workpad sheet to create your Workpad poster. Next, as indicated on the right, write down the KEY WORDS underneath this:

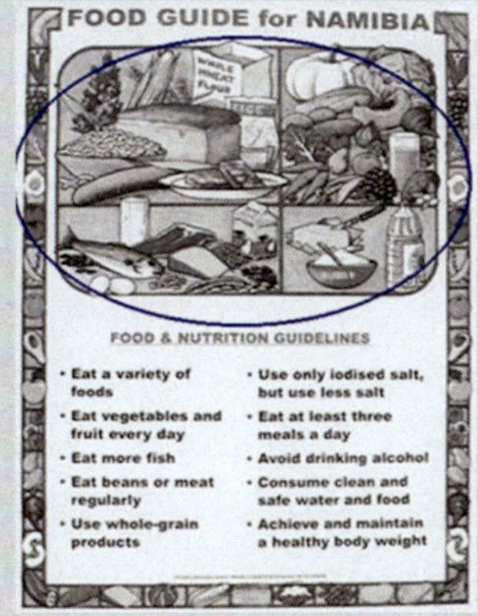

Functions of Food:

- growth & development
- muscles,
- bones and teeth
- energy
- healthy bodies
- fighting disease
- good brains
- healthy organs

If you think it would be beneficial, write these important terms in your native tongue so that group members can comprehend them more easily. Adding illustrations or cutting out images from magazines that best illustrate these key words and adhering them to each point will help your poster look more appealing. To demonstrate healthy teeth, you may, for example, draw a picture of a smiling mouth or an energetic person running or playing soccer. Keep this Work pad poster handy for future use in similar activities involving other youth groups.

Food Group Cards

Using a marker, write the names of each food group on one of four pieces of paper or cardboard: 1. Cereals, 2. Veggies and Fruit, 3. Beans, Fish, Meat, and Milk, 4. Sugars and Fats. Keep these in your Kit for later use.

Introduction

- Tell the people in your group that the purpose of this activity is to help everyone understand the four different food groups and the reasons the body needs a variety of foods.

Experience

- Ask everyone to take a seat so they can all see the Food Guide poster when it is displayed.
- It should be noted that despite their beliefs, all group members possess prior knowledge on the significance of food for our bodies. This understanding has been acquired through familial teachings (from parents, grandparents, and other relatives) and education received in school.
- Name the benefits that foods have for the body to help your participants recall and discuss what they have learned. Answers to expect are:
  - Growth and development
  - Develop strong muscles, bones and teeth;
  - Keep our bodies warm and give us energy;
  - Keep our bodies healthy by helping our immune systems to prevent or fight disease;
  - Repair or healing of injuries.
  - Helps our brains to grow and function properly
  - Development and proper working of our organs: heart, kidneys, lungs, liver, etc.
- All foods fit into these four basic groups, as you can see by pointing to the picture on the Food Guide poster. I want to know if anyone can identify these: 1. Cereals, 2. Veggies and Fruit, 3. Beans, Fish, Meat, and Milk, 4. Sugars and Fats.

Process

- Place your Workpad poster next to the poster for the Food Guide.
- Give each of the four groups that your participants have formed a card representing a food group.
- Indicate to the other groups that they have five minutes to consider the list of KEYWORDS on your Work pad poster, discuss among themselves, and then identify which of the bodily functions of food are provided by the Food Group on their Food Group card.

Everyone must understand that group members should be allowed to complete their explanations without being corrected or interrupted by others.

Apply

- After everyone has finished, gather them all back into a single, sizable group.
- Ask your participants to take another look at the Food Guide Poster. Explain that:
  - The largest group, grains, should be consumed in greater amounts since they contain nutrients that warm and give us energy.
  - Numerous vitamins and minerals found in fruits and vegetables support a healthy immune system.
  - Proteins found in meat, dairy products, nuts, seeds, and other foods aid in muscle growth.
  - Fats and sugars are dense sources of energy and should be consumed in moderation. Fats and oils in particular provide a significant amount of energy, more than twice as much as sugars. People requiring additional energy, such as those who are malnourished or engaged in strenuous physical activity, can benefit from including fats, oils, and sugar in their diet for added energy.

Overconsumption of fat, oil, and sugar, however, can exacerbate the condition of those who are already overweight (fat), which not only makes them uncomfortable but also impairs their heart and blood circulation and causes them to become seriously overweight.

SUMMARY

- A living organism conducts self-sustaining biological processes. A cell is the smallest and most basic form of life.
- Energy is required to build molecules into larger macromolecules, and to turn macromolecules into organelles and cells, and then turn those into tissues, organs, and organ systems, and finally into an organism.
- Energy is stored in a nutrient's chemical bonds. Energy comes from sunlight, which plants then capture and, via photosynthesis, use to transform carbon dioxide in the air into the molecule, glucose. When the glucose bonds are broken, energy is released.
- Digestion begins even before you put food into your mouth. When you feel hungry, your body sends a message to your brain that it is time to eat. Sights and smells influence your body's preparedness for food. Smeling food sends a message to your brain. Your brain then tells the mouth to get ready, and you start to salivate in preparation for a delicious meal.
- When the digestive system has broken down food to its nutrient components the body eagerly awaits delivery. The first stop of most absorbed nutrients is the liver. One of the liver's primary functions is to regulate metabolic homeostasis. Metabolic homeostasis may be defined as when the nutrients consumed and absorbed match the energy required to carry out life's biological processes. Simply put, nutrient energy intake equals energy output.
- The circulatory system is one of the 11 organ systems of the human body. Its main function is to transport nutrients to cells and wastes from cells. This system consists of the heart, blood, and blood vessels.
- The human brain (which weighs only about 3 pounds, or 1,300 kilograms) is estimated to contain over 100 billion neurons. Neurons form the core of the central nervous system, which consists of the brain, spinal cord, and other nerve bundles in the body. The main function of the central nervous system is to sense changes in the external environment and create a reaction to them.
- The muscular system allows the body to move voluntarily, but it also controls involuntary movements of other organ systems such as heartbeat in the circulatory system and peristaltic waves in the digestive system.
- The functions of the endocrine system are intricately connected to the body's nutrition. This organ system is responsible for regulating appetite, nutrient absorption, nutrient storage, and nutrient usage, in addition to other functions, such as reproduction.
- The immune system is comprised of several types of white blood cells that circulate in the blood and lymph. Their jobs are to seek, recruit, attack, and destroy foreign invaders, such as bacteria and viruses. Other less realized components of the immune system are the skin (which acts as a barricade), mucus (which traps and entangles microorganisms), and even the bacteria in the large intestine (which prevent the colonization of bad bacteria in the gut). Immune system functions are completely dependent on dietary nutrients.

- Metabolism of all nutrients occurs within the cells of our body. The final oxidation of nutrients into carbon dioxide and water occurs in the mitochondria.
- The human body can be thought of as an engine that releases the energy present in the foods that it digests. This energy is utilized partly for the mechanical work performed by the muscles and in the secretory processes and partly for the work necessary to maintain the body's structure and functions.

REVIEW QUESTIONS

1. Define digestion. Where does it occur?
2. Identify two organ systems that control the process of digestion by the digestive system.
3. What is mechanical digestion? Where does it occur?
4. Describe chemical digestion.
5. What is the role of enzymes in chemical digestion?
6. What is absorption? When does it occur?
7. Where does most absorption occur in the digestive system?
8. Why does most of the absorption occur in this organ and not earlier in the GI tract?
9. Name two digestive enzymes found in saliva and identify which type of molecule they digest.
10. Where is bile produced?
11. What are some functions of bile?
12. List the enzymes involved in the digestion of food and the functions of each enzyme.

REFERENCES

1. Aggarwal, R., Sentz, J., & Miller, M. A. (2007). Role of zinc administration in prevention of childhood diarrhea and respiratory illnesses: A meta-analysis. *Pediatrics, 119*(6), 1120–1130. https://doi.org/10.1542/peds.2006-3481.
2. Boettner, D. R., Huston, C. D., Linford, A. S., Buss, S. N., Houpt, E., Sherman, N. E., & Petri, W. A. (2008). Entamoeba histolytica phagocytosis of human erythrocytes involves PATMK, a member of the transmembrane kinase family. *PLOS Pathogens, 4*(1), e8. https://doi.org/10.1371/journal.ppat.0040008.
3. Cascales, E., & Christie, P. J. (2003). The versatile Type IV secretion systems. *Nature Reviews Microbiology, 1*(2), 137–149. https://doi.org/10.1038/nrmicro753.
4. Dusenbery, D. B. (2009). *Living at Micro Scale* (p. 280). Harvard University Press.
5. Kong, F., & Singh, R. P. (2008). Disintegration of solid foods in human stomach. *Journal of Food Science, 73*(5), R67–R80. https://doi.org/10.1111/j.1750-3841.2008.00766.x.

6. Kuehn, M. J., & Kesty, N. C. (2005). Bacterial outer membrane vesicles and the host-pathogen interaction. *Genes & Development, 19*(22), 2645–2655. https://doi.org/10.1101/gad.1299905.
7. Luz, M. R. M. P. (2008). Glucose as the sole metabolic fuel: A study on the possible influence of teachers' knowledge on the establishment of a misconception among Brazilian high school students. *Advances in Physiology Education, 32*, 225–230. https://doi.org/10.1152/advan.00050.2007.
8. Miserez, A., Li, Y., Waite, H., & Zok, F. (2007). Jumbo squid beaks: Inspiration for design of robust organic composites. *Acta Biomaterialia, 3*(1), 139–149. https://doi.org/10.1016/j.actbio.2006.09.004.
9. Oliveira, G. A., et al. (2003). Students' misconception about energy yielding metabolism: Glucose as the sole metabolic fuel. *Advances in Physiology Education, 27*, 97–101. https://doi.org/10.1152/advan.00009.2003.
10. Salyers, A. A., & Whitt, D. D. (2002). *Bacterial Pathogenesis: A Molecular Approach* (2nd ed.). ASM Press.
11. Wooldridge, K. (2009). *Bacterial Secreted Proteins: Secretory Mechanisms and Role in Pathogenesis*. Caister Academic Press.

CHAPTER 3

Essential Nutrients

LEARNING OBJECTIVES

After studying this chapter, you will be able to:

- Explain carbohydrates
- Identify fats and other lipids
- Discuss the role of proteins and amino acids
- Understand energy metabolism
- Focus on vitamins and minerals
- Describe the importance of water, fluids, electrolytes and acid-base balance
- Identify antioxidants in food.

KEY TERMS FROM THIS CHAPTER

Albumin	Antibodies	Blood cells
Blood vessels	Cholesterol	Collagen
Dietary fiber	Dietary protein	Disaccharides
Endocrine glands	Energy metabolism	Galactose
Glucose	Glycemic index	Hemoglobin
Hormones	Kreb's cycle	Lactose
Maltose	Monosaccharides	Monounsaturated fats
Neurons	Starch	Sterols

INTRODUCTION

An essential nutrient is one that the body needs for regular bodily functions but cannot produce on its own, or cannot produce it in sufficient quantities for optimal health. As a result, the nutrient must come from the diet. The six essential nutrients are traditionally thought to be water, vitamins, minerals, protein, fat, and carbohydrates. Consider carbohydrates to be the body's energy source. Glucose, which is produced from all carbohydrates, powers bodily functions and activities. Carbohydrates are found in all plant foods, including fruits, vegetables, grains, breads, and cereals; they are also found in dairy products like lactose; and they are found in sweets that include sucrose or table sugar. Humans can make a wide range of decisions for better health among the macronutrients—fat, protein, and carbs. For example, they can choose fruits over desserts or skim milk over full-fat dairy products.

The body's structural component is protein. All of the components are built upon a protein framework. Human skeleton is made up of a protein matrix that calcium has hardened. Protein makes up tissues and organs. Protein makes up every single component of the body, including hormones, blood cells, antibodies, neurons, and literally every single cell. Moreover, regular use wears out all of these components. Dietary protein provides the building blocks a person needs to assemble spare parts and swap them out as needed. All meats, seafood, eggs, and dairy products contain protein. Additionally, vegetables, grains, beans, nuts, seeds, and starches provide smaller amounts of protein.

With more than twice the number of calories as carbohydrates, fat is a concentrated form of dietary energy. When there was little to eat, ancient ancestors needed fat to survive. These days, eating foods high in fat can put at risk for obesity and all the long-term health issues that come with being overweight. Saturated fats, which clog blood vessels and raise the risk of heart disease, are found in meats, whole milk, cheese, ice cream, and other animal products. Omega-3 fats, which are present in salmon and tuna, as well as monounsaturated fats from nuts, seeds, avocados, and olive oil lower the risk of heart disease. Whether they are healthy or not, fats are concentrated sources of calories, so in order to maintain a healthy weight, it's critical to choose them carefully and control serving size. Walking for 60 minutes will burn the 300 calories in a small order of French fries if a person weigh 200 pounds.

Vitamins are minuscule molecules that serve vital purposes. The process of converting carbohydrates into energy involves the utilization of the B vitamins thiamin and niacin. A person wouldn't be able to utilize the carbohydrates they consumed without them. Vitamin D aids in the growth and development of cells as well as the absorption of calcium from the digestive system. Collagen, which gives blood vessels, bones, and ligaments structure, is made possible by vitamin C. Additionally tiny, minerals are listed on the periodic table in any chemistry classroom. Every single cell in the body receives oxygen from the lungs thanks to iron, a component found in red blood cells. Potassium prevents an excessive rise in blood pressure caused by an increase in sodium intake and maintains the fluid volume both inside and outside of cells. Vitamins and minerals are obtained from every food group. Whole grains are a good source of magnesium and

B vitamins. Dairy products contain calcium, potassium, and the B vitamin riboflavin. Meats contain thiamine and iron. A wide variety of vitamins and minerals can be found in abundance in fruits, vegetables, beans, nuts, and seeds. All plant-based foods (those that aren't fried) are generally regarded as being high in vitamins and minerals and low in calories, a quality known as nutrient density. This also applies to lean meats and dairy products, such as skim and low-fat varieties.

Eating a variety of foods from all the food groups is crucial for a balanced diet and to ensure a person receive all the necessary nutrients for good health. Opting for nutrient-rich, low-calorie foods and staying physically active are key elements in leading a healthy and fulfilling life.

3.1. CARBOHYDRATES

With four kilocalories per gram, carbohydrates—which are made up of carbon, hydrogen, and oxygen—are the body's main energy source. The elements hydrogen and oxygen are found in most carbohydrates in the same 2:1 ratio as in water; hence, the terms "hydrate" (for water) and "carbo" (for carbon).

3.1.1. Types

The ideal nutrient to satisfy the body's nutritional demands is a carbohydrate. They support the nervous system and brain, give all of the cells energy (as long as a person consumes them within reasonable calorie ranges), and keep the body in shape. In particular, while indigestible carbohydrates offer a good amount of fiber along with a host of other health advantages, digestible carbohydrates give food structure and absorb vitamins and minerals. Using the energy from the sun and carbon dioxide found in the air and water, plants create glucose, a fast-releasing carbohydrate. Plants use solar energy to transform it into chemical energy in the glucose molecule. Glucose is used by plants to synthesize larger, slower-releasing carbohydrates. Consuming plants allows us to obtain glucose energy, which is needed to sustain life's functions (Figure 3.1).

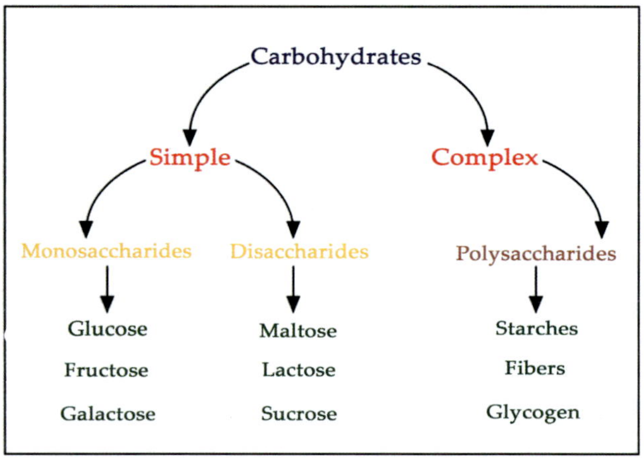

Figure 3.1. *Carbohydrate classification scheme. Carbohydrates are broken down into the subgroups of "fast-releasing" and "slow-releasing" carbohydrates. These subgroups are further categorized into mono-, di-, and polysaccharides.*

Source: https://2012books.lardbucket.org/books/an-introduction-to-nutrition/s08-carbohydrates.html.

One carbon atom, two hydrogen atoms, and one oxygen atom make up the group of organic compounds known as carbohydrates. They are hydrated carbons, basically. "Hydrate" refers to water, and "carbo" means carbon. 12 hydrogen atoms, six oxygen atoms, and six carbon atoms make up glucose, the most prevalent carbohydrate in the human body. Glucose has the chemical formula $C_6H_{12}O_6$ written down. The Greek word "saccharide," which means sugar, is synonymous with the word "carbohydrate." The monosaccharide is the most basic type of carbohydrate. "Slow-releasing" and "fast-releasing" are the two main subgroups into which carbohydrates are divided. Monosaccharides and disaccharides are two more categories into which fast-releasing carbohydrates are divided. Long chains of monosaccharides make up slowly-releasing carbohydrates.

3.1.1.1. Fast-Releasing Carbohydrates

Simpler terms for fast-releasing carbohydrates are "sugars." Monosaccharides and disaccharides are the two categories into which fast-releasing carbohydrates fall. Lactose, maltose, and sucrose are examples of disaccharides, whereas glucose, fructose, and galactose are examples of monosaccharides.

3.1.1.2. Monosaccharides

Glucose is the preferred fuel source for all organisms, from bacteria to plants to animals. The brain relies on glucose for energy, with the exception of extreme starvation conditions. Galactose, a monosaccharide similar to glucose but with a slight structural difference due to the arrangement of a hydroxyl group, is less stable than glucose. The liver quickly converts galactose to glucose because of this instability. Once converted, most of the absorbed galactose is used for energy production in cells. It is one of the simple sugars found in milk, released during the digestion process.

Although fructose and glucose share the same chemical formula, their chemical structures are different because fructose's ring structure only has five carbons instead of six. Unlike glucose, fructose does not provide energy to other body cells. One of the most prevalent monosaccharides in nature is fructose, which is mostly present in fruits, honey, and sugarcane. Moreover, it can be found in cereals, soft drinks, and other goods that have been sweetened with high fructose corn syrup. Pentoses are less common monosaccharides because they only contain five carbons rather than six. Pentoses are found in large quantities in both fiber and the nucleic acids DNA and RNA.

Finally, there are the industrially produced sugar alcohols, which are monosaccharide derivatives. Sugar alcohols include sorbitol, xylitol, and glycerol, to name a few. (Xylitol is about as sweet as regular table sugar. Since sugar alcohols are less calorie-dense due to incomplete digestion and absorption, they are frequently used to sweeten food instead of table sugar. Because the bacteria present in the mouth fight against them, sugar alcohols do not lead to tooth decay. It's interesting to note that chewing gum containing sugar alcohols causes a sensation of "coolness" because the gum dissolves in the mouth, a chemical reaction that needs heat from the inside of the mouth.

3.1.1.3. Disaccharides

Pairs of two monosaccharides joined together to form disaccharides. Lactose, maltose, and sucrose are examples of disaccharides. There is at least one glucose molecule in each disaccharide. Table sugar is another name for sucrose, which has both glucose and fructose molecules in it. In addition, sucrose can be found in a wide variety of fruits and vegetables. Sugar cane and sugar beets, which are used to make table sugar, contain large amounts of sucrose. One glucose and one galactose unit make up lactose, also referred to as milk sugar. Dairy products like milk, yogurt, and cheese often contain lactose. Two glucose molecules joined together to form maltose. It is frequently produced when plant starches break down and is infrequently present in food as a disaccharide.

3.1.1.4. Slow-Releasing Carbohydrates

Polysaccharides, which are extended chains of monosaccharides with or without branches, are slow-releasing carbohydrates. The two primary categories of polysaccharides are fibers and starches.

3.1.1.5. Starches

Grains, legumes, and root vegetables like potatoes are rich sources of starch molecules. Plant starch amylose is a linear chain made up of hundreds of glucose units. Another plant starch called amylopectin is a branched chain made up of thousands of glucose units.

These big starch molecules are what plants use to store energy in the form of crystals. Foods contain both of these starch molecules—amylose and amylopectin—but the smaller one, amylose, is more prevalent. Starchy foods are difficult for the digestive system to break down, so eating them raw gives very little energy. Starches become much easier to digest in the human body when they are cooked, as this breaks down their crystal structure. Resistant starches are those that hold their structure during the digestive process. Some of these can be broken down by gut bacteria, which may be advantageous to gastrointestinal health. In the food industry and during cooking, isolated and modified starches are frequently used as food thickeners. Animals and humans store glucose energy from starches as glycogen, a massive molecule. Its numerous branches enable it to decompose rapidly when the body's cells require energy. It is primarily present in animal muscle and liver tissue.

3.1.1.6. Dietary Fiber

The structural components of plants, or dietary fiber, are indigestible by humans due to a lack of digestive enzymes. These indigestible substances support optimal health even though they pass through the gut undigested (apart from a tiny portion that is fermented by bacteria in the large intestine).

Insoluble fiber gives roughage, or bulk, that promotes regularity in bowel movements and speeds up the body's removal of substances found in food that may be carcinogenic or otherwise harmful. It also does not dissolve in water. Cellulose, the majority of hemicelluloses, and lignin—a phenolic polymer rather than a carbohydrate—are examples of insoluble fiber types. Vegetables, wheat bran, and whole grain breads and cereals are important dietary sources of insoluble fiber. In addition to slowing down the amount of time food passes through the stomach (an undesirable effect), soluble fiber, which dissolves or swells in water, also lowers blood cholesterol levels (a desirable effect). Gums, pectins, some hemicelluloses, and mucilages are examples of soluble fiber types; fruits (particularly citrus fruits and apples), oats, barley, and legumes are important dietary sources of soluble fiber. Soluble and insoluble fiber contribute to the slowing down of glucose absorption, resulting in a more gradual and consistent release of blood glucose. It is believed that dietary fiber can protect against certain gastrointestinal disorders and lower the chance of developing other chronic illnesses.

3.1.2. Digestion, Absorption and Metabolism of Carbohydrates

The primary source of energy comes from carbohydrates, which are vital biomolecules. The majority of the body's daily energy needs are met by their consumption. They contain a lot of calories. But their consumption alone won't supply energy. Carbs must be properly ingested, absorbed, and metabolized to be converted into energy. It is important to remember that carbs are absorbed in the small intestine after beginning their digestion in the mouth. Based on how many saccharide units they contain, carbohydrates can be classified as simple or complex.

3.1.2.1. Digestion of Carbohydrates

Humans chew food as it is placed in their mouths. Mastication, also known as "chewing," is the process by which complex carbohydrates are broken down into simpler ones. Saliva (or salivary amylase) is secreted by salivary glands located in the mouth. Saliva breaks down starch, which is a complex carbohydrate, into simpler polysaccharides and maltose. In the mouth, maltose concentration rises. This explains why they perceive food as sweet. Food has a sweet taste because of the sugar maltose. Not all food, though, is digested in the mouth. Merely 5% are decomposed. It is beneficial because tooth decay can result from excessive breakdown of carbohydrates.

Because salivary amylase is not active in the stomach, digestion rarely occurs there. It travels to the duodenum, the first segment of the small intestine, from the stomach. Pancreatic amylase, which the pancreas secretes, hydrolyzes the polysaccharide's α-1,4 bonds. Nevertheless, it is unable to degrade terminal bonds or α-1,6 bonds. Currently, the villi are lined with enzymes called disaccharidase that are secreted by the small intestinal wall. The release of enzymes such as lactase, lactase hydrolyzes carbohydrates into monosaccharides. The small intestine absorbs these monosaccharides. Anything that is not absorbed and digested, like fiber, makes its way to the colon where the excretory processes get rid of it. It is known as excretion.

3.1.2.2. Absorption of Carbohydrates

After digestion, absorption comes next. Certain cells in the small intestine have transport proteins or transporters inside of them, which aid in transferring the carbohydrates into the bloodstream. The glucose transporter (GLUT5) transport protein aids in the absorption of fructose, while the sodium-glucose cotransporter (SGLT-1) transporter is in charge of absorbing glucose. The liver is the first organ in the body to receive these digested carbohydrates. Glucose is produced in the liver from galactose. The already-simple sugar fructose is further broken down into carbon units. Glycogen is a type of stored glucose. The blood receives extra glucose back. It is crucial to keep the body's glucose levels at the proper levels. A glucose level that is too low could cause issues like fatigue and difficulty concentrating, while a glucose level that is too high could have more dangerous consequences like diabetes. Fortunately for us, the body naturally uses a negative feedback mechanism to control glucose levels.

The human body's glucose level abruptly rises after eating a meal. In these situations, insulin is secreted by the pancreatic beta cells, and its purpose is to alert the body's cells to remove excess glucose from the blood. After that, extra glucose is either used to create macromolecules or stored in the cells. Glucagon, the hormone that opposes insulin, is released when the body's blood sugar levels fall. This is what occurs when a person engage in strenuous physical labor. It instructs the liver to release glucose into the bloodstream by breaking down glycogen and instructing the cells to stop storing glucose.

3.1.2.3. Carbohydrate Metabolism

The process through which body cells receive a steady supply of energy is known as carbohydrate metabolism. Carbohydrate metabolism is the process of breaking down glucose into smaller molecules through a process called glycolysis glucose joins Kreb's cycle, sometimes referred to as the citric acid cycle, or CAC. It happens within the mitochondria, the cell's power plant.

During the process of glycolysis, glucose is eventually reduced to two molecules of pyruvate (converting glucose to glucose-6-phosphate first, then fructose-6-phosphate, and finally pyruvate). There is a connection between the metabolism of carbohydrates and respiration. Humans breathe in oxygen from the surrounding air and exhale carbon dioxide when a person breathes. When glucose is converted into energy, oxygen is inhaled. Glycogen can occasionally be converted straight from glycogen to glucose-6-phosphate (without first having to be converted to glucose). Glycogenolysis is the direct conversion of glycogen to glucose-6-phosphate.

3.1.2.4. Glycemic Index

The glycemic index (GI) is a numerical value that is derived from measuring and ranking the glycemic responses of different foods relative to a reference food, which is typically a 50 g slice of white bread or just straight glucose. Foods with a lower GI don't cause blood sugar levels to rise as quickly as those with a higher GI. Studies involving epidemiology and clinical trials have demonstrated that a low-GI diet can promote weight loss and lower the risk of obesity, Type 2 diabetes, and cardiovascular disease (Brand-Miller et al., 2009).

Table 3.1. The Glycemic Index (GI): Foods in Comparison to Glucose: Representative Foods Low in GI (<55) and GL (<10)

Food	GI	GL	Food	GI	GL
Apples	40	6	Barley	28	13
Oranges	40	4	Rice noodles	54	22
Cherries	20	5	Full-fat milk	27	3
Raspberries	32	3	Low-fat milk	32	4
Grapefruits	25	3	Soy milk	55	6
Kiwis	53	9	Skim milk	32	4
Dates	54	21	Yogurt	15	1
Pear	33	3	Chocolate	43	7
Apricot	34	3	Fructose	23	2
Apple juice	40	12	Peanuts	14	1
Orange juice	55	14	Walnuts	20	1
Chickpeas	28	8	Almonds	10	<1
Kidney beans	28	7	Pecans	10	<1
Lentils	30	11	Hazelnuts	15	<1
Soybeans	16	9	Eggplant	15	2
Mushrooms	10	1	Celery	15	1
Zucchini	15	1	Spinach	15	1
Asparagus	32	2	Carrots	39	2
Cucumber	15	0	Tomatoes	15	1
Onion	15	<1	Lettuce	15	1

Source: Mavroeidi I, Manta A, Asimakopoulou A, Syrigos A, Paschou SA, Vlachaki E, Nastos C, Kalantaridou S, Peppa M. the role of the glycemic index and glycemic load in the dietary approach of gestational diabetes mellitus. Nutrients. 2024; 16(3):399. https://doi.org/10.3390/nu16030399.

A food's GI is influenced by the type of carbohydrates it contains, but it is also lowered by the amount of fat and fiber it contains. Foods higher in fat and fiber take longer to digest and take longer for the stomach to empty into the small intestine. A food's GI is also impacted by cooking and processing because these methods make food more digestible. The United States' high demand for convenient, precooked foods combined with advancements in food processing technologies have produced foods that, regardless of their fiber content, digest and absorb more quickly. Breads, pastas, and many prepared foods, as well as modern breakfast cereals, have a high GI. The majority of raw foods, however, have a lower GI. Nonetheless, a fruit or vegetables GI increases with ripeness. Table 3.1 lists the GI values for a variety of foods. Although it has certain limitations, the GI can be used as a guide to help make better carbohydrate choices. One is that the GI only considers the kind of carbohydrates, not their quantity, in a serving of food. Another is that the GI of the meal is altered when low- and high-GI foods are combined. Additionally, some foods high in nutrients have higher GIs than foods low in nutrients. For example, because chocolate contains more fat than oatmeal, oatmeal has a higher GI than chocolate. Lastly, since they don't contain carbs, meats and fats have no GI.

3.1.3. The Functions of Carbohydrates in the Body

Carbohydrates serve five main functions in the human body: producing energy, storing energy, building macromolecules, preserving protein, and aiding in lipid metabolism.

3.1.3.1. Energy Production

Carbohydrates are essential for providing energy to all of the body's cells. When it comes to energy sources, many cells favor glucose over other substances like fatty acids. Certain cells, like red blood cells, can only use glucose as a source of cellular energy. The brain exclusively requires glucose for energy production and proper operation, making it extremely vulnerable to low blood-glucose levels (unless it is experiencing severe starvation). The liver redistributes roughly 70% of the glucose that enters the body during digestion so that other tissues can use it. Because they have a transport protein in their membranes, energy-dependent cells take out glucose from the circulation. The carbon atoms' chemical bonds with one another provide glucose with its energy. During the process of photosynthesis, sunlight energy was needed to create these high-energy bonds. Human cells dissociate these bonds and absorb energy to carry out cellular respiration. In essence, cellular respiration is the difference between an uncontrolled and regulated burning of glucose. To more effectively capture the energy contained within the chemical bonds in glucose and to slow down the release of energy (no explosion), a cell employs numerous chemical reactions in multiple enzymatic steps.

Glycolysis is the initial phase of the breakdown of glucose. The process of splitting glucose, or glycolysis, involves 10 complex enzymatic reactions. The energy-producing organelles known as mitochondria are where the second stage of glucose breakdown takes place. More energy is produced by removing two oxygen atoms and one carbon atom. This carbon bond-derived energy is transferred to a different region of the mitochondria, where it is converted into a form that can be utilized by the cells.

3.1.3.2. Energy Storage

The extra glucose is stored as glycogen if the body already has enough energy to sustain its operations (the majority of which is stored in the muscle and liver). When glucose is required to produce cellular energy, it can spread quickly through the highly branched structure of glycogen, which can contain more than 50,000 single glucose units. The body contains roughly 4,000 kilocalories of glycogen at any given time, of which 3,000 are found in muscle tissue and 1,000 in the liver.

The glycogen energy reserve can be exhausted by prolonged muscle use, such as prolonged exercise lasting more than a few hours. Because it takes longer than glucose to convert the chemical energy in fatty acids and proteins into usable energy, muscles begin to weaken. Glycogen is depleted after extended exercise, forcing muscles to rely more on lipids and proteins for energy. Three to five days before an event, athletes can slightly increase their glycogen reserve by decreasing training intensity and increasing their carbohydrate intake to 60–70% of total calories. It is not necessary for someone who is not training hard to eat a large plate of pasta right before a 5-kilometer race because increased muscle glycogen will not be adapted without long-term, intense training.

Similar to muscle tissue, the liver can store glucose energy as glycogen; however, when blood glucose levels are low, the liver will give up its stored glucose energy to other body tissues. The liver contains about 25% of the body's total glycogen content, which is comparable to a four-hour supply of glucose, but this is highly dependent on one's level of activity. The glycogen reserve is utilized by the liver to maintain blood glucose levels within a specific range in between meals. To preserve metabolic homeostasis, glucose is produced from amino acids obtained from the breakdown of proteins when the liver's supply of glycogen is depleted.

3.1.3.3. Building Macromolecules

While the majority of absorbed glucose is used to produce energy, some glucose is also converted to the essential building blocks of important macromolecules like RNA, DNA, and ATP: ribose and deoxyribose. Moreover, glucose is needed to produce the molecule NADPH, which is critical for defense against oxidative stress and involved in a variety of other bodily chemical reactions. Excess glucose can be used to make fat if the body needs it for energy, building, and glycogen storage. This is the reason why a diet excessively rich in calories and carbs can contribute to weight gain—a subject that will be covered in a moment.

3.1.3.4. Sparing Protein

When the body doesn't have enough glucose to meet its needs, amino acids are used to synthesize glucose. Since amino acids do not have a storage molecule, this process necessitates the breakdown of proteins, mostly found in muscle tissue. When there is enough glucose available, the body doesn't need to break down proteins to produce the necessary amount of glucose.

3.1.3.5. Lipid Metabolism

When blood-glucose levels increase, the body's utilization of lipids for energy is suppressed. This means that glucose has a "fat-sparing" effect. The rise in blood glucose triggers the release of insulin, which instructs cells to use glucose (instead of lipids) for energy production. Maintaining adequate glucose levels in the blood helps to prevent the onset of ketosis, a metabolic condition caused by elevated levels of ketone bodies in the blood. Ketone bodies serve as an alternative energy source for cells when glucose is scarce, such as during fasting. Excessive ketone levels can lead to acidosis in the blood. Although rare in healthy individuals, acidosis can occur in alcoholics, malnourished individuals, and those with Type 1 diabetes. Adults require a minimum of 50 grams of carbohydrates per day in their diet to prevent the onset of ketosis.

The synthesis of energy, which is life's most fundamental function, depends on carbohydrates. None of the other biological processes could exist without energy. Human bodies can produce glucose, but doing so results in the breakdown of proteins. Carbohydrates should be taken in moderation, just like all other nutrients, as eating too much or too little of them can have negative health effects.

3.1.4. Food Rich in Carbohydrates

Below are top carbohydrate-rich foods one should consider incorporating into their diet:

- **Quinoa:** Quinoa is considered one of the most nutritious foods on the planet, as its seeds are classified as a pseudocereal and can be consumed as a grain. Quinoa is high in carbohydrates (with around 70% in cooked form) and is also a great source of fiber.
- **Oats:** Oats are a well-known fine-grain food that is high in carbohydrates and also contains vitamins and antioxidants. One cup of raw oats provides 54 grams of carbohydrates and a negligible amount of fiber. Oats are 70% carbohydrate. Studies have indicated that eating oats regularly may lower blood cholesterol levels, which may reduce the risk of heart disease. Additionally, oats are an effective way to lower blood glucose levels.
- **Buckwheat:** Similar to quinoa, buckwheat is regarded as a pseudocereal; however, it is unrelated to the wheat family and contains no gluten. Conversely, it is low in protein and high in fiber and nutrients. Carbohydrate-rich foods can control blood sugar levels and are beneficial to heart health.
- **Brown Rice:** Brown rice is a whole-grain food packed with carbohydrates. A single cup or 195 grams of cooked brown rice contains around 45 grams of carbs, providing a substantial source of energy.
- **Kidney Beans:** Popular as a legume, kidney beans offer an abundance of carbohydrates. About 45 grams of carbohydrates are present in 177 grams of cooked kidney beans. These beans' fiber improves gut health, aids in digestion, controls blood sugar, and may even serve as a viable protein alternative to animal meat.
- **Bananas:** A large banana (130 g) contains 31 g of carbohydrates in the form of starch or natural sugar. Along with potassium, vitamin B6, and vitamin C, this food is high in carbohydrates. In addition to supporting the digestive system, resistant starch, which is mostly present in green bananas, has the ability to lower blood pressure. Pectin enhances the functions of the digestive system by feeding the beneficial bacteria in the gut.
- **Sweet Potatoes:** Sweet potatoes are a well-liked and nutrient-dense root vegetable. 20 grams of carbohydrates, a small amount of natural sugar, starch, and a good amount of fiber are found in 100

grams of mashed sweet potatoes. Sweet potatoes are well-known for being an excellent source of antioxidants and vitamins, particularly B and C. Carbohydrate-containing foods contain antioxidants that help neutralize deadly radicals in body cells.

- **Beets:** As root vegetables, beets have 10 grams of carbohydrates per 100 grams. The healthiest foods for carbohydrates are undoubtedly rich in minerals, vitamins B and C, potent antioxidants, and plant-based compounds like beetroot (Brand-Miller et al., 2009).

3.2. FATS AND OTHER LIPIDS

A class of organic substances that are insoluble in water is referred to as lipids. One kind of lipid that occurs naturally as a triglyceride in oils and fats is dietary fat. Triglycerides come in various forms in most foods, including saturated, monounsaturated, and polyunsaturated ones. For instance, butter contains 34% unsaturated fat and 66% saturated fat. Sterols, cholesterol, and phospholipids are examples of additional lipids. Fat is broken down in the small intestine. The liver and gall bladder produce bile, which emulsifies fat and facilitates digestion. The pancreatic enzyme lipase converts triglycerides into monoglycerides and two fatty acids, which the small intestine's lining cells passively absorb. A chain of carbon atoms, some joined by single bonds and some by double bonds, makes up a fatty acid. Saturated fatty acids contain only one bond; unsaturated fatty acids contain two bonds. The level of saturation affects one's health. While monounsaturated and polyunsaturated fatty acids are linked to a lower risk of heart disease, saturated fatty acids are associated with an increased risk. The body can't produce two essential fatty acids, so a person has to get them through their diet. They are found in vegetable oils and are linoleic and alpha-linolenic acid. Triglycerides assemble into lipoproteins to travel through the bloodstream. An elevated level of low-density lipoprotein (LDL) has been linked to a higher risk of heart disease. On the other hand, there is a link between a lower risk of heart disease and having an adequate level of high-density lipoprotein (HDL). At room temperature, a fat that is solid is called a fat; an oil is a liquid. Fat contributes significantly to daily caloric intake, containing nine calories per gram. When hydrogenation of polyunsaturated fatty acids occurs during food manufacturing, trans fats are produced. Trans fat-containing diets have been linked to negative health outcomes. Even though food manufacturers have made efforts to eliminate trans fats from processed foods, it's crucial to check the trans fat content by reading the nutrition facts label. Reducing overall fat consumption is the best course of action. Even though fat is a vital nutrient, it should be consumed in moderation.

There are numerous roles that lipids play in the body.

- **Store Energy:** The body stores excess energy as adipose tissue, or fatty tissue, commonly referred to as fat. Lipids and carbohydrates supply the majority of the energy that the human body needs. Glucose is stored in the body as glycogen, as was covered in the unit on carbohydrates. Although glycogen is an easily accessible energy source, the body cannot store large amounts of it for extended periods due to its large bulk and high water content. Conversely, fats have the potential to function as a more substantial and enduring energy store. Without water, fats clump together tightly and can store much more energy in a smaller area. A gram of fat has more than twice as much energy as a gram of carbohydrate, making it highly concentrated.

When we're at rest or need to refuel the muscles for movement throughout the day, such as walking to class, playing with kids, dancing through dinner prep, or working a shift, a person uses the energy stored in fat. Storing energy as fat was essential to human survival during periods of famine in the past, when people relied on hunting and gathering wild foods or the success of crops. People still struggle with hunger throughout the world, and being able to store energy during good times can help them get through a period of food insecurity. In other situations, a person may be able to endure a protracted illness thanks to the energy stored in adipose tissue.

Fat cells are specialized for storing fat and have an almost infinite capacity to grow in size, in contrast to other body cells that have a limited capacity to do so. An excess of adipose tissue can be harmful to health because it puts the body under mechanical stress from weight gain as well as hormonal and metabolic changes. Numerous illnesses, such as type 2 diabetes, heart disease, stroke, kidney disease, and some types of cancer, can be made more likely by obesity. Moreover, it may disrupt mood, cognition, and reproduction. Therefore, while a certain amount of body fat is necessary for survival and general well-being, excessive amounts of it may be detrimental.

- **Regulate and Signal:** The human body's temperature is regulated by lipids to keep it constant—neither too hot nor too cold. Additionally, lipids aid in the body's production and control of hormones related to blood clotting, the reproductive system, and appetite. Lipids are essential to the structure and operation of the brain because they help transmit signals within the brain, form the membranes of nerve cells, and insulate neurons, which are the cables that carry messages throughout the body.
- **Insulate and Protect:** Fat cushions bodies, shielding from the daily grind. Men typically have 18–24% body fat, while women typically have 25–31% body fat. Even so, a person's degree of obesity can determine how much more of their body weight is made up of adipose tissue. Visceral fat is the type of fat that is stored inside the abdominal cavity, and subcutaneous fat is the type of fat that is stored directly beneath the skin. Visceral fat shields important organs, including the liver, kidneys, and heart. The subcutaneous fat layer acts as a blanket to protect the body from high and low temperatures and to regulate body temperature. Since hands and buttocks frequently come into contact with hard surfaces, it cushions these areas and reduces friction. Additionally, it provides the body with the extra cushioning needed for physically taxing activities like snowboarding, ice skating, and horseback riding. Adipose tissue stores two different types of fat: visceral fat and subcutaneous fat.
- **Aid Digestion and Increase Bioavailability**: Fat is released during the breakdown of food in the stomach and mixes with nutrients that are soluble in fat. The nutrients can be absorbed and digested more readily by the body when they are combined with fat. This

enhanced absorption is referred to as increased bioavailability. The bioavailability of substances referred to as phytochemicals—non-essential plant compounds thought to be advantageous to human health—can also be increased by dietary fats. Since many phytochemicals, like beta-carotene in carrots and lycopene in tomatoes, are fat-soluble, eating fat enhances their absorption in the digestive system.

Not only do fat-soluble vitamins increase in bioavailability, but foods high in fat are also among the best dietary sources of these vitamins. Nuts (including peanut butter and other nut butter), seeds, and plant oils—like those in salad dressings—are a few good sources of vitamin E. Eating a very low-fat diet makes it difficult to get enough vitamin E in the diet.

Even though vegetable oils are typically used to cook fried foods, high heat destroys vitamin E, so onion rings and French fries don't contain much of it. Whole foods with little to no processing are the best options. The four fat-soluble vitamins are A, D, K, and E. They can be found in a variety of foods like butternut squash, broccoli, and salmon. Vegetable oils also contain some vitamin K, and fatty fish and eggs are good sources of vitamins A and D. Consuming dietary fat, such as walnuts and olive oil dressing in a kale salad, aids in the absorption of these fat-soluble vitamins.

3.2.1. Types of Lipids and Where They Are Found

Phospholipids, sterols, and triacylglycerols—also referred to as triglycerides—are the three primary categories of lipids.

3.2.1.1. Phospholipids

Phospholipids form the outer layer of cells in animal and human bodies, creating a protective barrier around the cells. While most people do not need to be concerned about phospholipids, individuals with the rare autoimmune disorder known as antiphospholipid syndrome (APS) may experience problems. This condition, commonly referred to as sticky blood, often occurs in individuals with lupus, particularly women. APS causes proteins on phospholipids to be attacked, leading to damage to the protective layers around the cells. This can result in an increased risk of blood clots, potentially leading to heart attacks and strokes in affected individuals. Additionally, APS may also result in complications during pregnancy.

3.2.1.2. Sterols

One type of steroid is a sterol. A class of hormones produced by the body from fats is called steroids. It's possible that one has heard of phytosterols or plant sterols. These compounds are present in food and have a number of health advantages. Similar to cholesterol, which is the primary sterol in humans, are phytosterols. About 20% of the cholesterol in bodies is derived from diet, with the majority being produced by the liver and intestines. Every cell in the body contains cholesterol, which is essential for the synthesis of hormones and vitamin D. Bile salts, which aid in the body's absorption of vitamins and aid in the breakdown of fat, are also made largely from cholesterol (Macierzanka et al., 2019).

Because cholesterol is a fat and not soluble in water, it needs protein to be transported through the bloodstream. Lipoproteins are lipid and protein-based particles that carry cholesterol and triglycerides through the bloodstream. Although lipoproteins come in a variety of forms, high-density lipoprotein (HDL) and low-density lipoprotein (LDL) are the two main varieties. Because HDL absorbs cholesterol and returns it to the liver, it is referred to as "good" cholesterol. LDL is referred to as "bad" cholesterol due to its propensity to accumulate in blood vessels.

3.2.1.3. Triglycerides

The most prevalent form of lipid in bodies, triglycerides are derived from the fats and oils a person eats. A person needs triglycerides because they provide them with energy. Eat saturated or unsaturated fats as lipids in the diet. At room temperature, saturated fats remain solid and are commonly found in animal-based foods like cheese, butter, and meat. Unsaturated fats are typically liquid at room temperature and come from plants, though they don't always. Although fat and lipids are sometimes used synonymously, fat (also known as fatty acid) is just one kind of lipid.

Nuts, seeds, avocados, tuna, salmon, and leafy vegetables are examples of foods high in unsaturated fats, which are also known as omega-3 fatty acids. These unsaturated fatty acids have the potential to lower blood pressure, triglycerides, and inflammation in the body.

They also lessen the chance of a heart attack ending suddenly and stop blood clots from forming. Trans fats, also known as partially hydrogenated oils, are manufactured artificially to give processed foods a specific texture. They also extend baked goods' shelf life. Consuming foods high in trans fats can raise LDL (bad) cholesterol levels while lowering HDL (good) cholesterol levels. What foods, then, cause elevated levels of triglycerides? In reality, the excess calories a person takes in from food and beverages—particularly from alcohol and carbohydrates—are what cause the body to store excess fat, or adipose tissue, as triglycerides. If a person doesn't burn off the excess calories they're consuming, they may eventually experience high triglycerides and fat storage, which can result in several metabolic disorders like obesity, diabetes, insulin resistance, and even acute pancreatitis (Yuan, Al-Shali, Hegele, 2007).

3.2.2. Digestion

Because (a) fats and other lipids are insoluble in water and (b) the solubility of lipolytic enzymes in aqueous media is limited, the digestion of fats and other lipids presents unique challenges. But emulsification of fats solves this issue in the gut, especially by phospholipids and bile salts, which are found in bile. The dietary lipid intended for digestion has a significantly larger surface area as a result of this emulsification. Therefore, the action of bile salts significantly increases the fat's accessibility to digestive enzymes.

The lipid component of the Indian diet is composed of phospholipids, sterols (primarily cholesterol), and triacylglycerols (TAG). Triacylglycerols, also referred to as fats or triglycerides, are among them. The average daily consumption rate of TAG is approximately 150 g. In contrast, the daily intake of cholesterol is normally between 300 and 600 mg, contingent upon the amount of animal products consumed. Esterases are digestive enzymes that aid in the gastrointestinal tract's breakdown of dietary lipids. They break the ester bonds in phospholipids (phospholipases), triacylglycerols (lipase), and cholesteryl esters (cholesterol esterase).

3.2.2.1. Triacylglycerol Digestion

In the small intestine's lumen, the majority of dietary triacylglycerol digestion is finished (SI). The process starts in the stomach where the main stomach cells produce gastric lipase and the serous gland releases lingual lipase. These lipases secrete basally continuously. These are triggered by dietary, mechanical, and neurological factors. The majority of the limited digestion (10%–30%) of TAG that takes place in the stomach is caused by these lipases. The enzymes at the low pH of the gastric juices enable the lipase activity. Medium- and short-chain fatty acid-containing triacylglycerols are the target of both gastric and lingual lipases. They hydrolyze fatty acids, yielding 1,2-diacylglycerols and a fatty acid as byproducts.

Lissification, gastric lipases, and lingual hydrolyze dietary fat in the stomach. Shearing forces and muscular contractions squirt the fat's surface, causing emulsification. In addition to the muscle contraction, phospholipids, complex polysaccharides, and peptic digests of dietary proteins are some of the emulsifiers in the acid medium. The stomach's natural emptying rate is slowed down when there is undigested fat present. Because of this inhibition of stomach motility, fats have a "high satiety value."

The small intestine is where most TAG digestion takes place. Long-chain fatty acid hydrolysis requires less acidity, the right lipases, and more potent emulsifiers (bile salts). The upper small intestine lumen provides these conditions. Fine lipid droplets made of partially hydrolyzed lipid emulsion exit the stomach and enter the duodenum. Continued mechanical shearing and bile lead to effective emulsification. Bile salts and products of triacylglycerol breakdown function as emulsifying agents. When pancreatic lipase breaks down ingested triacylglycerols, it produces a complex mixture of free fatty acids or FFAs, monoacylglycerols, and diacylglycerols. Thus, triacylglycerols → 2,3-diacylglycerols → 2-monoacylglycerols is the main pathway of this digestion. Merely a minor portion of the triacylglycerols undergo complete hydrolysis, releasing glycerol.

3.2.2.2. Cholesterol and Phospholipid Digestion

A particular process hydrolyzes phospholipids and cholesterol esters. A fatty acid and cholesterol are released during the hydrolysis of esterified cholesterol. The cholesterol esterase enzyme is the catalyst for this reaction. The products of the partial digestion of lipids, mainly 2-monoacylglycerols, lysolecithin, cholesterol, and fatty acids, combine with bile salts to form micelles. Phospholipase, a specific esterase, hydrolyzes lecithin to produce lysolecithin and a FFA.

3.2.3. Absorption and Utilization of Lipids

The bloodstream absorbs small molecules of broken-down triglycerides, such as glycerol and short- and medium-chain fatty acids, straight away. They travel straight to the liver after binding with albumin. Larger molecules combine to form spherical complexes called micelles, as was previously mentioned. The intestinal cells absorb the lipids from the micelles through diffusion. Micelles are inserted into chylomicrons, which are transport vehicles, within the intestinal cells. Chylomicrons are subsequently released into the lymphatic system by the intestinal cells. The chylomicrons are released into the bloodstream by the lymphatic circulation. Lipids are carried by the blood to the other parts of the body, where they are absorbed and used by cells as fuel. This process of breaking down fat to produce energy is called "lipolysis." The majority of lipids travel to the liver through the lymphatic system, where they combine with proteins (triglycerides, cholesterol) to form lipoproteins. There are four different kinds of lipoproteins, as seen in the classification. They are HDL, LDL, VLDL, and chylomicrons. Lipids are transported and deposited from the intestine and liver to the tissues for absorption by chylomicrons, VLDL, and LDL. LDL has the largest percentage of cholesterol. LDL is known as "bad" cholesterol because it promotes lipid deposition in tissues, including blood vessels. The lipids are taken out of the tissues by HDL cholesterol, which then returns them to the liver for elimination. It is therefore referred to as "good cholesterol." An increased risk of cardiovascular disease is indicated by high LDL cholesterol levels.

3.2.4. Food Sources of Lipids

The two main types of dietary lipids are fats (solid) and oils (liquid). Canola, corn, olive, peanut, safflower, soy, and sunflower oils are among the oils that are frequently consumed. Nuts, seeds, avocados, peanut butter, olives, salad dressing, and certain types of fish are foods high in oils. Meat from animals, dairy products, and cocoa butter all contain fats.

3.2.5. Diet Recommendations

Fat is a primary source of energy for infants, with an adequate intake recommended for different age groups. For infants from 0 to 6 months, the recommended intake is 31 g/day based on exclusively breast-fed infants. For older infants (7–12 months), the recommended intake is 30 g/day based on human milk and complementary foods. For adults, there is no established requirement for total fat intake, with recommendations falling within the acceptable macronutrient distribution range (AMDR) of 20–35% of energy. Saturated fatty acids (SFAs) and monounsaturated fatty acids (MUFAs) are not considered essential, therefore there are no specific recommendations for their intake across the life cycle. There are also no upper limits established for SFAs, MUFAs, or polyunsaturated fatty acids (PUFAs) at this time. Trans fatty acids do not provide any health benefits, therefore there is no dietary reference intake (DRI) established for them. The current recommendation is to consume as little trans fatty acids as possible in the diet.

Because the body cannot produce n-3 and n-6 PUFA on its own, unlike SFAs or MUFAs, dietary intake is necessary. AI has been set for n–6 and n–3 fatty acids, corresponding to linoleic and α-linolenic acid, the main forms in the diet. The AI (19 to >70 y) for men and women is 1.6 g/d and 1.1 g/d for α-linolenic acid and 17 g/d and 12 g/d for linoleic acid, respectively. Both of these fatty acids have an increase in AI during lactation and pregnancy.

While there exists some proof that linoleic and α-linolenic acid play a functional role in the body, it is widely acknowledged that the 20-carbon and longer elongated and desaturated PUFAs are the metabolically essential forms of these PUFAs. The energy content of n–6 (linoleic acid) has an AMDR of 5–10%, whereas n–3 has an AMDR of 0–11.2%. Of this, ≤10% is absorbed as eicosapentaenoic acid and/or docosahexaenoic acid. The following advice is provided by the 2015–2020 Dietary Guidelines for Americans: stay away from trans-fat, keep SFA intake to less than 10% of daily energy (calories), and swap out SFAs for healthier MUFAs and PUFAs.

3.2.6. Food and Clinical Product Sources

North Americans eat a lot of fat, and there are many different types of fatty acids in all food sources. Deep-fried and bakery goods, as well as animal fats, are the main sources of SFAs. More than 90% of the MUFAs in food come from oleic acid, which is found in olive oil, animal fats (which contain 40% of MUFAs), and oils from the more recent high MUFA cultivars of canola, soy, and sunflower. PUFAs are mostly found in nuts, plant and seed oils, and food items prepared with these. The main sources of long-chain omega-3 PUFAs in the diet are seafood and other marine foods. Plant oils that have undergone partial hydrogenation are the primary source of trans fats in the diet. Meat and milk from ruminant animals account for a minor fraction of trans fats. Conjugated linoleic acid, the main fatty acid derived from ruminants, is not included in the calculation of trans fat on food labels because the detrimental health effects of rumen-derived trans fatty acids have not been proven. For people who are unable to eat enough food to meet their energy needs, lipids, mainly TGs, makeup 20–40% of the energy in enteral products. In the context of clinical practice, lipids serve as the primary energy source for parenteral (intravenous) feeding of patients unable to swallow food.

3.2.7. Deficiencies

A person will enter a negative energy balance, lose weight, and/or not grow if they consume insufficient amounts of dietary fat, along with insufficient amounts of protein and carbohydrates. A scaly rash, poor growth, and neurological abnormalities are examples of classic deficiencies of n–3 and/or n–6 PUFAs. Only those suffering from severe malnutrition/starvation and/or chronic fat malabsorption, such as those with cystic fibrosis, pancreatic insufficiencies, or damaged intestines, are at this level of deficiency.

3.2.8. Toxicity/Health Risks

People are capable of adjusting to a variety of fat intakes. A higher risk of obesity, type 2 diabetes, coronary heart disease, and cancer has been linked to higher intake of fat, particularly saturated fat (SFA).

Some of these worries regarding high fat and high SFA intakes and disease risk, however, are being called into question by new research, which is being spurred by the popularity of low-carb, high-fat diets. Significantly, it has been shown that an imbalanced or low intake of omega-3 and omega-6 fatty acids is linked to a higher risk of developing several chronic illnesses, including cancer and cardiovascular disease (CVD).

A class of naturally occurring substances known as "lipids" are defined by their solubility in "fat solvents" and insoluble in water. The three components of fats are carbon, hydrogen, and oxygen, just like in carbohydrates. An essential part of Indian cuisine is fats and oils. They serve a variety of purposes in baked goods, including as leavening and shortening agents. Foods with fats have better flavor, taste, and palatability.

There are three categories for lipids: derived, compound, and simple lipids. Because lipids are insoluble in water, digestion of lipids presents unique challenges. Bile salts and phospholipids can emulsify fats to solve the solubility problem. The digestive enzymes esterases, phospholipases, and cholesterol esterase are involved in the digestion of dietary lipids in the gastrointestinal tract. It first produces diacylglycerols and subsequently monoacylglycerols during the digestion of lipids. SCFAs enter the bloodstream right away after absorption. Diffuse into the intestinal cells are micelles produced through partial digestion of lipids. Micelles aid in the absorption of lysolecithin, cholesterol, 2-monoacylglycerols, FFA, and cholesterol esters, among other digested lipid contents. Lipoproteins in the blood carry lipids.

3.3. PROTEINS AND AMINO ACIDS

In an attempt to reduce their waist sizes, some dieters make protein bars a staple of their diets. Many customers of exercise cafes order protein shakes to aid in muscle growth and improve post-workout recuperation. Some people no longer eat meat and feel that taking protein supplements is necessary to make sure they are getting the daily recommended amount of protein. After all, protein is an essential component of every organ in the body and is needed for the synthesis of numerous molecules, including hormones and enzymes. It makes sense that maximizing dietary protein intake is a major concern for so many people. Dieters, athletes, physically active people, and vegetarians may be concerned that their diets aren't high enough in protein and that, to perform better and maintain optimal health, they should consume more protein in the form of shakes, protein bars, or supplements. Let's start by examining vegetarian diets.

While there are many varieties of vegetarians, all of them abstain from eating meat. There are four popular types of vegetarianism:

1. Lacto-ovo vegetarian. This is the most typical version. This kind of vegetarian consumes dairy and eggs.
2. Lacto-vegetarian. This kind of vegetarian consumes dairy products but abstains from eggs.
3. Ovo-vegetarian. This kind of vegetarian consumes eggs but abstains from dairy.
4. Vegan. This kind of vegetarian avoids all animal products, including dairy and eggs.

A vegetarian diet can be chosen for a variety of reasons, such as a dislike of meat, health issues, ecological and animal welfare concerns, or religious beliefs. For centuries, people have been practicing vegetarianism. Prominent intellectuals like Pythagoras and Plato encouraged vegetarian diets in their natural philosophies during the fourth century BC. One month before the Olympic Games, the ancient Olympians were ordered to eat only vegetarianism.

Strong men known as gladiators engaged in bloody combat in gladiatorial contests supported by the Roman Empire's aristocratic elite. Not far from the Temple of Artemis, archaeologists discovered a gladiator burial ground in 1993. They were shocked to learn that the gladiators ate a vegetarian diet after forensic examination of their bones. This data is consistent with other historical narratives describing the diet of gladiators as being high in barley and dried fruits. Hulled barley is an incredibly nutrient-dense whole grain that offers over 20 grams of protein and all nine essential amino acids in a single cup serving, making it a complete protein source.

The complete nutrient package of the food is crucial to take into account when deciding whether to consume protein from plant- or animal-derived sources. The protein source may include additional fats, nutrients, additives, or preservatives. Red meat is a common source of protein, but it is high in saturated fat. Another excellent source of protein is fish, which also has a higher content of healthy fats and substantially less saturated fat than other meats. Certain plant-based protein sources have good levels of healthy fats along with high protein content—less than one gram of less-than-ideal fat per serving.

3.3.1. Amino Acids

When proteins, like albumin in egg whites, casein in dairy products, and gluten in wheat, are broken down into their constituent amino acids during digestion, the resulting amino acids are absorbed and added to the body's metabolic pool. Peptide linkages are then used to connect amino acids to form distinct proteins, which are assembled following the genetic code and the body's current demands. Every gene produces one or more proteins, each of which has a distinct amino acid sequence and exact three-dimensional arrangement. Other significant nonprotein substances like creatine, certain neurotransmitters, and peptide hormones require amino acids to be synthesized.

There are about 20 common amino acids found in food, nine of which are regarded as indispensable, meaning that humans cannot synthesize them or not in sufficient amounts, so they must be obtained through diet. Histidine, isoleucine, leucine, lysine, methionine, phenylalanine, threonine, tryptophan, and valine are the amino acids that are necessary for human health. Arginine, cysteine, and tyrosine are conditionally indispensable amino acids that may require special provision due to impaired conversion from precursors, as in the case of premature infants or individuals with liver disease.

Different foods have different relative amounts of the different amino acids. Meat, fish, eggs, and dairy products are examples of foods of animal origin that are good sources of complete protein, meaning that their essential amino acid patterns match those needed by humans. (An exception is gelatin, which does not contain the amino acid tryptophan. Except for soybeans, individual plant-based foods are either inferior or insufficient sources of protein.

The main limiting amino acids are tryptophan, methionine, and lysine; as a result, the amount of protein that can be synthesized is restricted due to their limited availability. However, if the foods that contain protein are balanced so that their essential amino acids complement each other, a varied vegetarian diet can easily meet human protein requirements. Legumes, like beans, have a high lysine content and a low methionine content, whereas grains have complementary advantages and disadvantages.

As a result, eating rice and beans together throughout the day will enhance each other's amino acid patterns and result in a higher-quality protein than consuming either food by itself. Native cultures have made good use of protein complementarity in their traditional food patterns. On the other hand, people whose intake of protein is insufficient or insufficient do not require careful plant protein balancing. Getting enough high-quality protein is typically only an issue for young children who do not receive animal proteins in wealthy populations where protein intake is significantly above needs.

3.3.2. Protein Intake

For adults of both sexes, the World Health Organization suggests consuming 0.75 grams of high-quality protein per kilogram of body weight each day. Therefore, 52.55 grams of protein would be required for a 70-kg (154-pound) man and approximately 41 grams for a 55-kg (121-pound) woman. This suggestion is predicated on research on nitrogen balance and requires a sufficient intake of energy. Extra protein is required by infants, kids, pregnant women, and nursing mothers in order to support the synthesis of new tissue or the production of milk. Although bodybuilders and endurance athletes may have slightly higher protein requirements than sedentary people, this is not really significant because athletes usually consume far more protein than they require.

When the body consumes more protein than it needs, it breaks down the protein, excretes nitrogen as urea, and uses the remaining keto acids for energy (each gram provides 4 kilocalories) or converts them to fat or carbohydrates. Lean tissue is broken down in times of fasting, starvation, or inadequate protein intake to provide amino acids needed for essential bodily processes. Insufficient amounts of protein in the body cause poor metabolic performance and a higher risk of illness and infection.

3.3.3. Supplementing Diets with Amino Acids

Adding the right amounts of the amino acids required in realistic diets to the protein is one way to make use of proteins that are relatively low in one or more amino acids. It appears that fish use free amino acids in different ways. It has been demonstrated that young carp, *Cyprinus carpio*, cannot develop on diets where the protein (casein, gelatin) is swapped out for a combination of amino acids that are similar in overall composition. Casein hydrolyzed with trypsin proved to be equally ineffectual. However, some growth of young carp does occur if a diet containing free amino acids as the protein component is carefully neutralized with NaOH to pH 6.5–6.7. Under the same conditions, this growth was noticeably less than that on a comparable casein diet. Additionally, free amino acids added to inadequate proteins cannot be utilized by channel catfish. When menhaden meal was replaced with soybean meal isonitrogenously, channel catfish growth and feed efficiency were significantly decreased. The most limiting amino acids—free methionine, cystine, or lysine—did not increase weight gain when added to these soy-substituted diets. Increased weight gain was obtained by isonitrogenously substituting gelatin for casein, raising the arginine level in catfish diets from 11 to 17 g/kg. However, growth and food conversion were not affected much by adding free arginine, cystine, tryptophan, or methionine to casein. Free amino acids can be used by salmonids to fuel their growth. When weight gain and protein utilization were taken into consideration, a zein-gelatin diet supplemented with lysine and tryptophan was found to be significantly better for rainbow trout than an unsupplemented zein-gelatin diet. Numerous researchers have shown that limiting the amount of amino acids in salmonid diets can be an effective way to supplement proteins that are low in amino acids. When six amino acids were added to casein, Atlantic salmon showed feed conversion ratios that were comparable to what happened when an isolated fish protein was used as the main source of protein in the diet. For rainbow trout, soybean meal combined with five or more amino acids—such as lysine and methionine—was a better source of protein than soybean meal alone. Soybean meal's value was not enhanced by the addition of methionine or lysine alone. These findings imply that the amino acid spectrum of the fish protein isolate they utilized might be close to what rainbow trout require in terms of amino acids. The first limiting amino acid, i.e., could be added to soy protein isolate to increase its nutritional value methionine. Supplementing with cystine (10 g/kg) and tryptophan (5 g/kg) together can improve diets that include fishmeal, meat and bone meal, yeast, and soybean meal as protein sources. In rainbow trout diets, fishmeal can be completely substituted with a combination of poultry by-product meal and feather meal along with 17 g lysine HCL/kg, 4.8 g DL-methionine/kg, and 1.44 g DL-tryptophan/kg without lowering the food conversion rate.

3.3.4. Protein Digestion and Absorption

When a person eats, the body's digestive system breaks down the protein into individual amino acids, which are then absorbed and used by cells to build other proteins and a few other macromolecules, like DNA. But how do the proteins from foods, denatured or not, get processed into amino acids that cells can use to make new proteins? We'll use eggs as an example to explain how proteins go through the digestion and absorption processes because they're a good dietary source of protein. Regardless of its preparation—raw, hardboiled, scrambled, or fried—one egg provides roughly six grams of protein.

3.3.4.1. From the Mouth to the Stomach

Chewing is the first step in the digestion of an egg, or any other protein-rich food, unless it is consumed raw. The big egg pieces are mechanically broken down by the teeth into smaller pieces that can be ingested. Saliva is produced by the salivary glands to help with swallowing and the partially mashed egg's passage through the esophagus. The esophageal sphincter allows the broken pieces of egg to pass into the stomach. The hydrochloric acid and pepsin-containing gastric juices that the stomach secretes start the protein's breakdown. The stomach's acidity aids in the breakdown of protein aggregates created during cooking as well as the unfolding of proteins that have preserved some portion of their three-dimensional structure after cooking. The stomach's lining cells secrete a substance called pepsin, which breaks down protein chains into ever-tinier pieces. Because egg proteins are big, globular molecules, it takes time and mixing to break them down chemically. The partially digested protein is agitated by the strong mechanical stomach contractions into a more homogenous mixture known as chyme. In the stomach, digesting protein takes longer than digesting carbohydrates, but it takes less time to digest fat. A high-protein meal lengthens the time needed in the stomach for the food to be adequately broken down. Food stays longer in the stomach, extending the feeling of fullness (Figure 3.2).

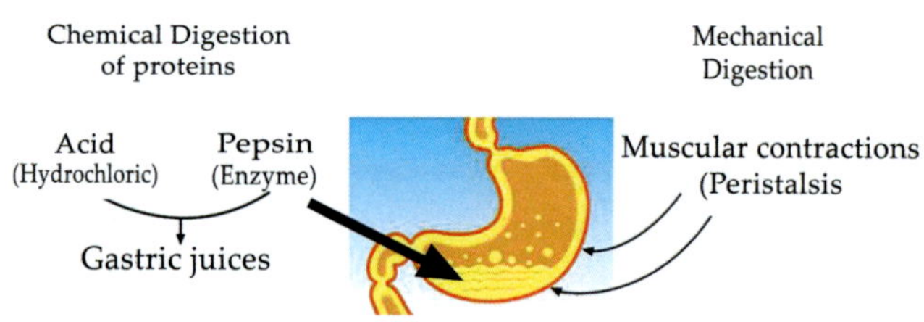

Figure 3.2. *Protein digestion requires the chemical actions of gastric juice and the mechanical actions of the stomach.*

Source: https://2012books.lardbucket.org/books/an-introduction-to-nutrition/s10-proteins.html.

3.3.4.2. From the Stomach to the Small Intestine

The small intestine is where most protein digestion takes place after the stomach empties the chyme containing the broken-down egg pieces. More enzymes found in the digestive juice secreted by the pancreas help break down the protein fragments even more. Chymotrypsin and trypsin are the two main pancreatic enzymes involved in protein digestion. Further enzymes are released by the cells lining the small intestine, which ultimately disassembles the smaller protein fragments into their constituent amino acids. The digested proteins are mixed and propelled to the absorption sites by the small intestine's muscular contractions. Amino acids are transferred from the intestinal lumen to the bloodstream via the intestinal cells in the lower portions of the small intestine. Adenosine triphosphate (ATP), a molecule that provides energy to cells, and specific transport proteins are needed for the movement of individual amino acids. The amino acids are carried to the liver by the blood after they enter it. The liver serves as the distribution checkpoint for amino acids and the site of very little additional amino acid breakdown, just like it does for other macronutrients. Because nitrogen is a component of amino acids, ammonia-containing nitrogen is released during further amino acid breakdown. Ammonia is poisonous, so the liver converts it to urea, which is subsequently sent to the kidney and eliminated in the urine. Two nitrogens make up the molecule urea, which is very soluble in water. It is therefore a wise option for eliminating extra nitrogen from the body. More than 90% of the protein consumed is not broken down any further than the amino acid monomers because amino acids are building blocks that the body stores to synthesize other proteins.

3.3.4.3. Amino Acids Are Recycled

Amino acids are recycled to create new proteins, much like certain plastics can be recycled to create new goods. The process known as protein turnover occurs continuously in all body cells as old proteins are broken down and replaced with new ones. The human body breaks down more than 250 grams of protein every day and builds up another 250 grams of protein every day. Amino acids from food and those from protein degradation are combined to create a "pool" that creates these new proteins. The extra amino acids that the body contains can provide an amino acid when it is needed to assemble another protein, even though it is not a literal pool. In addition to building proteins, amino acids are also needed to build other nitrogen-containing biological molecules like DNA and RNA and, to a lesser extent, to produce energy.

It is imperative to sustain the levels of amino acids in this cellular pool through the consumption of high-quality proteins through diet, or else the amino acids required for the synthesis of new proteins will be acquired by increasing the breakdown of proteins in other bodily tissues, particularly muscle. Less than 1% of the body's total protein content is made up of this pool of amino acids. As a result, unlike fats and carbohydrates (stored as triglycerides in adipose tissue and glycogen in the muscles and liver), the body does not store protein.

3.3.5. Functions of Proteins in the Body

The "workhorses" of the body, proteins are involved in a wide range of physiological processes. Proteins have various shapes and sizes, and each is uniquely structured for a specific function.

3.3.5.1. Structure and Motion

Although the human body contains over a 100 distinct structural proteins, collagen is the most prevalent and accounts for approximately 6% of the total body weight. 30% of bone tissue is composed of collagen, which is also abundant in skin, muscle, cartilage, ligaments, and tendons. The amino acids glycine and proline make up the majority of the robust, fibrous protein collagen.

Three protein strands form a rope-like structure within its quaternary structure, and these collagen ropes overlap with one another. Compared to steel fibers of the same size, this highly ordered structure is even stronger. Strong yet flexible bones are made of collagen. The skin's dermis contains collagen fibers that give it structure, and the elastin protein fibrils that go with it give it flexibility.

The skin on the hand can return to its natural shape when pinched and then release it due to the presence of collagen and elastin proteins. Blood vessels are surrounded by smooth-muscle cells that release the proteins collagen and elastin, which give the vessels structure and the capacity to expand back after blood is pumped through them. Keratin, the protein found in skin, hair, and nails, is another robust and fibrous protein.

The synchronous mechanical movements of bones and muscles as well as the tissues' capacity to rebound after a movement is finished are made possible by the densely packed collagen fibrils found in tendons and ligaments. Observe the synchronization of knuckle movements as a person moves their fingers. It takes a contraction of the muscles to move. The proteins myosin and actin make up the contractile components of muscles. These proteins slide past one another in response to nerve impulse stimulation, shortening the muscle cell. Muscle contraction results from the simultaneous shortening of several muscle cells in response to stimulation.

3.3.5.2. Enzymes

While connective tissues like bone contain the highest concentrations of proteins, their most remarkable role is that of enzymes. Proteins called enzymes carry out particular chemical reactions. A chemical reaction needs an enzyme to provide a site for it as well as to reduce the energy and time required for the reaction to occur (this process is called "catalysis"). Cells go through over a 100 chemical reactions per second on average, and the majority of these require enzymes. There are over a 1000 different enzyme systems in the liver alone. Enzymes are unique in that they can only utilize specific substrates that fit into their active site; this is comparable to how a lock can only be opened with a specific key. An enzyme is needed for almost all chemical reactions. Thankfully, an enzyme can repeatedly act as a catalyst even though it eventually gets destroyed and rebuilt. Enzymes are involved in every physiological process, such as the digestion of food in the stomach and small intestine, the conversion of nutrients into molecules that cells can use, and the synthesis of all macromolecules, including protein.

3.3.5.3. Hormones

Hormone synthesis is carried out by proteins. The chemical signals that the endocrine glands produce are called hormones. An endocrine gland releases a hormone upon stimulation. After that, the hormone travels through the blood to its intended cell, where it sends a signal to start a particular cellular process or reaction. For example, blood glucose levels increase following a meal. The pancreas secretes the hormone insulin in response to elevated blood glucose levels. Insulin alerts the body's cells to the presence of glucose in the blood, directing them to either store it for later use or use it to create macromolecules or energy. Certain proteins can even control the activities of other proteins because hormones play a major role in the on-and-off regulation of enzymes. Although many hormones are not derived from proteins, some are.

3.3.5.4. Fluid and Acid-Base Balance

The body's fundamental biological functions can preserve the status quo in a changing environment when the right amount of protein is consumed. Maintaining the body's water distribution is referred to as fluid balance. A tissue may swell and possibly even die if there is an abrupt influx of excessive blood water. Water always moves from a highly concentrated area to a lowly concentrated area. Water therefore gravitates toward regions where other solute concentrations are higher, such as those of proteins and glucose. Proteins constantly circulate in the blood at high concentrations to maintain the water's equal distribution throughout the blood and cells. The protein with a butterfly shape called albumin is the most prevalent type of protein in the blood. The blood's protein concentration is comparable to that of cells due to the presence of albumin. To maintain the status quo, fluid exchange between the blood and cells is therefore minimized rather than increased.

Additionally, protein is necessary to keep the blood's pH balance—which indicates how basic or acidic a substance is. Blood pH is kept in the slightly basic range of 7.35–7.45. The body's processes can be impacted by even a small shift in blood pH. Remember that proteins can become denaturated in acidic environments, rendering them non-functional. To stop this from happening, the body has many systems that keep the blood pH within the normal range. The albumin in circulation is one of these. Because albumin is negatively charged and somewhat acidic, it balances out the many positively charged molecules found in blood, including calcium, potassium, magnesium, and hydrogen protons (H^+). By serving as a buffer against sudden variations in these molecules' concentrations, albumin helps to keep blood pH stable and in check. By binding hydrogen protons, the protein hemoglobin contributes to the maintenance of the acid-base balance.

3.3.5.5. Transport

Hemoglobin and albumin are also involved in molecular transport. Albumin carries substances throughout the circulatory system by chemically binding to hormones, fatty acids, certain vitamins, important minerals, and medications. Millions of hemoglobin molecules are found in each red blood cell, which binds oxygen in the lungs and carries it to every tissue in the body. Large polar molecules typically cannot pass through a cell's plasma membrane, therefore the cell membrane contains numerous transport proteins that allow nutrients and other molecules to enter the cell. A portion of these proteins function as channels, permitting specific molecules to enter and exit cells. Others need energy to run and serve as one-way taxis.

3.3.5.6. Protection

Skin's robust collagen fibers give it support and structure. The thick collagen fiber network of the skin acts as a barrier against dangerous substances. Enzymes and antibodies are also proteins that are necessary for the immune system's ability to attack and destroy targets. Saliva secretes an enzyme called lysozyme, which targets bacterial walls and tears them apart. Blood circulating proteins can be directed to form a molecular knife that pierces foreign invaders' cellular membranes. White blood cells secrete antibodies that scour the circulatory system for pathogenic bacteria and viruses to encircle and eradicate. To locate and eliminate undesired invaders, antibodies also stimulate other immune system components.

3.3.5.7. Wound Healing and Tissue Regeneration

Wound healing occurs in three phases: inflammatory, proliferative, and remodeling. All aspects of this process involve proteins. For instance, a person's flesh would become red and inflamed if they were sewing and accidentally pricked their finger with a needle. It would stop bleeding in a matter of seconds. Bradykinin and other proteins cause blood vessels at the site of injury to dilate, which is the first step in the healing process. To stabilize platelets and create a clot to stop the bleeding, another protein known as fibrin is needed. Afterward, during the proliferative phase, newly formed collagen fibers are inserted by the cells to repair the damaged tissue. Collagen fibers aid in drawing the edges of the wound together. More collagen is deposited during the remodeling phase, creating a scar. Scar tissue functions only 80% as well as healthy, uninjured tissue. A diet low in protein causes a noticeable slowdown in the healing of wounds.

Although wound healing occurs only following an injury, the body is continuously going through a process known as tissue regeneration. Tissue regeneration is primarily distinguished from wound healing by the process of creating a precise anatomical and functional replica of the lost tissue. As a result, fresh, fully functional tissue replaces aging, dying tissue rather than scar tissue. Certain cells, like those found in the skin, hair, nails, and intestines, regenerate extremely quickly, whereas other types of cells, like those found in the heart and muscles, do not regenerate at all. The process of tissue regeneration involves cell division or the synthesis of new cells. This requires a variety of proteins, including transport proteins, collagen, hormones, and enzymes that synthesize RNA and proteins. Cells divide and a hair grows longer in a hair follicle. Nails grow about one centimeter every 100 days, and hair grows one centimeter on average every month. Every three to five days, the intestinal lining cells renew themselves. Diets lacking in protein hinder the regeneration of tissue, which can lead to various health issues such as impaired absorption and digestion of nutrients and, most obviously, stunted growth of hair and nails.

3.3.5.8. Energy Production

Proteins contain amino acids that can be broken down and utilized to produce energy. Merely 10% of the proteins consumed are catabolized daily to produce energy for cells. Amino acids can be converted by the liver into the carbon skeleton, which can subsequently be supplied to the citric acid cycle. This is comparable to how ATP is produced using glucose. A person's body will use more amino acids to produce energy if their diet is low in fats and carbohydrates, which will impair the synthesis of new proteins and damage muscle proteins. Conversely, if a person consumes more protein than their body requires through diet, the excess amino acids will be metabolized and turned into fat.

3.4. ENERGY METABOLISM

The biochemical reactions that occur in living cells to convert nutrients from food into energy that the body can use for various physiological functions is known as energy metabolism. These nutrients, which are broken down into smaller molecules by the body's enzymes, include proteins, lipids, and carbohydrates.

The main output of energy metabolism is adenosine triphosphate, the body's basic energy currency. ATP provides energy for essential cellular processes like metabolic synthesis, neuronal signaling, and muscle contraction. The cycle of citric acid production, oxidative phosphorylation, and glycolysis are a few of the interconnected mechanisms that comprise energy metabolism. Different enzymes and metabolic intermediates regulate these pathways.

Energy metabolism must be managed in order to maintain the balance between energy intake and expenditure, which is essential for an organism's overall health and wellbeing. Hormonal and signaling pathways like glucagon, adiponectin, and insulin have a major impact on fat storage and utilization, as well as the uptake and utilization of nutrients. A disruption in energy metabolism can lead to a number of metabolic diseases, including obesity, type 2 diabetes, and cardiovascular disease. Therefore, in order to develop effective preventative and therapeutic measures, it is imperative to understand the mechanisms governing energy metabolism. Energy metabolism occurs in multiple ways within the body, utilizing different biochemical pathways and processes. Some significant examples of energy metabolism are as follows:

- **Glycolysis**: The process that converts glucose, a carbohydrate, into pyruvate is known as glycolysis. In the cell cytoplasm, glycolysis produces a tiny quantity of ATP through substrate-level phosphorylation.
- **The Citric Acid Cycle**: The Krebs cycle, also known as the TCA cycle or the citric acid cycle, is a metabolic process that produces $CO2$ and highly energetic compounds like NADH and FADH2 from pyruvate and other molecules. In the mitochondria, the citric acid cycle produces ATP through oxidative phosphorylation.
- **Oxidative Phosphorylation**: Oxidative phosphorylation is the process by which oxygen is produced in the mitochondria by transferring electrons from NADH and FADH2. This process involves many protein complexes and electron transporters, such as the electron transport chain.
- **Beta-Oxidation**: Acetyl-CoA, which is produced when fatty acids undergo beta-oxidation, can then enter the citric acid cycle to produce ATP. The role of the mitochondria as an energy-generating pathway is essential for people who

fast or participate in high-intense physical activity.

- **Gluconeogenesis**: The process by which glucose is produced from non-carbohydrate building blocks like fatty acids and amino acids is known as gluconeogenesis. Because gluconeogenesis mostly occurs in the liver, blood glucose levels must be maintained during fasting periods.

These examples demonstrate the many pathways the body uses to produce ATP and how complex and intertwined energy metabolism is.

3.4.1. Steps to Energy Metabolism

The body uses a network of connected pathways and reactions called energy metabolism to create ATP (adenosine triphosphate), which is its main source of energy. The following are the four basic steps in energy metabolism:

- **Digestion and Absorption:** The first step in the metabolism of energy is the digestion and absorption of nutrients from meals. This means breaking down fats, proteins, and carbohydrates into smaller molecules that the body can absorb and deliver to the cells.
- **Glycolysis and the Citric Acid Cycle:** Following absorption, nutrients are metabolized in the cells by a sequence of biochemical reactions known as the citric acid cycle and glycolysis. In the cytoplasm of cells, glucose is broken down into pyruvate via the process of glycolysis before being transferred to the mitochondria. There, as part of the citric acid cycle, pyruvate is further metabolized to create energy-dense molecules like NADH and FADH2.
- **Oxidative Phosphorylation:** The energy-dense molecules generated by the citric acid cycle and glycolysis are oxidatively phosphorylated to produce ATP. This process needs the transfer of electrons from NADH and FADH2 to a variety of protein complexes and electron carriers in order to produce ATP.
- **Utilization of ATP:** The final stage of energy metabolism is when cells use ATP to carry out different physiological processes. ATP provides energy for a variety of processes, such as nerve transmission, muscular contraction, and metabolic synthesis.

In general, the energy metabolism system is a complex and tightly regulated system that is essential to the survival and proper functioning of living organisms. Because defects in energy metabolism can lead to a variety of metabolic diseases, it is important to understand the systems involved in energy metabolism.

3.4.2. Functions of Energy Metabolism

Energy metabolism is the fundamental activity of all living organisms, including bacteria and humans. Converting the nutrients in food into energy that the body can use for various physiological functions is the primary objective of energy metabolism.

Adenosine triphosphate, or ATP, is the body's main energy source. It powers essential cellular processes like muscle contraction, nerve transmission, and metabolic synthesis.

Energy metabolism in the body does many vital tasks, one of which is producing ATP. For example, energy metabolism contributes to maintaining cellular homeostasis, regulating body temperature, and enhancing immunological response. Energy metabolism affects the synthesis of important substances like lipids, nucleotides, and amino acids. Moreover, energy metabolism is intimately linked to the metabolism of fats and carbohydrates as well as other metabolic processes in the body. A disruption in energy metabolism can lead to several metabolic diseases, including obesity, type 2 diabetes, and cardiovascular disease. Thus, an understanding of energy metabolism is necessary to preserve general health and prevent metabolic disorders.

3.4.2.1. Production of Energy for Cellular Functions

One of the primary byproducts of energy metabolism is the production of adenosine triphosphate, the body's primary energy currency. The process by which ATP is produced in cells is called cellular respiration. It entails the breakdown of glucose, fatty acids, and other molecules via a variety of metabolic pathways, such as glycolysis, the citric acid cycle, and oxidative phosphorylation. The ATP produced during cellular respiration provides energy for numerous cellular functions, such as muscular contraction, neuron transmission, protein synthesis, and DNA replication. To sustain cellular homeostasis, ATP is also necessary for sustaining ion gradients across cell membranes and facilitating active transport processes.

Different energy needs for different types of cells are determined by their activity level and mode of operation. For example, the activity of cells requiring a lot of energy, like muscle cells, depends on ATP. On the other hand, adipocytes and other cells with lower energy needs might require less ATP. In general, the production of energy for cellular functions is a significant aspect of energy metabolism. Disturbances in energy metabolism may lead to a range of metabolic illnesses, such as mitochondrial diseases, which are characterized by reduced energy generation and can impact various body organs and tissues. Therefore, knowledge of the processes underlying energy production is necessary to maintain overall health and prevent metabolic disorders.

3.4.2.2. Glucose Production

The process of glucose synthesis is an essential component of energy metabolism. The body uses simple sugars like glucose as its primary energy source, particularly for the brain and nervous system. The process by which the body produces glucose from food or other sources is called gluconeogenesis. The primary location of glucose synthesis is the liver, with the kidneys and small intestine following in order of importance. During gluconeogenesis, non-carbohydrate precursors such as amino acids and glycerol are converted into glucose through a series of biochemical processes. This procedure is essential for maintaining stable blood glucose levels during exercise or fasting, when glucose intake may be limited.

Glucose is not only a fuel source but also necessary for maintaining cellular homeostasis. For example, glucose is required for the synthesis of NADPH, an essential chemical involved in various cellular processes, including antioxidant defense and the synthesis of fatty acids and cholesterol. Disturbances in glucose metabolism can cause a range of metabolic diseases, including type 2 diabetes, which is characterized by a decreased absorption and utilization of glucose. Understanding the mechanisms underlying glucose synthesis and utilization is essential for preserving normal blood glucose levels and preventing metabolic disorders.

3.4.2.3. Production of Fats

Another function of energy metabolism is the synthesis of lipids, an essential component of the body's energy storage system. Fat that has been stored in adipose tissue can be used as fuel during prolonged fasts or periods of low blood glucose. The body produces fats through a process called lipogenesis, which includes changing glucose and other metabolites into fatty acids. Liposomal tissue, the liver, and the mammary glands are where the majority of lipogenesis occurs during lactation.

Triglycerides formed through lipogenesis can either be stored in adipose tissue or used for energy production in a process called beta-oxidation. During beta-oxidation, fatty acids are converted into acetyl-CoA, which then enters the citric acid cycle and oxidative phosphorylation to produce ATP. Fats also play a crucial role in insulation, immunity, hormone synthesis, and cell membrane formation in the body. However, disruptions in fat metabolism can lead to various metabolic disorders such as obesity and dyslipidemia, which are associated with an increased risk of cardiovascular disease, type 2 diabetes, and other health problems.

3.4.2.4. Protein Production

Proteins play a vital role in numerous cellular functions, such as enzymatic reactions, providing structural support, and facilitating the movement of molecules across cell membranes. Energy metabolism is essential for protein synthesis. Translation is the mechanism by which amino acids are assembled into polypeptide chains to form proteins. This intricate metabolic process requires the hydrolysis of ATP, which is energetically demanding.

The amino acids required for protein synthesis can either be produced by the body through a process called amino acid metabolism, or they can be obtained through diet. A variety of metabolic pathways, including the citric acid cycle and oxidative phosphorylation, can use amino acids for energy production. Alternatively, gluconeogenesis can convert amino acids into glucose. Protein metabolism problems can lead to inherited metabolic diseases that change how specific amino acids are synthesized or used. Phenylalanine hydroxylase, for example, is an enzyme required for the conversion of the amino acid phenylalanine to tyrosine; its absence causes the genetic disorder phenylketonuria (PKU). PKU can cause seizures, cognitive decline, and other neurological problems if left untreated.

3.4.2.5. Production of Oxidative Stress

Life depends on energy metabolism, but it can also lead to oxidative stress, a state where the body's antioxidant defense systems fail to adequately counteract an excess of reactive oxygen species (ROS) produced in comparison to the quantity of ROS that can be neutralized. ROS are extremely reactive substances that are produced as a byproduct of cellular respiration and other metabolic processes. High levels of ROS can harm DNA, proteins, and lipids and accelerate the development of several diseases, such as cancer, cardiovascular disease, and neurodegenerative disorders, even though low levels are essential for immunological response and cellular signaling.

Oxidative stress can result from many processes, including inflammation, exposure to chemicals and pollutants in the environment, and incomplete reduction of oxygen during the electron transport chain. The body has numerous antioxidant defense systems in place to fight oxidative stress, including non-enzymatic antioxidants like glutathione and vitamin E as well as enzymes like catalase and superoxide dismutase. Oxidative stress, on the other hand, is caused by the body's inability to neutralize ROS due to either excessive production of them or malfunctions in antioxidant defense mechanisms.

3.4.2.6. Detoxification of Chemicals

Energy metabolism also plays a major role in the detoxification of substances that are exposed to through the environment and diet. Several of these substances, such as medicines, heavy metals, and pesticides, can accumulate in the body over time and result in a number of health problems. Detoxification of these compounds primarily occurs in the liver, where they are processed and eliminated from the body through bile or urine. The detoxification process consists of two main stages: Phase I and Phase II.

Phase I entails the metabolism of the materials by a range of enzymatic procedures, including hydrolysis, reduction, and oxidation, which raises their reactivity and speeds up their excretion. But if the reactive intermediates created by this process are not removed, they may lead to the production of ROS and free radicals, both of which are harmful to cells. Phase II involves the conjugation of endogenous compounds, such as glutathione, amino acids, or glucuronic acid, with reactive intermediates generated during phase I to render them less toxic and easier for the body to eliminate. Since many of the phase I and phase II enzymatic processes require the hydrolysis of ATP, energy is required for the detoxification process. Therefore, abnormalities in energy metabolism may hinder the body's ability to properly detoxify substances, leading to the accumulation of toxins and the emergence of various health problems.

3.4.2.7. Hormone Production

Energy metabolism has a major impact on hormone synthesis, which is essential for maintaining homeostasis and managing many physiological processes in the body. Hormones are chemical messengers produced by endocrine glands that are released into the bloodstream and target specific tissues and organs to regulate their functions. Hormones are involved in many processes, such as growth and development, metabolism, reproduction, and stress response. Because many of the enzymatic processes involved in the synthesis and secretion of hormones require the hydrolysis of ATP, hormone production requires energy. For example, during the synthesis of steroid hormones such as cortisol and testosterone, cholesterol is converted by a variety of enzyme processes into distinct steroid intermediates. Disturbances in the metabolism of energy can lead to a variety of hormonal disorders by affecting the production of hormones. For example, diabetes affects the synthesis and secretion of insulin, which leads to hyperglycemia and other abnormalities in metabolism. In diseases like hypothyroidism, the synthesis of thyroid hormones is reduced, leading to a slowdown in metabolism and a variety of symptoms like fatigue, weight gain, and sensitivity to cold.

3.4.2.8. Cellular Growth and Replication

Energy metabolism is necessary for cellular growth and reproduction, which are essential functions for maintaining tissue and organ function and repairing cell damage. The energy needed for the synthesis of macromolecules like proteins, nucleic acids, and lipids increases with an increase in cell mass and size. One of the signaling pathways that regulates cell development is the mammalian target of rapamycin (mTOR) pathway, which combines signals from growth factors, nutrition, and energy state.

Cell replication, also referred to as cell division, is the process by which a single parent cell divides into two identical daughter cells. Mitosis and cytokinesis are the two main stages of cell division. Each of these phases requires energy for several enzymatic functions, such as chromosomal segregation, membrane trafficking, and DNA replication. Energy metabolism issues can impede the growth and replication of cells, leading to a range of cellular diseases, such as cancer, in which cells proliferate and divide uncontrollably. Cancer cells' altered energy metabolism allows them to meet their increased energy requirements and encourages their faster growth and division.

3.4.2.9. Production of Adenosine Triphosphate (ATP)

The primary objective of energy metabolism is the synthesis of ATP, the cell's energy currency. ATP is essential for almost all biological processes, such as muscle contraction, cell division, and the synthesis of big molecules like proteins, nucleic acids, and lipids. The process of ATP synthesis is carried out by the mitochondria, the organelles of the cell that produce energy. By means of oxidative phosphorylation, mitochondria generate ATP. It involves the movement of electrons via several electron transport chain (ETC) complexes from oxygen to electron donors such as NADH and FADH2. Using the energy produced during this process, protons are pumped across the mitochondrial membrane to create a proton gradient, which in turn drives the ATP synthase enzyme to produce ATP.

Apart from oxidative phosphorylation, ATP can also be produced by the oxygen-free cytoplasmic reaction known as glycolysis. For every glucose molecule, two ATP molecules are produced on average during the breakdown of glucose into pyruvate. Disturbances in energy metabolism can impact the synthesis of ATP and lead to a range of metabolic illnesses such as metabolic syndrome and mitochondrial diseases. The functioning of the mitochondria is hampered in mitochondrial diseases, which reduces ATP production and results in a range of symptoms like fatigue, neurological problems, and muscle weakness. Metabolic syndrome impairs the body's ability to produce and use ATP, which leads to several metabolic abnormalities such as insulin resistance, obesity, and cardiovascular disease.

3.4.2.10. Regulation of Body Energy Levels

Energy metabolism is responsible for controlling body energy levels, which is essential for preserving overall health and preventing metabolic disorders.

The body's energy levels are regulated by a multitude of hormones and signaling pathways, including insulin, glucagon, leptin, and the AMP-activated protein kinase (AMPK) pathway. Insulin is a hormone secreted by the pancreas that regulates glucose metabolism and increases energy storage. Insulin promotes the synthesis of lipids and glycogen, which are subsequently stored for use at a later date, as well as the cell's ability to absorb glucose. Moreover, the pancreas secretes the hormone glucagon, which regulates glucose metabolism and enhances energy release.

In low-glucose environments, glucagon promotes the energy-producing process known as gluconeogenesis, which breaks down glycogen and produces glucose. The hormone leptin, which regulates appetite and energy balance, is produced by adipose tissue. Leptin helps people maintain a healthy weight and avoid obesity by decreasing hunger and increasing energy expenditure.

The AMPK system is a signaling mechanism that responds to changes in energy levels to regulate energy metabolism. When cellular energy levels are low, AMPK is triggered, which promotes energy synthesis while preventing energy-consuming processes like protein synthesis. Issues with energy metabolism can cause disturbances in the regulation of the body's energy levels, leading to various metabolic disorders such as obesity, diabetes, and the metabolic syndrome. Diabetes alters the regulation of glucose metabolism, leading to hyperglycemia and additional issues. Obesity is the outcome of an imbalance in the regulation of energy balance, which occurs when there is an accumulation of body fat and an unbalanced energy intake and expenditure.

3.4.3. Regulatory Mechanisms of Energy Metabolism

Energy metabolism is a complex process that involves the regulation of multiple metabolic pathways and enzymes to ensure that the body has an adequate supply of energy for its cellular processes. Numerous signaling pathways, hormones, and enzymes are all involved in the regulatory processes of energy metabolism to maintain the body's energy homeostasis. These regulatory systems are essential for maintaining overall health and preventing metabolic disorders such as diabetes, obesity, and metabolic syndrome, which are characterized by abnormalities in energy metabolism.

It is necessary to comprehend the regulatory mechanisms regulating energy metabolism in order to develop treatments and preventions for metabolic diseases. It can also help identify possible therapeutic targets and develop customized treatments based on an individual's metabolic profile. This section will examine the various mechanisms that regulate energy metabolism, such as the effects of hormones, signaling pathways, and enzymes on the maintenance of energy balance and homeostasis. We'll also discuss potential prophylactic and curative actions for metabolic disorders as well as the consequences of disruptions in energy metabolism.

3.4.3.1. Signaling through Hormones and Receptors

One of the primary methods for regulating energy metabolism is signaling through hormones and receptors. Hormones are chemical messengers produced by the body's many organs and tissues. These messengers regulate many physiological functions, including energy metabolism. Hormones such as insulin, glucagon, leptin, and ghrelin all have a major impact on the regulation of energy metabolism. Through their specific receptor binding on target cell surfaces, these hormones either stimulate or inhibit metabolic processes such as glucose absorption, glycogen synthesis, lipolysis, and energy expenditure.

For example, the pancreas secretes insulin, which acts on target cells to improve the absorption of glucose, stimulate the synthesis of glycogen, and stimulate the process of lipogenesis, which leads to the storage of energy. Insulin signaling is mediated by the transmembrane protein insulin receptor, which activates downstream signaling pathways such as the phosphatidylinositol 3-kinase (PI3K)/Akt pathway. This promotes the synthesis of glycogen and lipogenesis and causes the translocation of glucose transporters (GLUT4) to the cell surface. On the other hand, the hormone glucagon, which is produced by the pancreas, causes target cells to increase glycogenolysis and gluconeogenesis, which releases glucose into the bloodstream and releases energy. By triggering downstream signaling pathways like the cyclic AMP/PKA pathway via the glucagon receptor, a G protein-coupled receptor, glucagon promotes glycogenolysis and gluconeogenesis.

The hormone leptin, which is produced by adipose tissue, works with the brain to increase energy expenditure and decrease food intake, ultimately regulating the energy balance.

The transmembrane protein known as the leptin receptor is the site of leptin signaling, which activates downstream signaling pathways such as the Janus kinase (JAK)/signal transducer and activator of transcription (STAT) pathway to promote energy expenditure and inhibit food intake. It's critical to comprehend how hormones and receptors regulate energy metabolism because problems with hormone signaling can lead to metabolic diseases like diabetes, obesity, and metabolic syndrome.

3.4.3.2. Reactions in the Krebs Cycle and Oxidative Phosphorylation

Oxidative phosphorylation and the Krebs cycle comprise yet another essential energy metabolism regulatory mechanism. Often called the citric acid cycle, the Krebs cycle is a series of enzymatic reactions that occur in the mitochondria of cells and convert acetyl-CoA into carbon dioxide while also producing energy-dense molecules like FADH2 and NADH. These energy-dense molecules enter the electron transport chain and then donate their electrons to various protein complexes that span the mitochondrial membrane to produce a proton gradient. ATP synthase then uses this proton gradient to produce ATP, the basic energy unit of the cell.

The Krebs cycle and oxidative phosphorylation are closely regulated processes that ensure the production of ATP is balanced with the energy needs of the cell. This regulation is influenced by a multitude of mechanisms, including substrate availability, enzyme allostery, and feedback inhibition.

For example, the activity of the enzyme isocitrate dehydrogenase, which catalyzes an important step in the Krebs cycle, is allosterically controlled by the amounts of ATP and NADH. The rate of the Krebs cycle and ATP synthesis is regulated by the concentrations of ATP and NADH. Low concentrations of these molecules enhance the enzyme's activity, while high concentrations inhibit it.

In a similar vein, the electron transport chain's activity is regulated by the availability of electron donors such as NADH and FADH2, which are produced by the Krebs cycle. The rate of ATP synthesis by ATP synthase is regulated by the proton gradient, which is upheld by the pace of electron transport and the activity of proton transporters. Reduced energy synthesis and effects on multiple organ systems are hallmarks of mitochondrial diseases, which can be caused by problems with the Krebs cycle and oxidative phosphorylation regulation. Thus, knowledge of these metabolic pathways' regulatory mechanisms is necessary to preserve general health and prevent metabolic disorders.

3.4.3.3. Glucose Metabolism and Production of Glucose

Glucose metabolism, another important regulatory mechanism of energy metabolism, governs the body's production and utilization of glucose. Since glucose is the primary fuel source for the majority of body cells, maintaining healthy glucose levels is essential for proper cellular function. Glucose metabolism is controlled by several hormones, including insulin, glucagon, and cortisol, which also affect how much glucose is produced, absorbed, and used by different organs and tissues. The liver produces glucose through a process called gluconeogenesis, which converts non-carbohydrate substrates like amino acids and glycerol into glucose. This process is necessary to maintain glucose levels during periods of reduced carbohydrate consumption, such as during fasting.

Glycolysis is an additional step in the metabolism of glucose, in addition to gluconeogenesis. This process produces ATP and changes glucose to pyruvate inside the cytoplasm of cells. Pyruvate can then enter the Krebs cycle and oxidative phosphorylation to produce more ATP. The complex regulation of glucose metabolism involves numerous pathways and feedback loops. The pancreatic hormone insulin reduces blood glucose levels by promoting cell uptake and utilization of glucose and blocking gluconeogenesis in the liver. Glucagon, another hormone produced by the pancreas, promotes gluconeogenesis and glycogenolysis in the liver, which raises blood sugar. Other hormones, such as growth hormone and cortisol, can also affect glucose metabolism by inhibiting cell absorption of glucose and encouraging gluconeogenesis. Disturbances in glucose metabolism can lead to metabolic diseases such as diabetes mellitus, which is typified by elevated blood glucose levels and inadequate glucose regulation. Therefore, knowledge of the regulatory mechanisms of glucose metabolism is necessary to maintain healthy glucose homeostasis and prevent metabolic diseases.

3.4.3.4. Fatty acid Breakdown and Synthesis

Fatty acid synthesis and breakdown are important regulatory processes of energy metabolism that govern the production and distribution of fats in the body. Fats are the body's main energy source and are essential for the synthesis of membranes, signaling, and energy storage. Fatty acid breakdown, also referred to as beta-oxidation, is the process that converts fatty acids into acetyl-CoA, which can then participate in the Krebs cycle and oxidative phosphorylation to produce ATP. This process occurs in the mitochondria of cells. A few factors that regulate beta-oxidation are the availability of fatty acids, the activity of the enzymes involved in the process, and hormonal regulation.

On the other hand, acetyl-CoA undergoes a series of enzymatic conversions into fatty acids during fatty acid synthesis, which occurs in the cytoplasm of cells. A few factors that govern the synthesis of fatty acids are the availability of substrates, the activity of the enzymes involved, and hormonal regulation. Insulin is essential for controlling fatty acid metabolism during high blood glucose levels because it promotes the synthesis and storage of lipids. On the other hand, glucagon and other hormones such as growth hormone and cortisol promote the breakdown of fats, increasing the availability of energy substrates, during periods of fasting or stress. Disruptions in the metabolism of fatty acids can lead to metabolic diseases such as obesity, dyslipidemia, and fatty liver disease. Therefore, knowledge of the regulatory mechanisms governing fatty acid synthesis and breakdown is necessary to maintain healthy lipid homeostasis and prevent metabolic diseases.

3.4.3.5. Intermediary Metabolism

Intermediary metabolism is the intricate web of chemical processes that the body uses to produce and regulate energy substrates like glucose, fatty acids, and amino acids. During these reactions, a variety of metabolites are broken down and synthesized, including sugars, lipids, and amino acids, which are the building blocks for the synthesis of energy. Intermediary metabolism encompasses a wide range of metabolic activities, including amino acid metabolism, oxidative phosphorylation, glycolysis, and the Krebs cycle. Energy substrates can be produced and regulated efficiently in response to changes in nutrient availability and energy demands because of the interconnection and strict regulation of these pathways.

Intermediary metabolism is regulated by a variety of factors, such as transcriptional factors, hormones, and enzymes. Hormones such as insulin, glucagon, and cortisol are important regulators of intermediary metabolism because they control the absorption, use, and storage of energy substrates. Enzymes are essential for intermediate metabolism and are responsible for catalyzing many of the chemical reactions involved in energy production and regulation. These enzymes are strictly regulated; their level of activity is affected by the availability of substrate, pH, and allostery. Transcription factors are proteins that regulate gene expression and are crucial for the regulation of intermediate metabolism because they control the expression of genes involved in metabolic pathways. These transcription factors allow the adaptive control of intermediate metabolism by responding to changes in nutrition availability and energy requirements.

Disruptions in intermediary metabolism can lead to various metabolic disorders such as diabetes, obesity, and metabolic syndrome. To prevent such disorders, it is important to have an understanding of the complex network of metabolic pathways involved in intermediary metabolism, as well as the regulatory mechanisms that govern these pathways. Maintaining normal metabolic function is crucial in avoiding these conditions.

3.4.3.6. Amino Acid Metabolism

Amino acid metabolism is the process by which the body breaks down digested proteins and repurposes amino acids to make new proteins, synthesize hormones and neurotransmitters and produce energy. Amino acids are the building blocks of proteins and are necessary to maintain the balance of metabolism and cellular activity. Since amino acids are deaminated and transformed into intermediates in the liver, which can subsequently be utilized in the Krebs cycle and oxidative phosphorylation to produce ATP, the liver is the main location of amino acid catabolism. The nitrogen released during deamination is converted by the kidneys into urea, which they then excrete.

Amino acid anabolism, the process by which all cells synthesize new proteins, depends on amino acids. Many factors, such as the availability of nutrients, hormones, and signaling pathways, tightly control the activity of the enzymes involved in protein synthesis. In addition to being involved in the synthesis of energy and proteins, amino acids are also required for the synthesis of hormones and neurotransmitters. For example, tyrosine is the precursor for the synthesis of dopamine, adrenaline, and norepinephrine, while tryptophan is the precursor for the synthesis of serotonin. Disruptions in amino acid metabolism can lead to amino acid disorders such as homocystinuria, which is characterized by the inability to metabolize methionine, and phenylketonuria, which is defined by the inability to metabolize phenylalanine. Therefore, an understanding of the regulation of amino acid metabolism is necessary to maintain healthy metabolic function and prevent metabolic disorders.

3.4.3.7. Energy Expenditure and Intake

Energy expenditure and intake refers to the equilibrium between the body's energy needs and its energy intake from food and drink. Total energy expenditure is the total amount of energy that the body needs for physical activity, thermogenesis, and its resting metabolic rate. The basal metabolic rate is the quantity of energy required to maintain vital bodily functions such as breathing, heart rate, and body temperature. Walking, standing, and fidgeting are examples of non-exercise behaviors that are included in the energy consumption of physical activity. Thermogenesis is the process of using energy to produce heat in response to changes in temperature or food intake. Energy intake is the total amount of calories consumed through food and drink. Depending on factors like age, sex, weight, height, and level of exercise, the body requires varying amounts of energy. Consuming more calories than the body needs leads to weight gain, while consuming fewer calories than the body needs results in weight loss.

The balance of energy intake and expenditure is complex and is influenced by various factors such as hormones, genes, and the environment. Hormones like leptin and ghrelin regulate hunger and food intake, while genetic factors can affect metabolism and appetite. Environmental factors like diet, exercise, and stress also play a role in maintaining energy balance. A diet high in calories but low in nutrients can lead to overeating and weight gain, while a nutrient-dense, low-calorie diet can support healthy weight management. Stress can impact appetite, while physical activity can increase energy expenditure. Imbalances in energy intake and expenditure can contribute to metabolic disorders like obesity and the metabolic syndrome. To maintain metabolic health, it is important to understand how energy balance is regulated and the factors that can disrupt it.

3.4.3.8. How Does Exercise Affect Energy Metabolism?

Exercise has a big impact on energy metabolism. Exercise causes an increase in energy expenditure because it requires more energy for the body to perform physical tasks. This increase in energy use is caused by both increased basal metabolic rate and physical activity. During exercise, the body uses fats and carbohydrates as energy sources. During high-intensity exercise, carbohydrates are the primary energy source; they are transformed into glucose and used to produce ATP. The body stores glucose in the muscles and liver as glycogen, which can be broken down and used as energy during exercise. Nonetheless, the body uses fats as an energy source for prolonged activity. Fatty acids from fats are broken down and supplied to the mitochondria to make ATP.

Additionally, exercise increases insulin sensitivity, which makes it easier for the body to use glucose more efficiently. Insulin is a hormone that regulates blood sugar levels and facilitates the uptake of glucose by cells. Exercise raises insulin sensitivity, which reduces the amount of insulin required to enter cells and deliver glucose. Improving cells' capacity to take up and utilize glucose lowers blood sugar levels and supports healthy metabolism. Frequent exercise can also lead to changes in the composition of the body, such as an increase in muscle mass and a decrease in body fat. This may help energy metabolism because muscle tissue requires more energy to maintain than fat tissue does. Because of this, increasing muscle mass may raise basal metabolic rate, which raises the body's energy expenditure during rest (Kristinsson, 2014; Kumar and Bagchi, 2021).

3.5. VITAMINS

Vitamins are organic substances that are present in food in very small amounts and are necessary for survival as well as normal functioning. To a certain extent, humans can synthesize some vitamins. For instance, sunlight exposure generates vitamin D; tryptophan, an amino acid, can be used to make niacin; and bacteria in the stomach can produce vitamin K and biotin. But generally speaking, people get their vitamins from their food. A specific deficiency syndrome arises when a vitamin is either insufficient or cannot be used effectively. The signs and symptoms are reversed if the deficient vitamin is replenished before irreversible damage happens. The units of measurement for vitamin content in food and daily requirements are milligrams and micrograms.

Vitamins are not used by the body as an energy source or as building blocks for tissue, in contrast to macronutrients. Instead, they support metabolic and physiologic processes that occur throughout the body and aid in reactions that produce energy. For example, in addition to its role in vision, vitamin A is necessary for proper immunological function, growth, reproduction, embryonic development, and the integrity of epithelial cells. One of the B vitamins, folic acid (folate), helps prevent birth defects in the early stages of pregnancy. The B vitamins work as coenzymes to aid in energy metabolism. In addition to being an antioxidant that helps guard against damage from reactive molecules (free radicals), vitamin C is involved in the formation of connective tissue. Vitamin D is involved in calcium and phosphorus homeostasis as well as bone metabolism and is now thought to be a hormone. Another antioxidant, vitamin E, guards lipid systems from the damaging effects of free radicals, and vitamin K is essential for blood clotting. Even though vitamins are frequently discussed separately, many of their roles are connected, and a lack of one can affect how well another works.

Vitamin nomenclature is a bit complicated, with chemical names progressively taking the place of the initial letter designations developed during the first half of the 20th century when vitamins were first discovered. The knowledge that vitamins belong to families that may have several active forms complicates nomenclature even more. Certain vitamins are present in food in precursor forms that the body needs to activate for them to perform their intended functions. For instance, the body transforms the plant-based compound beta(⊠)-carotene into vitamin A. Based on their solubility, the 13 vitamins that are known to be necessary for human health are divided into two categories. Vitamins A, D, E, and K are the four fat-soluble vitamins (soluble in nonpolar solvents). The activated form of vitamin D, called vitamin D hormone (calcitriol), is still included in the category of vitamins even though it is now understood to function as a hormone. The eight B-complex vitamins—thiamin, riboflavin, niacin, vitamin B6, folic acid, vitamin B_{12}, pantothenic acid, and biotin—as well as vitamin C are the nine water-soluble vitamins (soluble in polar solvents). Similar to a vitamin, choline is a dietary element that the body can synthesize but is obviously necessary for proper metabolism. While some premature infants and people with specific medical conditions may require choline in their diet, it has not been proven to be a necessary component of a human's diet over the course of their lifetime.

Certain vitamins are more or less vulnerable to being destroyed by chemicals and the environment. For instance, riboflavin is particularly susceptible to UV or fluorescent light, thiamin to extended heating, and vitamin C to oxidation (as occurs when a piece of fruit is cut open and the vitamin is exposed to air). Generally speaking, cooking can more readily destroy water-soluble vitamins than fat-soluble ones. A vitamin's solubility affects not only where it is found in food but also how the body absorbs, transports, stores, and excretes it. The water-soluble vitamins are produced by plants and can be found in both plant and animal-based diets, except vitamin B_{12}, which can only be obtained from foods derived from animals. As a result, strict vegetarians and vegans who abstain from all animal products run the risk of developing a vitamin B_{12} deficit. Conversely, fat-soluble vitamins are associated with fats and oils in food and the body, and they usually need protein carriers to get through the body's water-filled tissues.

Except for vitamin B_{12}, water-soluble vitamins cannot be significantly stored in the body and must be taken regularly through diet. Although even water-soluble vitamins have the potential to be toxic, if taken in excess, they are easily eliminated through the urine. Vitamin B_6 is particularly notable in this respect. As long as average intakes over time—weeks, months, or even years—meet the body's needs, fat-soluble vitamins don't necessarily need to be taken daily because they are stored in the liver and fatty tissue. The ability of these vitamins to be stored, however, raises the risk of toxicity if extremely high dosages are taken. This is especially concerning in the case of vitamins A and D, which when taken in excess can be toxic. Certain vitamins have recognized medical uses when they are present in pharmacological ("megadose") levels, which are many times greater than the amounts normally found in food. For example, vitamin D is used to treat psoriasis; niacin lowers blood cholesterol; pharmacological derivatives of vitamin A reduce skin wrinkling and treat acne and other skin conditions. It is not recommended to take large amounts of vitamins or other dietary supplements beyond recommended limits without medical supervision.

The body cannot tell the difference between vitamins that are synthesized in a lab and vitamins that are obtained from food. Different vitamin forms aren't always interchangeable, though. As an example, supplements labeled d-⊠-tocopherol (or natural) typically contain higher levels of vitamin E activity than supplements labeled dl-⊠-tocopherol. Vitamins found in food have a clear benefit over vitamins found in supplements because they are linked to other potentially advantageous substances and have a lower risk of toxicity. Supplements with nutrients cannot take the place of a balanced diet.

3.6. MINERALS

The human body needs minerals to remain healthy. Minerals are needed by a person's body for a variety of functions, such as maintaining healthy bones, muscles, heart, and brain. The synthesis of hormones and enzymes depends on minerals as well. Macrominerals and trace minerals are the two categories of minerals. Greater quantities of macrominerals are required. These consist of sulfur, sodium, potassium, calcium, phosphorus, magnesium, and chloride. Tiny amounts of trace minerals are all that are required. Iron, manganese, copper, iodine, zinc, cobalt, fluoride, and selenium are some of them. The majority of people eat a wide range of foods to get the amount of minerals they need. A doctor might suggest a mineral supplement in specific circumstances. Individuals who take certain medications or have certain medical conditions might need to consume less of one of the minerals. People who suffer from chronic kidney disease, for instance, should limit their intake of foods high in potassium.

3.6.1. Classification of Minerals

Depending on the quantity required by the body, minerals are categorized as major or trace minerals. Major minerals are those that need to be consumed in food in quantities greater than 100 milligrams per day. These consist of calcium, phosphorus, magnesium, sulfur, sodium, potassium, and chloride. A wide variety of foods contain these important minerals. Although mineral deficiencies can occur, eating a varied diet greatly increases a person's capacity to meet their nutrient needs.

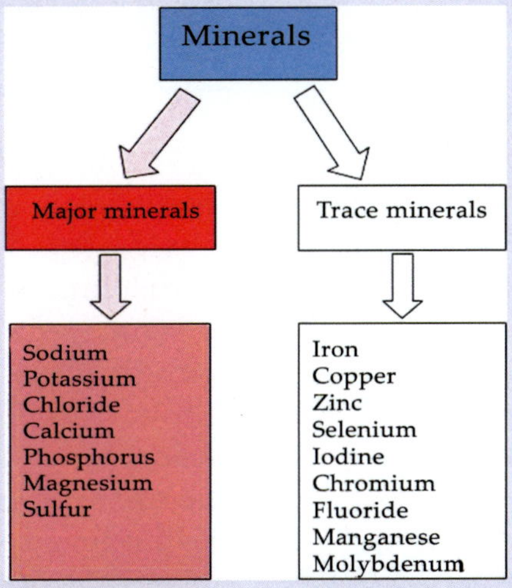

Figure 3.3. *The classification of minerals as either major minerals or trace minerals.*

Source: By Allison Calabrese, CC BY 4.0, https://openoregon.pressbooks.pub/nutritionscience/chapter/8a-classification-vitamins-minerals/.

Trace minerals are defined as those that must be consumed in smaller doses—100 milligrams or fewer per day—in the diet. They consist of manganese, molybdenum, iron, copper, zinc, selenium, iodine, chromium, and fluoride.

Even though trace minerals are required in smaller quantities, a trace mineral deficiency can have an equally negative impact on the health as a major mineral deficiency (Figure 3.3).

Minerals do not require enzymatic digestion because they are soluble in water. While certain minerals require the help of transport proteins to be absorbed and transported in the blood, most minerals are directly absorbed into the bloodstream.

The majority of vitamins are more readily absorbed than minerals, and a variety of factors affect how bioavailable minerals are.

- In general, foods derived from animals provide better absorption of minerals. Compounds found in plant-based diets, such as phytates and oxalates, can bind to minerals and prevent their absorption.
- Generally speaking, absorption of a given mineral will decrease with increased dietary intake.
- Certain minerals affect how others are absorbed. For example, an excess of zinc in the diet can hinder the absorption of copper and iron. On the other hand, some vitamins improve the absorption of minerals. For instance, vitamin D increases the absorption of calcium and magnesium, while vitamin C increases the absorption of iron.
- Just like vitamins, mineral absorption can be hindered by various gastrointestinal disorders and diseases like Crohn's disease, kidney disease, and the aging process. This puts individuals with malabsorption issues and older adults at increased risk for mineral deficiencies.

3.6.1.1. Calcium

A vital role for calcium in the body. The majority—99%—is kept in bone tissue. The most well-known benefit of calcium is that it helps to strengthen and develop teeth and bones. In the body, calcium is essential for the mineralization of the skeleton (bone), the contraction of muscles, and the formation of blood clots. In addition, calcium is required for the nervous system to transmit messages and for the secretion of hormones and enzymes.

Blood calcium levels are meticulously regulated so that in the event of a decrease, the body will quickly react by inducing bone resorption and releasing calcium that has been stored in the body. Calcium levels are regulated by a variety of hormones; calcitriol and parathyroid hormone combine to raise blood calcium levels, whereas calcitonin has the opposite effect and lowers blood calcium levels.

Bone fractures and osteoporosis can result from low calcium levels. To maintain levels, the body requires 1000–1200 mg of calcium per day. Dairy products, leafy greens, legumes, sardines, and fortified orange juice are good sources of calcium. A few things can affect how well calcium is absorbed. For instance, vitamin D enhances the absorption of calcium. Its absorption can be hindered by additional food ingredients like oxalic acid and phytic acid, which are present in spinach, collard greens, and beans. These variables don't really affect the nutritional status of individuals who eat a wide range of foods.

3.6.1.2. Phosphorus

The second most common mineral in the human body is phosphorus. The skeleton contains roughly 85% of it. Phosphorus in the form of phosphate is a primary mineral in the skeleton and also forms the backbones of RNA and DNA, phospholipids, and adenosine triphosphate (ATP), a molecule that stores energy in cells. A phosphorus shortage is uncommon. Organic phosphorus, also known as phosphorus in food, is naturally present in foods high in protein, including meats, poultry, fish, nuts, beans, and dairy products. Fast food, canned and bottled drinks, ready-to-eat foods, and the majority of processed foods all contain phosphorus that has been added to them as an additive or preservative (inorganic phosphorus). Different dietary sources absorb phosphorus differently; food additives absorb phosphorus more easily. 700 milligrams of phosphorus should be consumed daily. The recommended daily allowance (UL) for phosphorus is 4,000 mg for adults and 3,000 mg for those over 70.

3.6.1.3. Magnesium

1% of mineralized bone tissue is made up of the skeleton, which contains 60% of the magnesium present in the human body. Numerous bodily processes, such as calcium regulation, enzyme reactions, platelet activity, and bone mineralization, are influenced by magnesium. Damage to bone remodeling is linked to a diet low in magnesium, which also lowers parathyroid hormone levels and activates vitamin D. The majority of foods contain magnesium. Wholegrain cereals, nuts, and seeds are among the foods high in magnesium. Food processing, like whole wheat refinement, significantly lowers the magnesium content of the food (by more than 75%). A substantial amount of magnesium is also found in leafy green vegetables, which also contain chlorophyll. Dairy products, seafood, spices, legumes, and legumes are additional dietary sources of magnesium. For adults, the recommended daily intake of magnesium is 420 mg for males and 320 mg for females.

3.6.1.4. Sulfur

Although sulfur is a significant mineral, the body does not use it as a nutrition on its own. The body uses sulfur in the structures of several different substances, including lipoic acid, methionine, taurine, and cysteine, as well as vitamins like biotin and thiamin. Sulfur is frequently found in foods that contain protein and in body proteins, particularly in the skin, hair, and nails. This is because sulfur is present in amino acids. When protein requirements are satisfied, there is no recommended intake or deficiency for sulfur. Dairy products, eggs, and animal proteins like meat are sources of sulfur.

3.6.2. Essential Trace and Ultratrace Minerals

3.6.2.1. Iron

Iron serves a variety of purposes in the body. A protein called hemoglobin transports oxygen from the lungs throughout the body. It is also a component of myoglobin, a protein that gives muscles oxygen. Certain hormones are also made possible by iron. Hemoglobin makes up more than 65% of the body's iron, myoglobin can make up to 10% of it, enzymes can contain between 1% and 5% of it, and the rest is stored in the body and found in the blood. There are two types of iron in food: heme and nonheme. Heme iron is derived from myoglobin and hemoglobin and is present in meat, poultry, and fish products. The predominant type of iron found in plant-based diets is nonheme iron, which is found in whole grains, nuts, legumes, peas, fruits, vegetables, and tofu. Meat, poultry, fish, and foods high in vitamin C, like citrus fruits, strawberries, sweet peppers, tomatoes, and broccoli, enhance the absorption of iron derived from plant sources. Oxalates, phytates, polyphenols (tannins in tea and coffee), and other minerals are among the substances that prevent the body from absorbing iron. For men, the recommended daily iron intake is 8 mg, while for women it is 18 mg.

3.6.2.2. Zinc

Zinc is present in every organ, tissue (particularly bone and muscle), bodily fluid, and cell throughout the body. It supports the immune system's defense against bacterial and viral invasions. The synthesis of DNA and proteins requires zinc as well. It is also necessary for healthy taste and smell perception and aids in the healing of wounds. Seafoods like oysters and red meats, particularly organ meats, are excellent providers of zinc. Dairy products, pork, and poultry are some additional reasonably good animal sources of zinc. It is estimated that 40–70% of the zinc that most Americans consume comes from animal products. Legumes and whole grains also contain modest levels of zinc. Fortified cereals are thought to provide about 30% of the zinc in the U.S. diet. Fruits contain little amounts of zinc. Zinc content in fruits is relatively low. Plant-based zinc is less absorbed than zinc obtained from animal sources, such as meat. For men, the recommended daily intake is 11 mg, and for women, it is 8 mg.

3.6.2.3. Copper

All bodily tissues and the majority of secretions contain copper. Four parts of the body contain copper: 40% is found in bone, 24% is found in muscle, 9% is found in the liver, and 6% is found in the brain. Copper is essential for the health of bones, nerves, collagen, and skin pigment expression, among other things. It helps the body absorb iron as well. Similar to iron, copper aids in the production of red blood cells. The highest dietary sources of copper are shellfish and meats, particularly organ meats. Rich sources of plant food that are high in copper are legumes, nuts, seeds, and dried fruits. Whole grains, potatoes, and cocoa (chocolate) are additional excellent sources of the mineral. The recommended daily intake of copper is 900 micrograms.

3.6.2.4. Selenium

Selenium is necessary for DNA synthesis, thyroid gland function, reproduction, and defense against infection and free radical damage to the body. Generally speaking, plants, seafood, and muscle meats are the best providers of selenium. Organic forms of selenium are primarily found as selenocysteine in animal-based diets and as selenomethionine primarily in plant-based diets. Vegetables (such as beets and cabbage) and livestock that have received inorganic selenium supplements contain inorganic forms of selenium. Due to the wide variations in selenium levels in soil, foods grown on these soils have varying selenium contents. 55 mcg of selenium should be consumed daily.

3.6.2.5. Chromium

Water, soil, and the air all contain the metal chromium. It can be found in a range of oxidation states, including Cr^{2}–Cr^{6+}. Nonetheless, the trivalent form, Cr^{3+}, is the most significant in humans and is the most stable oxidation state. Though the exact mechanism is still unknown, chromium is thought to improve the action of insulin. The impact of chromium supplements on improvements in glucose control measures in people with type 2 diabetes is still being investigated in numerous studies. Meats and whole grains are generally regarded as excellent dietary sources of chromium, in addition to fruits and vegetables. Few foods, nevertheless, have undergone precise chromium content analysis. For males, 35 micrograms per day and for females, 25 micrograms per day is the recommended daily intake of chromium. The average adult in the US takes in between 23 and 54 micrograms of chromium per day.

3.6.2.6. Iodine

The thyroid gland uses iodine extensively in the synthesis of thyroid hormones. Along with many other vital processes, the body's metabolism is regulated by these hormones. Thyroid hormones are also necessary for healthy bone and brain development in the womb and during infancy. Foods' iodine content varies widely and is mostly influenced by soil concentrations. Eggs, dairy, meats, and seafood are all excellent sources of iodine. Iodine intake guidelines call for 150 mcg daily. An enlarged thyroid gland, or goiter, is frequently the initial outward sign of an iodine deficiency.

3.6.2.7. Fluoride

The mineral known to prevent tooth decay is fluoride. It supports the growth and maintenance of teeth and bones. Fluoride works against tooth decay in three ways: by preventing bacteria from producing acid, by keeping teeth from demineralizing, and by helping damaged enamel to remineralize. Foods that are solid have a low fluoride content. This fluoride content varies depending on whether they were cooked in fluoridated water or grown in fluoridated soil and water. Fluoridated water provides over 70% of the fluoride that an individual has in their body. The majority of toothpastes and mouthwashes contain fluoride, which can marginally increase the daily fluoride intake even though they shouldn't be consumed. The majority of Americans obtain adequate fluoride from their diet, beverages, and dental products (Kristinsson, 2014; Kumar and Bagchi, 2021) (Table 3.2).

Table 3.2. Certain Micronutrients Have Key Roles in the Immune System

Immune Function Roles	Micronutrient	Comments
Physical and Biochemical Barriers		
Maintenance of structural and functional integrity of mucosal cells in innate barriers (e.g., skin, respiratory tract)	Vitamin A	Normal differentiation of epithelial tissue; retinoic acid essential to imprint T and B cells with gut-homing specificity and array T cells and IgA^+ cells into intestinal tissues (Macierzanka, Torcello-Gómez, Jungnickel, & Maldonado-Valderrama, 2019). important for intestinal immune response, thus supporting the gut barrier (Schrauzer, 2002; Soliman, 2018; Warren, & Kauffman, 2003); carotenoids (either provitamin A or nonprovitamin A carotenoids) have immunoregulatory actions including reducing the toxic effects of ROS and regulating membrane fluidity and gap-junctional communication (Yuan, Al-Shali, & Hegele, 2007).

	Vitamin D	Calcitriol regulates antimicrobial proteins (cathelicidin and ⊠-defensin), responsible for modifying intestinal microbiota to a healthier composition and supporting the gut barrier (Schrauzer, 2002; Zarse, et al. 2011), as well as protecting the lungs against infection [Gombart, A.F.2009]; increases tight junction protein expression, E-cadherin, and connection 43 in the gut [Clairmont, A., et al, 1996; Gniadecki, R., et al. 1997; Palmer, H.G. et al. 2001]; maintains renal epithelial barrier function [Mihajlovic, M., et al. 2017]; enhances corneal epithelial barrier function [Yin, Z., et al. 2011]
	Vitamin C	Promotes collagen synthesis and protects cell membranes from damage caused by free radicals, thus supporting integrity of epithelial barriers (Macierzanka, Torcello-Gómez, Jungnickel, & Maldonado-Valderrama, 2019); enhances keratinocyte differentiation and lipid synthesis as well as fibroblast proliferation and migration [Carr, A.; Maggini, S., 2017]
	Vitamin E	Protects cell membranes from damage caused by free radicals and support the integrity of epithelial barriers (Kumar, & Bagchi, 2020; Macierzanka, Torcello-Gómez, Jungnickel, & Maldonado-Valderrama, 2019)
	Vitamins B6, B12 and folate	All involved in intestinal immune regulation (e.g., by mediating lymphocyte migration into the intestine in the case of vitamin B6, while folate is essential for the survival of regulatory T cells in the small intestine, and human gut microbes use vitamin B12 as a cofactor for metabolic pathways), thus supporting the gut barrier (Schrauzer, 2002) [Yoshii, K.; Hosomi, K.; Sawane, K.; Kunisawa, J., 2019]
	Iron	Essential for differentiation and growth of epithelial tissue (Brand-Miller, et al. 2009)
	Zinc	Helps maintain integrity of skin and mucosal membrane (e.g., cofactor for metalloenzymes required for cell membrane repair [Lin, P.H.; Sermersheim, M.; Li, H.; Lee, P.H.U.; Steinberg, S.M.; Ma, J., 2017])
Immune Cells		
Differentiation, proliferation, functioning, and movement of innate immune cells	Vitamin A	For example, regulates number and function of NK cells (Azam et al. 1983; Macierzanka, Torcello-Gómez, Jungnickel, & Maldonado-Valderrama, 2019), contributes to phagocytic and oxidative burst activity of macrophages (Macierzanka, Torcello-Gómez, Jungnickel, & Maldonado-Valderrama, 2019)
	Vitamin D	Vitamin D receptor found in, e.g., monocytes, macrophages, and DCs (Kumar, & Bagchi, 2020); increases differentiation of monocytes to macrophages (Brand-Miller et al., 2009); calcitriol promotes movement and phagocytic ability of macrophages (Macierzanka, Torcello-Gómez, Jungnickel, & Maldonado-Valderrama, 2019; Wishart, K., et al. 2017 Sly, L.M. et al. 2001; Tanaka, H. et al. 1991]

	Vitamin C	Involved in proliferation, function, and movement of neutrophils, monocytes, phagocytes (Azam et al. 1983); maintains or enhances NK cell activities and chemotaxis (Azam et al. 1983; Brand-Miller, et al. 2009; Macierzanka, Torcello-Gómez, Jungnickel, & Maldonado-Valderrama, 2019; [Wu, D.; Lewis, E.D.; Pae, M.; Meydani, S.N., 2019]; enhances phagocytosis and ROS generation, enhances microbial killing [Carr, A.; Maggini, S. 2017]; involved in apoptosis and clearance of spent neutrophils from sites of infection by macrophages [Carr, A.; Maggini, S. 2017] attenuates extracellular trap (NET) formation, thus reducing associated tissue damage [Bozonet, S.M.; Carr, A.C.2019].
	Vitamin E	Maintains or enhances NK cell cytotoxic activity (Azam et al. 1983; Brand-Miller, et al. 2009; Macierzanka, Torcello-Gómez, Jungnickel, & Maldonado-Valderrama, 2019; Wu, D.; Lewis, E.D.; Pae, M.; Meydani, S.N., 2019] inhibits PGE2 production by macrophages (thus indirectly protecting T-cell function) [Wu, D.; Lewis, E.D.; Pae, M.; Meydani, S.N., 2019; Wu, D.; Meydani, S.N. 2014]
	Vitamin B6	Maintains or enhances NK cell cytotoxic activity (Azam et al. 1983; Macierzanka, Torcello-Gómez, Jungnickel, & Maldonado-Valderrama, 2019; [Wu, D.; Lewis, E.D.; Pae, M.; Meydani, S.N., 2019]
	Vitamin B12	May act as immunomodulator for cellular immunity, effects on cytotoxic cells (e.g., NK cells, cytotoxic T cells) (Brand-Miller, et al. 2009)
	Folate	Maintains or enhances NK cell cytotoxic activity ((Azam et al. 1983; Macierzanka, Torcello-Gómez, Jungnickel, & Maldonado-Valderrama, 2019; Wu, D.; Lewis, E.D.; Pae, M.; Meydani, S.N., 2019)
	Zinc	Maintains or enhances NK cell cytotoxic activity (Azam et al. 1983; Macierzanka, Torcello-Gómez, Jungnickel, & Maldonado-Valderrama, 2019; Wu, D.; Lewis, E.D.; Pae, M.; Meydani, S.N., 2019); central role in cellular growth and differentiation of immune cells that have a rapid differentiation and turnover [World Health Organization; Food and Agricultural Organization of the United Nations. 2006]; enhances the phagocytic activity of peritoneal macrophages for E. coli and S. aureus [Gao, H.; Dai, W.; Zhao, L.; Min, J.; Wang, F. 2018]; improves phagocytic capacity of monocytes [Wu, D.; Lewis, E.D.; Pae, M.; Meydani, S.N., 2019; Sheikh, A., et al. 2010]
	Iron	Forms highly-toxic hydroxyl radicals, thus involved in killing of bacteria by neutrophils; component of enzymes critical for functioning of immune cells (e.g., ribonucleotide reductase involved in DNA synthesis); involved in regulation of cytokine production and action (Brand-Miller, et al. 2009); iron-rich status promotes M2-like macrophage phenotype and negatively regulates M1 pro-inflammatory response [Agoro, R.; Taleb, M.; Quesniaux, V.F.J.; Mura, C. 2018]

	Copper	Role in functions of macrophages (e.g., copper accumulates in phagolysosomes of macrophages to combat certain infectious agents [Besold, A.N.; Culbertson, E.M.; Culotta, V.C., 2016]), neutrophils and monocytes [Saeed, F.; Nadeem, M.; Ahmed, R.; Nadeem, M.; Arshad, M.; Ullah, A., 2016]; enhances NK cell activity [Beck, M.A. 1999]
	Selenium	Selenoproteins important for antioxidant host defense system, affecting leukocyte and NK cell function [Saeed, F.; Nadeem, M.; Ahmed, R.; Nadeem, M.; Arshad, M.; Ullah, A., 2016]
	Magnesium	Cofactor of enzymes of nucleic acid metabolism and stabilizes structure of nucleic acids; involved in DNA replication and repair [Petrovic, J., et al., 2016]; roles in antigen binding to macrophages [Laires, M.J.; Monteiro, C. 2008]; regulates leukocyte activation [Laires, M.J.; Monteiro, C. 2008]; involved in the regulation of apoptosis [Petrovic, J., et al., 2016]
Antimicrobial effects	Vitamin A	Downregulates IFNγ production (Brand-Miller, et al. 2009; Carr, A.; Maggini, S., 2017)
	Vitamin D	Calcitriol regulates antimicrobial protein expression (cathelicidin and defensin), which directly kill pathogens, especially bacteria (Kumar, & Bagchi, 2020; Wang, T.T. et al., 2004; Gombart, A.F. et al., 2005; Weber, G. et al., 200); inhibits IFNγ production [Wu, D.; Lewis, E.D.; Pae, M.; Meydani, S.N. , 2019; Matsui, T. et al., 1986; Reichel, H. et al., 1987; Rigby, W.F. et al., 1987; Inoue, M. et al., 1998]
	Vitamin C	High levels can improve antimicrobial effects; increases serum levels of complement proteins (Kumar, & Bagchi, 2020); role in IFN⊠ production (Brand-Miller, et al., 2009; Carr, A.; Maggini, S., 2017)
	Zinc	Involved in complement activity; role in IFNγ production (Brand-Miller, et al., 2009; Carr, A.; Maggini, S., 2017)
	Iron	Role in IFNγ production (Brand-Miller, et al., 2009; Carr, A.; Maggini, S., 2017)
	Copper	Intrinsic antimicrobial properties (Kumar, & Bagchi, 2020)
	Selenium	Increases IFNγ production (Brand-Miller, et al., 2009; Carr, A.; Maggini, S., 2017)
Roles in inflammation, antioxidant effects, and effects in oxidative burst	Vitamin A	Helps to regulate the production of IL-2 and the proinflammatory TNF-α, which activates the microbial action of macrophages; involved in phagocytic and oxidative burst activity of macrophages activated during inflammation (Macierzanka, Torcello-Gómez, Jungnickel, & Maldonado-Valderrama, 2019)
	Vitamin D	Calcitriol increases the oxidative burst potential of macrophages [Wishart, K., et al. 2017 Sly, L.M. et al. 2001; Tanaka, H. et al. 1991; Tanaka, H. et al., 1991] increases superoxide synthesis (Macierzanka, Torcello-Gómez, Jungnickel, & Maldonado-Valderrama, 2019); reduces the expression of pro-inflammatory cytokines and increases the expression of anti-inflammatory cytokines by macrophages [Wishart, K. 2017; Lin, Z.; Li, W. 2016; Zhang, Y. et al., 2012; Topilski, I. 2004]

	Vitamin C	Maintains redox homeostasis within cells and protects against ROS and RNS during oxidative burst (Macierzanka, Torcello-Gómez, Jungnickel, & Maldonado-Valderrama, 2019); regenerates other important antioxidants, such as glutathione and vitamin E, to their active state [Wintergerst, E.; Maggini, S.; Hornig, D. 2006]; modulates cytokine production and decreases histamine levels [Carr, A.; Maggini, S., 2017]
	Vitamin E	Important fat-soluble antioxidant that hinders the chain reaction induced by free radicals (chain-breaking effect) and protects cells against them [Haryanto, B.; Suksmasari, T.; Wintergerst, E.; Maggini, S. 2015; Micronutrient Information Center. 2016]; enhances IL-2 production [Haryanto, B.; Suksmasari, T.; Wintergerst, E.; Maggini, S. 2015]; decreases production of PGE2 (indirectly protecting T-cell function) [Lee, G.Y.; Han, S.N. 2018]
	Vitamin B6	Required in endogenous synthesis and metabolism of amino acids, the building blocks of cytokines (Kumar, & Bagchi, 2020); helps to regulate inflammation (higher levels of the active form result in lower rates of inflammation) [Saeed, F.; Nadeem, M.; Ahmed, R.; Nadeem, M.; Arshad, M.; Ullah, A. 2016; Sakakeeny, L.; Roubenoff, R. et al., 2012; Ueland, P.M. 2017]
	Zinc	Anti-inflammatory agent [Jarosz, M.; Olbert, M.; Wyszogrodzka, G.; Młyniec, K.; Librowski, T. 2017]; helps to modulate cytokine release (Brand-Miller, et al. 2009; Wintergerst, E.; Maggini, S.; Hornig, D. 2006] by dampening the development pro-inflammatory Th17 and Th9 cells [Wu, D.; Lewis, E.D.; Pae, M.; Meydani, S.N. 2019; Kitabayashi, C. et al., 2010; Maywald, M.; Wang, F.; Rink, L. 2018] and influencing the generation of cytokines such as IL-2, IL-6, and TNF-⊠ [Foster, M.; Samman, S. 2012; Wessels, I.; Rink, L. 2019]; has antioxidant effects that protect against ROS and reactive nitrogen species [Wintergerst, E.; Maggini, S.; Hornig, D. 2006]; influences activity of antioxidant proteins (Macierzanka, Torcello-Gómez, Jungnickel, & Maldonado-Valderrama, 2019)
	Iron	Involved in regulation of cytokine production and action (Brand-Miller, et al. 2009); required for generation of pathogen-killing ROS by neutrophils during oxidative burst (Kumar, & Bagchi, 2020)
	Copper	Accumulates at sites of inflammation (Kumar, & Bagchi, 2020; Saeed, F.; Nadeem, M.; Ahmed, R.; Nadeem, M.; Arshad, M.; Ullah, A. 2016); part of copper/zinc-superoxide dismutase, a key enzyme in defense against ROS (Macierzanka, Torcello-Gómez, Jungnickel, & Maldonado-Valderrama, 2019); free-radical scavenger [Alpert, P. 2017]; changes in copper homeostasis a crucial component of respiratory burst (Macierzanka, Torcello-Gómez, Jungnickel, & Maldonado-Valderrama, 2019); important for IL-2 production and response (Kumar, & Bagchi, 2020; Saeed, F.; Nadeem, M.; Ahmed, R.; Nadeem, M.; Arshad, M.; Ullah, A. 2016); maintains intracellular antioxidant balance, suggesting important role in inflammatory response (Macierzanka, Torcello-Gómez, Jungnickel, & Maldonado-Valderrama, 2019)

	Selenium	Essential for function of selenoproteins that act as redox regulators and cellular antioxidants, potentially counteracting ROS produced during oxidative stress (Azam et al., 1983)
	Magnesium	Can help to protect DNA against oxidative damage [Petrovic, J.; Stanic, D.; Dmitrasinovic, G.; Plecas-Solarovic, B.; Ignjatovic, S.; Batinic, B.; Popovic, D.; Pesic, V. 2016]; high concentrations reduce superoxide anion production [Bussiere, F.I.; Mazur, A.; Fauquert, J.L.; Labbe, A.; Rayssiguier, Y.; Tridon, A. 2002]
Differentiation, proliferation and normal functioning of T cells	Vitamin A	Involved in development and differentiation of Th1 and Th2 cells [Ross, A.C. 2012]; enhances TGF-⊠-dependent conversion of naïve T cells into regulatory T cells (Macierzanka, Torcello-Gómez, Jungnickel, & Maldonado-Valderrama, 2019); plays a role in acquisition of mucosal-homing properties by T and B cells (Macierzanka, Torcello-Gómez, Jungnickel, & Maldonado-Valderrama, 2019)
	Vitamin D	Homing of T cells to the skin [igmundsdottir, H.; Pan, J.; Debes, G.F.; Alt, C.; Habtezion, A.; Soler, D.; Butcher, E.C. 2007]; calcitriol inhibits T-cell proliferation [Micronutrient Information Center. 2016]; inhibitory effects mainly in adaptive immunity (e.g., Th1-cell activity) (Kumar, & Bagchi, 2020); stimulatory effects in innate immunity (Kumar, & Bagchi, 2020); inhibits the effector functions of T helper cells and cytotoxic T cells [Wu, D.; Lewis, E.D.; Pae, M.; Meydani, S.N. 2019; Sassi, F.; Tamone, C.; D'Amelio, P. 2018], but promotes the production of Tregs [Wu, D.; Lewis, E.D.; Pae, M.; Meydani, S.N. 2019; Sassi, F.; Tamone, C.; D'Amelio, P. 2018; Cantorna, M.T.; Snyder, L.; Lin, Y.D.; Yang, L. 2015; inhibitory effect on the differentiation and maturation of the antigen-presenting DCs, and helps program DCs for tolerance [Wu, D.; Lewis, E.D.; Pae, M.; Meydani, S.N. 2019; Penna, G.; Adorini, L. 2000; Piemonti, L.et al., 2000; Bscheider, M.; Butcher, E.C. 2016]
	Vitamin C	Roles in production, differentiation, and proliferation of T cells, particularly cytotoxic T cells (Brand-Miller, et al., 2009; Carr, A.; Maggini, S., 2017]
	Vitamin E	Enhances lymphocyte proliferation and T-cell-mediated functions (Brand-Miller, et al., 2009); optimizes and enhances Th1 response (Brand-Miller, et al., 2009)
	Vitamin B6	Involved in lymphocyte proliferation, differentiation, maturation, and activity (Kumar, & Bagchi, 2020); maintains Th1 immune response (Brand-Miller, et al., 2009)
	Vitamin B12	Involved in one-carbon metabolism (interactions with folate) [Saeed, F.; Nadeem, M.; Ahmed, R.; Nadeem, M.; Arshad, M.; Ullah, A. 2016]; facilitates production of T cells [Saeed, F.; Nadeem, M.; Ahmed, R.; Nadeem, M.; Arshad, M.; Ullah, A. 2016], such as cytotoxic T cells (Brand-Miller, et al., 2009; Macierzanka, Torcello-Gómez, Jungnickel, & Maldonado-Valderrama, 2019) helps to regulate ratio between T helper cells and cytotoxic T cells (Macierzanka, Torcello-Gómez, Jungnickel, & Maldonado-Valderrama, 2019)
	Folate	Supports Th 1-mediated immune response [Saeed, F.; Nadeem, M.; Ahmed, R.; Nadeem, M.; Arshad, M.; Ullah, A. 2016]

	Zinc	Induces proliferation of cytotoxic T cells [Wintergerst, E.; Maggini, S.; Hornig, D. 2007]; involved in Th1 cytokine production and thus supports Th1 response [Haryanto, B.; Suksmasari, T.; Wintergerst, E.; Maggini, S. 2015]; essential for intracellular binding of tyrosine kinase to T cell receptors, required for T cell development, differentiation, and activation [Wintergerst, E.; Maggini, S.; Hornig, D. 2006; induces development of Treg cells and is thus important in maintaining immune tolerance [Wu, D.; Lewis, E.D.; Pae, M.; Meydani, S.N. 2019; Kitabayashi, C. et al., 2010; Maywald, M.; Wang, F.; Rink, L. 2018]
	Iron	Important in differentiation and proliferation of T cells (Kumar, & Bagchi, 2020); helps to regulate ratio between T helper cells and cytotoxic T cells (Brand-Miller, et al., 2009)
	Copper	Roles in differentiation and proliferation of T cells [Saeed, F.; Nadeem, M.; Ahmed, R.; Nadeem, M.; Arshad, M.; Ullah, A. 2016]
	Selenium	Roles in differentiation and proliferation of T cells [Saeed, F.; Nadeem, M.; Ahmed, R.; Nadeem, M.; Arshad, M.; Ullah, A. 2016; Alpert, P. 2017]; helps to improve Th cell counts [Hurwitz, B.E.; Klaus, J.R.; Llabre, M.M.; Gonzalez, A.; Lawrence, P.J.; Maher, K.J.; Greeson, J.M.; Baum, M.K.; Shor-Posner, G.; Skyler, J.S.; et al. 2007]
Antibodies		
Antibody production and development	Vitamin A	Development and differentiation of Th 1 and Th2 cells (Macierzanka, Torcello-Gómez, Jungnickel, & Maldonado-Valderrama, 2019); maintains normal antibody-mediated Th2 response by suppressing IL-12, TNF-⊠, and IFN-⊠ production of Th1 cells (Kumar, & Bagchi, 2020)
	Vitamin D	Calcitriol suppresses antibody production by B cells (Kumar, & Bagchi, 2020)
	Vitamin C	Promotes proliferation of lymphocytes, resulting in increased generation of antibodies [Carr, A.; Maggini, S., 2017]
	Vitamin E	Suppresses Th2 response (Brand-Miller, et al., 2009)
	Vitamin B6	Required in endogenous synthesis and metabolism of amino acids, the building blocks of antibodies (Kumar, & Bagchi, 2020); inhibits Th2 cytokine-mediated activity (Macierzanka, Torcello-Gómez, Jungnickel, & Maldonado-Valderrama, 2019)
	Vitamin B12	Important for antibody production and metabolism, via folate mechanism (Kumar, & Bagchi, 2020; Macierzanka, Torcello-Gómez, Jungnickel, & Maldonado-Valderrama, 2019; Saeed, F.; Nadeem, M.; Ahmed, R.; Nadeem, M.; Arshad, M.; Ullah, A. 2016]; required for optimal clonal expansion (Macierzanka, Torcello-Gómez, Jungnickel, & Maldonado-Valderrama, 2019)
	Folate	Important for antibody production and metabolism (Kumar, & Bagchi, 2020; Macierzanka, Torcello-Gómez, Jungnickel, & Maldonado-Valderrama, 2019, Saeed, F.; Nadeem, M.; Ahmed, R.; Nadeem, M.; Arshad, M.; Ullah, A. 2016]
	Zinc	Involved in antibody production, particularly IgG [Shankar, A.H.; Prasad, A.S. 1998; Ibs, K.-H.; Rink, L. 2003]
	Selenium	Helps to maintain antibody levels [Saeed, F.; Nadeem, M.; Ahmed, R.; Nadeem, M.; Arshad, M.; Ullah, A. 2016]

	Magnesium	Cofactor in antibody synthesis, role in antibody-dependent cytolysis and IgM lymphocyte binding [Laires, M.J.; Monteiro, C. 2008]
Responses to antigen	Vitamin A	Normal functioning of B cells, necessary for generation of antibody responses to antigen (Kumar, & Bagchi, 2020); required for B cell-mediated IgA antibody responses to bacterial polysaccharide antigens (Macierzanka, Torcello-Gómez, Jungnickel, & Maldonado-Valderrama, 2019)
	Vitamin D	Promotes antigen processing (Macierzanka, Torcello-Gómez, Jungnickel, & Maldonado-Valderrama, 2019); role in the down-regulation of MHC-II [Saeed, F.; Nadeem, M.; Ahmed, R.; Nadeem, M.; Arshad, M.; Ullah, A. 2016]
	Vitamin E	Helps to form effective immune synapses between and Th cells [Wu, D.; Lewis, E.D.; Pae, M.; Meydani, S.N. 2019]; increases proportion of antigen-experienced memory T cells [Han, S.N.; Adolfsson, O.; Lee, C.K.; Prolla, T.A.; Ordovas, J.; Meydani, S.N. 2004]
	Folate	Important for sufficient antibody response to antigens [Saeed, F.; Nadeem, M.; Ahmed, R.; Nadeem, M.; Arshad, M.; Ullah, A. 2016]
	Zinc	Involved in antibody response (Macierzanka, Torcello-Gómez, Jungnickel, & Maldonado-Valderrama, 2019); important in maintaining immune tolerance (i.e., the ability to recognize "self" from "non-self") [Wu, D.; Lewis, E.D.; Pae, M.; Meydani, S.N. 2019]
	Magnesium	Key role in antigen binding to macrophage RNA [Laires, M.J.; Monteiro, C. 2008]; involved in antibody-dependent cytolysis [Laires, M.J.; Monteiro, C. 2008]

Source: Gombart AF, Pierre A, Maggini S. A review of micronutrients and the immune system–working in harmony to reduce the risk of infection. Nutrients. 2020; 12(1):236. https://doi.org/10.3390/nu12010236.

Calcitriol = 1,25-dihydroxyvitamin D_3, the active form of vitamin D. Selenoproteins are selenium-dependent enzymes. APC, antigen-presenting cell; DC, dendritic cells; IFN, interferon; IL, interleukin; MHC, major histocompatibility complex; NK, natural killer; PGE2, prostaglandin E2; RNS, reaction nitrogen species; ROS, reactive oxygen species; Th, helper T cell; TGF, transforming growth factor; TNF, tumor-necrosis factor; Tregs, regulatory T cells.

3.7. WATER, FLUIDS, ELECTROLYTES AND ACID-BASE BALANCE

The human body helps to maintain homeostasis and appropriate functioning by keeping a careful balance of fluids and electrolytes. An imbalance in fluids or electrolytes puts people at risk for organ system dysfunction. Undiagnosed and untreated imbalances impair the ability of organ systems to function correctly, which leads to death. To act quickly, nurses must be able to spot subtle changes in their patients' fluid or electrolyte balances. Prompt evaluation and action avert difficulties and preserve lives (Kristinsson, 2014; Kumar and Bagchi, 2021).

3.7.1. Water

Water (H_2O) is the most important nutrient of all, despite being disregarded as a nutrient quite a bit. People can go for weeks without eating, but they can only last a few days without water. Water serves as a medium for the various biochemical reactions that take place during metabolism and the movement of nutrients and waste products throughout the body. Water is necessary for controlling body temperature, blood pressure, blood volume, large molecule structure, and tissue rigidity. Moreover, it serves as a protective cushion, a lubricant (in joints), a solvent, and an amniotic fluid and spinal fluid inside the eyes. By adjusting the concentration of electrolytes on either side of the cell membrane, the water flow into and out of cells can be precisely regulated. The main extracellular electrolytes are sodium and chloride, while the principal intracellular electrolytes are potassium, magnesium, phosphate, and sulfate.

About 50–70% of body weight is made up of water, with healthy adults making up roughly 60–70% and children making up even more. Body composition, and specifically the amount of fat, dictates the percentage of body water because lean tissue is approximately three-quarters water and fatty tissue is only approximately one-fifth water. Men typically have a higher percentage of water in their bodies than women because they have more lean tissue. In addition to being a component of other drinks and water, water is also consumed as a major ingredient in a variety of foods, especially fruits and vegetables, which can have an 85–95% water content. The body also produces water as a byproduct of metabolism. Every day, approximately 2 and a half liters (about 2.6 quarts) of water are turned over, and water excretion (mainly in the form of urine, vaporized water from the lungs, sweat from the skin, and feces) balances water intake from all sources. There is no universally accepted amount of water to drink each day because requirements differ depending on the climate, degree of activity, dietary composition, and other factors. Adults, however, normally require two liters (8 cups) or more of water per day, from all sources. When it comes to dehydration, which usually happens before the body is signaled to replace fluid, thirst is not a good indicator. Water consumption is therefore recommended throughout the day, particularly in hotter climates or during intense physical activity, during illness, or in situations that dehydrate, like flying.

3.7.2. Body Fluids and Fluid Compartments

Living things undergo chemical reactions in aqueous solutions. Solutes are the materials that dissolve in a solution. Solutes in the human body can vary depending on the area of the body, but they can include proteins, which are involved in the transportation of lipids, carbohydrates, and—most importantly—electrolytes.

An electrolyte is a term used frequently in medicine to describe a mineral that has separated from a salt and contains an ion, or electrical charge. Electrolytes, for example, include sodium ions (Na^+) and chloride ions (Cl^-).

Osmosis is the process by which water travels throughout the body's compartments and through semi-permeable cell membranes. Osmosis is essentially the diffusion of water through a semi-permeable membrane along an osmotic gradient from areas of higher concentration to regions of lower concentration.

Water will therefore enter and exit cells and tissues based on the relative concentrations of solutes and water present in each. To guarantee proper function, the solutes inside and outside of cells must be kept in the proper balance.

3.7.2.1. Body Water Content

The majority of human body mass is made up of water, which ranges from roughly 75% in newborns to between 50% and 60% in adults for both sexes, and as low as 45% in old age. Because the percentage of the body devoted to each organ, as well as to muscles, fat, bone, and other tissues, varies from infancy to adulthood, the percentage of body water also varies with development. 80–85% of the mass of kidneys and brain is made up of water. Teeth, on the other hand, contain the least amount of water—between 8–10% (Figure 3.4).

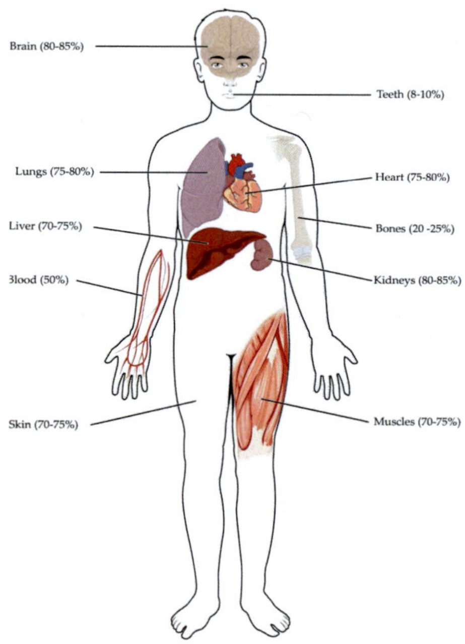

Figure 3.4. *Water content varies in different body organs and tissues, from as little as 8% in the teeth to as much as 85% in the brain.*

Source: https://courses.lumenlearning.com/suny-ap2/chapter/body-fluids-and-fluid-compartments-no-content/.

3.7.3. Fluid Compartments

Body fluids are referred to in terms of their own fluid compartment, which is primarily isolated from other compartments by a physical barrier. All fluid that is encased in cells by their plasma membranes is part of the intracellular fluid (ICF) compartment. Every cell in the body is surrounded by extracellular fluid or ECF. There are two main components of extracellular fluid: the interstitial fluid (IF), which surrounds all cells that are not in the blood, and the fluid component of blood called plasma.

- **Intracellular Fluid:** The ICF is the main part of the cytosol and cytoplasm and is found inside cells. About 60% of the water in the human body is found in the interstitial cyst (ICF), which holds approximately 25 liters (seven gallons) of fluid in an adult male of average size. The fluid volume in living cells is usually highly stable due to the tight regulation of water content. A cell's cytosol becomes too concentrated with solutes to support regular cellular functions when the volume of water inside the cell drops too low; conversely, when a cell absorbs too much water, it may burst and perish.
- **Extracellular Fluid:** The remaining one-third of the body's water content is made up of the ECF. The plasma contains about 20% of the ECF. A variety of substances, including blood cells, proteins (such as clotting factors and antibodies), electrolytes, nutrients, gases, and waste products, are transported by plasma as it passes through blood vessels in the body. Through the IF, gases, nutrients, and waste products move between capillaries and cells. A selectively permeable cell membrane that keeps materials from passing through between the IF and the inside of the cell divides cells from the IF. There are more water-based ECF in the body. These consist of the aqueous humor of the eye, lymph, pleural fluid in the pleural cavities, pericardial fluid in the cardiac sac, lymph, the cerebrospinal fluid that surrounds the brain and spinal cord, and synovial fluid in joints. These fluids are also regarded as elements of the ECF compartment because they are not contained within cells.

3.7.3.1. Composition of Body Fluids

Both the IF and the plasma have compositions that are more similar to one another than they are to the ICF. High levels of sodium, chloride, bicarbonate, and protein are found in blood plasma.

The IF has comparatively low protein concentrations but high sodium, chloride, and bicarbonate concentrations. On the other hand, the ICF has higher than normal levels of magnesium, potassium, phosphate, and protein. All things considered, the ICF has high potassium and phosphate (HPO_4^{2-}) concentrations, while the plasma and the ECF both have high sodium and chloride concentrations. The majority of bodily fluids have no charge. As a result, the balance of positively charged ions, or cations, and negatively charged ions, or anions, exists in fluids. The preceding graph illustrates that potassium (K^+) ions are concentrated inside cells, while sodium (Na^+) and chloride (Cl^-) ions are concentrated in the body's extracellular fluid (ECF). The high potassium and low sodium levels in the ICF are sustained by sodium-potassium pumps in the cell membranes, although both potassium and sodium can leak through pores into and out of cells, respectively. These pumps move potassium into the cell and sodium out of the cell using the energy provided by ATP.

3.7.3.2. Fluid Movement Between Compartments

The force that a fluid applies to a wall, known as hydrostatic pressure, is what moves the fluid between compartments. The pressure that blood exerts against blood vessel walls as a result of the heart's pumping action is known as the hydrostatic pressure of blood. At the arteriolar end of the capillary, hydrostatic pressure, also referred to as capillary blood pressure, is greater than the opposing "colloid osmotic pressure" in blood, which is a "constant" pressure mostly generated by circulating albumin. Plasma and nutrients are forced into the surrounding tissues from the capillaries by this pressure. At the venule end, where the hydrostatic pressure is lower than the osmotic pressure within the vessel, fluid and the waste products from the tissues enter the capillaries. Fluid from the blood's plasma is squeezed into the IF that surrounds the tissue cells by filtration pressure. The lymphatic system removes extra fluid from tissues in the interstitial space that isn't immediately returned to the capillaries. The extra fluid then re-enters the vascular system through the subclavian veins.

For the kidneys to properly filter blood and produce urine, the flow of water in their nephrons is primarily controlled by hydrostatic pressure. More urine filtrate is produced when the kidneys' hydrostatic pressure rises because more water is expelled from the capillaries. Dehydration can cause the hydrostatic pressure in the kidneys to drop too low, which can impair kidney function and reduce the amount of nitrogenous waste that is eliminated from the blood. Severe dehydration may lead to renal failure.

Additionally, fluid travels along an osmotic gradient between compartments. Remember that the concentration of all solutes varies on either side of a semi-permeable membrane to create an osmotic gradient. The difference between the concentration of solutes on one side of the cell membrane and the other side determines the size of the osmotic gradient. Utilizing osmosis, water will flow from the side of the membrane where its concentration is high (and the solute concentration is low) to the side where its concentration is low (and the solute concentration is high). Water travels through osmosis in the body from plasma to the IF (and vice versa) and from the IF to the ICF (and vice versa). Water continuously enters and exits fluid compartments in the body as various bodily parts experience changes in conditions. For instance, sweating causes the loss of water through the skin. Sweating causes tissues to lose water and their solute concentration to rise. As a result of the osmotic gradient, water diffuses from blood into the surrounding skin tissues and sweat glands, causing them to become dehydrated. Furthermore, the water in other tissues throughout the body that are not dehydrated replaces the water that leaves the blood. Dehydration spreads throughout the body if this keeps up. The water is redistributed by the same gradient, but in the opposite direction, when a dehydrated person drinks water and rehydrates, replenishing water in all of the tissues.

3.7.3.3. Solute Movement between Compartments

While some solutes move passively between compartments without requiring energy, other solutes move actively, requiring energy consumption during the transport process. With the help of a membrane protein and ATP energy, cells can move a particular material against its concentration gradient through a process known as active transport. The sodium-potassium pump, for instance, uses active transport to push potassium into cells and sodium out of them while both substances are traveling against concentration gradients. The ability of a molecule or ion to diffuse from a region of higher concentration to one of lower concentration is dependent upon both the molecule's ability to pass through the membrane and the presence of a concentration gradient. Certain molecules pass through the cell membrane relatively easily, such as gases, lipids, and water itself (which also makes use of water channels in the membrane called aquaporins); polar molecules, such as glucose, amino acids, and ions, do not. Some of these molecules use facilitated transport to enter and exit cells. This process involves the molecules passing through particular protein channels in the membrane and moving along a concentration gradient. There is no energy needed for this process. For instance, glucose transporters that use facilitated transport are responsible for transferring glucose into cells.

3.7.4. Water Balance

An adult will typically consume about 2,500 mL, or nearly 3 quarts, of aqueous fluid in a single day. About 230 mL (8 ounces) are produced metabolically each day during the final stages of aerobic respiration, even though the majority of intake passes through the digestive tract. Furthermore, the body loses roughly the same amount of water (2,500 mL) every day through a variety of pathways; the majority of this water is eliminated as urine. The kidneys' mechanisms for removing water from the filtrate and urine allow them to also regulate blood volume. The kidneys have the ability to control the body's water balance. If you are dehydrated, they can help retain water by diluting urine, which can help eliminate extra water when needed. Water is lost through the skin as a result of air being exhaled from the lungs and evaporating from the skin's surface without obvious sweating. Because a person is typically unaware of this type of water loss, it is referred to as insensible water loss.

3.7.4.1. Regulation of Water Intake

The ratio of solutes to solvent volume in a solution is known as osmolality. The ratio of solutes to water in blood plasma is therefore known as plasma osmolality. A person's level of hydration is indicated by their plasma osmolality value. A healthy body uses many processes to control the intake and output of water in order to keep plasma osmolality within a specific range. It is thought that drinking water is optional. How then does the body regulate how much water it takes in? Take the example of a person who is dehydrated—a person who has lost a net amount of water, leaving their blood and other tissues with inadequate levels of water. Blood plasma is ultimately where the water that is expelled from the body as sweat, urine, or exhaled air—is ultimately found. A series of physiological processes known as the thirst response are set off when the blood becomes more concentrated. Osmoreceptors are sensory receptors in the hypothalamic thirst center that track the blood's solute concentration, or osmolality. A sensation of thirst is brought on by signals sent by the hypothalamus when blood osmolality rises above optimal levels. Drinking water is the appropriate (and typical) response for the individual. Antidiuretic hormone (ADH) is also released by the hypothalamus of a dehydrated individual through the posterior pituitary gland. By instructing the kidneys to extract water from urine, ADH essentially dilutes blood plasma. A dehydrated person's hypothalamus also signals the salivary glands in the mouth to conserve water through the sympathetic nervous system. The stickier, thicker mucus output increases and watery, serous output decreases as a result of the signals. A "dry mouth" and a feeling of thirst are brought on by these modifications in secretions.

Two more effects result from decreased blood volume caused by water loss. First, a drop in blood volume causes a decrease in blood pressure, which is detected by baroreceptors, blood-pressure receptors in the carotid arteries in the neck and the arch of the aorta. Eventually, a signal is sent to the heart to accelerate and/or deepen its contractions to make up for the dropped blood pressure. Second, the kidneys' renin-angiotensin hormonal system raises the body's production of angiotensin II's active form, which not only helps to increase thirst but also causes the adrenal glands to release more aldosterone. In the kidneys, aldosterone promotes the reabsorption of sodium in the distal tubules of the nephrons, and water is subsequently reabsorbed and returned to the bloodstream.

Dehydration occurs when a person does not drink enough fluids, leaving their body with insufficient water to perform as intended. Infants, whose body mass is so low, can quickly become dangerously dehydrated. A person who vomits or has diarrhea on a regular basis may also become dehydrated. Distance runners and other endurance athletes frequently experience dehydration during lengthy races. If the body is not quickly rehydrated, a dehydrated person may lose consciousness, go into a comatose state, or even pass away. Dehydration can be a medical emergency.

3.7.4.2. Regulation of Water Output

The kidneys are the main organ through which the body loses water. Urine production is approximately 10.5 liters (10.6 quarts) per day on average for an individual. There is a minimum volume of urine production necessary for normal body functions, even though the volume of urine changes in response to hydration levels. To get rid of various excess salts and other water-soluble chemical wastes, the kidney excretes 100 to 1200 milliosmoles of solutes daily, the most common ones being creatinine, urea, and uric acid. Insufficient urine production can lead to ineffective removal of metabolic wastes from the body, potentially compromising organ function. About 0:47 liters (0.5 quarts) of urine production per day is the minimal amount required to sustain normal function. In the event of consuming excessive fluid, the kidneys also need to adapt. The production of urine above normal levels, or diuresis, starts approximately half an hour after consuming a large amount of fluid. After roughly an hour, diuresis peaks, and after roughly three hours, urine production returns to normal.

3.7.4.3. Role of ADH

Vasopressin, another name for antidiuretic hormone (ADH), regulates the volume of water that the kidney's tubules and collecting ducts reabsorb. The posterior pituitary receives this hormone from the hypothalamus, where it is stored and released. The hypothalamus signals the posterior pituitary's release of ADH into the blood when its osmoreceptors identify an increase in blood plasma concentration. ADH mainly affects two things. By narrowing the peripheral circulation's arterioles, it increases blood flow to the body's core while decreasing blood flow to the extremities. Aquaporins are water channel proteins that are moved from the inside of the cell to the apical surface of the cell, where they are inserted into the cell membrane, by the epithelial cells that line the renal collecting tubules as a result of

ADH. The end effect is an increase in these cells' water permeability, which increases the amount of water that passes from the urine through the collecting tubule walls and back into the circulation.

The removal of aquaporins from collecting tubule cell membranes and a decrease in the amount of water that passes from urine into the blood occur when blood plasma concentration falls and ADH levels drop.A diuretic is a substance that causes the body to produce more urine, which reduces its ability to conserve water. Diuretics are prescribed to treat menstrual-related fluid retention, hypertension, and congestive heart failure. Because alcohol prevents the release of ADH, it has a diuretic effect. Moreover, caffeine has a diuretic effect when taken in large quantities.

3.7.5. Electrolyte Balance

Numerous ions, also known as electrolytes, are found throughout the body and serve a range of purposes.

Certain ions help electrical impulses in neurons and muscles travel along their cell membranes. Enzyme protein structures are stabilized with the aid of other ions. Others still help endocrine glands release hormones.

Every ion present in plasma plays a role in maintaining the osmotic balance, which regulates the flow of water between cells and their surroundings. Sodium, potassium, chloride, bicarbonate, calcium, phosphate, magnesium, copper, zinc, iron, manganese, molybdenum, copper, and chromium are examples of electrolytes found in living systems. The six electrolytes that are most crucial to bodily function are sodium, potassium, chloride, bicarbonate, calcium, and phosphate.

3.7.5.1. Roles of Electrolytes

These six ions support the functions of nerve excitability, membrane permeability, endocrine secretion, bodily fluid buffering, and fluid transport between compartments. The digestive system allows these ions to enter the body. Bone serves as a mineral reserve for calcium and phosphate ions, with over 90% of these ions entering the body through teeth and bones. Bone tissue can be broken down to supply the blood and other tissues with calcium and phosphate if these minerals are required for other purposes. Since phosphate is a normal component of nucleic acids, whenever nucleic acids are broken down, blood phosphate levels will rise.

Ions are mostly excreted by the kidneys, with smaller amounts excreted through perspiration and feces. Sweating excessively can result in significant losses, particularly of sodium and chloride. Chloride and bicarbonate ions will be lost as a result of severe vomiting or diarrhea. The body can adjust renal and respiratory functions to control the concentrations of these ions in the extracellular fluid.

The reference values for blood plasma, urine, and cerebrospinal fluid (CSF) for the six ions covered in this section are shown in the Table 3.3. Sodium, potassium, and chloride are usually measured from a standard urine sample in a clinical setting. On the other hand, since the output of these ions can vary greatly throughout the day, urine must be collected for calcium and phosphate analysis for a full day. The rates of excretion of these ions are reflected in urine values. The one ion that is normally excreted in urine is bicarbonate, which the kidneys preserve for use in the body's buffering systems.

Table 3.3. Electrolyte and Ion Reference Values

Name	Chemical Symbol	Plasma	CSF	Urine
Sodium	Na^+	136.00–146.00 (mM)	138.00–150.00 (mM)	40.00–220.00 (mM)
Potassium	K^+	3.50–5.00 (mM)	0.35–3.5 (mM)	25.00–125.00 (mM)
Chloride	Cl^-	98.00–107.00 (mM)	118.00–132.00 (mM)	110.00–250.00 (mM)
Bicarbonate	HCO_3^-	22.00–29.00 (mM)	——	——
Calcium	Ca^{++}	2.15–2.55 (mmol/day)	——	Up to 7.49 (mmol/day)
Phosphate	HPO_4^{2-}	0.81–1.45 (mmol/day)	——	12.90–42.00 (mmol/day)

Source: https://courses.lumenlearning.com/suny-ap2/chapter/electrolyte-balance-no-content/.

3.7.5.1.1. Sodium

The main cation in the extracellular fluid is sodium. It is in charge of half of the gradient in osmotic pressure that occurs between the inside of cells and their external surroundings. Humans only need 1 to 2 milligrams of sodium per day, yet those following a typical Western diet—which is very high in NaCl—typically consume 130 to 160 milligrams per day. This overabundance of sodium seems to play a significant role in hypertension in certain individuals. Sodium excretion is mainly carried out by the kidneys. The kidneys' glomerular capillaries allow sodium to pass freely through them. While the proximal convoluted tubule reabsorbs most of the filtered sodium, some amount is left in the filtrate and urine and is typically eliminated.

A lower-than-normal sodium concentration is known as hyponatremia, and it is typically brought on by the body accumulating too much water, which dilutes the sodium.

Reduced sodium intake combined with ongoing excretion of the ion in the urine may result in an absolute loss of sodium. Several conditions can cause an abnormal loss of sodium from the body: acidosis, which can be either metabolic acidosis or diabetic ketoacidosis; diuretics; excessive urine production, which can occur in diabetes; and excessive sweating, vomiting, or diarrhea.

A dilution of sodium due to water retention associated with edema or congestive heart failure, or an imbalance of sodium in one of the body's other fluid compartments, such as IF, can cause a relative decrease in blood sodium levels. Hyponatremia causes an increase in the osmotic uptake of water into cells because the concentration of solutes in the cell is greater than the concentration of solutes in the diluted extracellular fluid (ECF).

The extra water makes the cells swell; this can lead to brain damage or even death. Red blood cells swell, reducing their ability to carry oxygen and possibly becoming too big to pass through capillaries. Brain cells swell as well.

An unnatural rise in blood sodium levels is called hypernatremia. It may be brought on by the blood losing water, which would cause all of the blood's components to become hemoconcentrated. Elevated sodium levels can also be caused by ADH and aldosterone-related hormonal imbalances.

3.7.5.1.2. Potassium

The main intracellular cation is potassium. After membrane depolarization and action potentials, it aids in the establishment of the resting membrane potential in neurons and muscle fibers. Potassium has a negligible impact on osmotic pressure in comparison to sodium. The sodium-potassium pumps in cell membranes, which preserve the typical potassium concentration gradients between the ICF and ECF, are the cause of the low potassium levels in blood and CSF. A daily intake or consumption of 4700 mg of potassium is advised. Both actively and passively, potassium is eliminated from the body through the renal tubules, particularly the collecting ducts and the distal convoluted tubule. Under the influence of aldosterone, potassium takes part in the sodium-potassium exchange in the renal tubules, which also depends on basolateral sodium-potassium pumps.

A blood potassium level that is unusually low is called hypokalemia. Like hyponatremia, hypokalemia can result from either a total decrease in potassium levels in the body or a relative decrease in potassium levels in the blood as a result of potassium redistribution. Intake can be reduced to the point of absolute loss, often as a result of starvation. Alkalosis, diarrhea, and vomiting can also cause it. Potassium redistribution causes a relative drop in blood potassium levels in certain insulin-dependent diabetic patients. Potassium enters the cell membrane with glucose when insulin is given, lowering the amount of potassium in the blood and intracellular potassium (IF). This can lead to hyperpolarization of the cell membranes of neurons, which lessens their reactivity to stimuli.

An increased potassium blood level known as hyperkalemia can also affect how well the heart, nervous system, and skeletal muscles work. Increased consumption of potassium in the diet may lead to hyperkalemia. Potassium from the blood ends up in the ECF in abnormally high concentrations under such circumstances. This may cause neurons, cardiac cells in the heart, and skeletal muscle fibers to partially depolarize (excite), as well as impair the ability of the affected cells to repolarize. This implies that the heart will not recover from a contraction; instead, it will essentially "seize" and cease pumping blood, which can be fatal in a matter of minutes. A person with hyperkalemia may also experience weakened respiratory muscles, numbness, and mental confusion as a result of these effects on the nervous system.

3.7.5.1.3. Chloride

The most common extracellular anion is chloride. The osmotic pressure gradient between the ICF and ECF is primarily caused by chloride, which is also crucial for preserving appropriate hydration. Chloride keeps the ECF electrically neutral by balancing the cations in the fluid. In the renal system, the routes taken by chloride ions during secretion and reabsorption correspond to those of sodium ions. Lowered-than-normal blood chloride levels, or hypochloremia, can be brought on by impaired renal tubular absorption. Hypochloremia can also result from vomiting, diarrhea, and metabolic acidosis. Higher-than-normal blood chloride levels, or hyperchloremia, can be brought on by aspirin overdose, excessive dietary salt (NaCl) intake, ingestion of seawater, congestive heart failure, and the inherited chronic lung disease cystic fibrosis. Sweat analysis is frequently used to diagnose cystic fibrosis because the disease's chloride levels are two to five times higher in patients with the condition than in healthy people.

3.7.5.1.4. Bicarbonate

The second most common anion in blood is bicarbonate. Its main job as a component of buffer systems is to keep the body's acid-base balance intact. We'll talk about this role in another section. Two molecules produced at the end of aerobic metabolism, carbon dioxide (CO_2) and water, initiate a chemical reaction that yields bicarbonate ions. Body fluids can only dissolve a small amount of CO_2. Therefore, through the following reactions, more than 90% of the 2 is transformed into bicarbonate ions or HCO_3^-.

$$CO_2 + H\ _2 \boxtimes H_2 + CO_3 \boxtimes H_2 + CO_3^- + H^+$$

The bidirectional arrows show that, depending on the reactant and product concentrations, the reactions can proceed in either way. Large amounts of carbon dioxide are produced by tissues with fast metabolic rates. An enzyme known as carbonic anhydrase works in the cytoplasm of red blood cells to convert carbon dioxide into bicarbonate. Blood carries bicarbonate throughout its journey. After entering the lungs, the processes change, generating CO2 from bicarbonate for exhalation as metabolic waste.

3.7.5.1.5. Calcium

The human body contains about two pounds of calcium in bone, which gives the bone its hardness and acts as a mineral reserve for calcium and its salts in the tissues that surround it. Moreover, teeth contain a large amount of calcium. Just over 50% of blood calcium is incorporated into proteins, with the remaining portion existing in its ionized state. Blood coagulation, enzyme activity, and muscle contraction all depend on calcium ions or Ca^{2+}. Furthermore, calcium is necessary for the release of hormones from endocrine glands and neurotransmitters from neurons, as well as for the stabilization of cell membranes.

When activated vitamin D is present, the intestines absorb calcium. A lack of vitamin D can result in reduced calcium absorption, leading to a depletion of calcium from the skeletal system. This can result in conditions such as rickets in children and osteomalacia in adults, contributing to osteoporosis. Hypocalcemia, or low calcium levels in the blood, can occur in cases of hypoparathyroidism, often seen after thyroid gland removal due to the presence of the parathyroid glands within it. Hypercalcemia, or high calcium levels in the blood, is associated with primary hyperparathyroidism and certain malignancies.

3.7.5.1.6. Phosphate

Phosphate is present in the body in three ionic forms: H_2PO_{4-}, HPO_4^{2-}, and PO_4^{3-}. The HPO_4^{2-}– form is the most prevalent. 85% of the phosphate in the body is bound up in calcium-phosphate salts found in bone and teeth. Phospholipids, which include the phospholipids in cell membranes, as well as ATP, nucleotides, and buffers, contain phosphate.

Low phosphate levels, known as hypophosphatemia, can be caused by excessive use of antacids, alcohol withdrawal, and malnutrition.

Normally, the kidneys conserve phosphate when levels are low but this conservation is compromised during starvation. On the other hand, high levels of phosphates, or hyperphosphatemia, can occur with decreased kidney function or in cases of acute lymphocytic leukemia. Phosphate is a major component of the ICF, so significant cell destruction can lead to phosphate being released into the ECF.

3.7.5.2. Regulation of Sodium and Potassium

In the renal collecting tubule, potassium is excreted into the filtrate and sodium is reabsorbed from it. The two hormones angiotensin II and aldosterone primarily regulate this exchange.

- **Aldosterone:** Remember that in the distal tubule, aldosterone promotes the reabsorption of sodium and the excretion of potassium. When potassium levels rise, when sodium levels sharply fall, or when blood pressure falls, aldosterone is released. Overall, it lowers the kidneys' excretion of sodium and water, which preserves and raises the amount of water in plasma. Increased osmolality of the ECF, which comes after aldosterone-stimulated sodium absorption, inhibits the release of the hormone in a negative feedback loop.
- **Angiotensin II:** Vasoconstriction and elevated systemic blood pressure are brought on by angiotensin II. More material is filtered into Bowman's capsule from the glomerular capillaries as a result of this action, which raises the glomerular filtration rate. An increase in the adrenal cortex's release of aldosterone is also signaled by angiotensin II. The sodium-potassium pump is synthesized and activated in the kidneys' collecting ducts and distal convoluted tubules by aldosterone. Sodium travels from the filtrate into the ECF, via the tubule and duct cells, and into the capillaries. Because of osmosis, water follows the sodium. As a result, aldosterone raises blood volume and sodium levels. Similar to how sodium affects potassium, aldosterone causes excess potassium to be pumped into the renal filtrate for the body's excretion.

3.7.5.3. Regulation of Calcium and Phosphate

Three hormones—calcitonin, dihydroxyvitamin D (calcitriol), and parathyroid hormone (PTH)—control the levels of calcium and phosphate. The blood levels of calcium determine the release or synthesis of all three. When blood calcium levels fall, the parathyroid gland releases parathyroid hormone (PTH). Inorganic calcium-phosphate salts are released and the bone matrix is broken down by osteoclasts when the hormone is activated. By transforming vitamin D into dihydroxyvitamin D (calcitriol), an active form of vitamin D that intestinal epithelial cells need to absorb calcium, PTH also improves the gastrointestinal absorption of dietary calcium.

PTH prevents calcium from being lost through the kidneys, which increases blood calcium levels. PTH also causes the kidneys to lose more phosphate. The thyroid gland releases calcitonin in reaction to high blood calcium levels. The hormone stimulates osteoblasts, which take calcium out of the bloodstream and add it to the bone matrix. Electrolytes perform a variety of functions, including stabilizing the structures of enzymes, releasing hormones from endocrine glands, and assisting in the conduct of electrical impulses along cell membranes in neurons and muscles. The osmotic balance, which regulates the flow of water between cells and their surroundings, is also influenced by the ions present in plasma. The body can experience several issues due to imbalances in these ions, and the concentrations of these ions are strictly controlled. The exchange of potassium and sodium between the renal collecting tubule and the renal filtrate is regulated by angiotensin II and aldosterone. PTH, calcitriol, and calcitonin control the levels of calcium and phosphate.

3.7.6. Acid-Base Balance

An extremely delicate equilibrium between the blood's acidic and basic concentrations is necessary for normal physiological functioning. The pH scale is used to measure acid-base balance, as demonstrated in. Blood and other body fluids can maintain a narrow pH range even in the face of disturbances thanks to a variety of buffering systems. A buffer is a chemical system that, in the event of an excess of acid or base, dampens the change in hydrogen ion concentrations, preventing a drastic change in fluid pH. Most frequently, weak bases or acids that absorb hydrogen ions or hydroxyl ions, respectively, are the substances that take up the ions.

3.7.6.1. Buffer Systems in the Body

The human body's buffer systems are incredibly effective, and they operate at various speeds. The blood's chemical buffers can quickly adjust pH in a matter of seconds. By exhaling CO_2, the respiratory system can raise the blood pH in a matter of minutes. By excreting hydrogen ions (H^+) and conserving bicarbonate, the renal system can also modify blood pH, though it may take hours or even days for this process to take effect. Phosphate, bicarbonate and carbonic acid buffers, and plasma proteins are among the buffer systems that are active in blood plasma. By excreting hydrogen ions and producing bicarbonate, which helps keep blood plasma pH within a normal range, the kidneys contribute to the regulation of acid-base balance. Systems for buffering proteins are mostly found inside cells.

- **Protein Buffers in Blood Plasma and Cells:** Proteins of almost any kind can serve as buffers. Amino acids, which comprise positively charged amino groups and negatively charged carboxyl groups, are the building blocks of proteins. These charged regions of the molecules can bind hydrogen and hydroxyl ions, making them useful as buffers. Most of the buffering within cells and two-thirds of the blood's buffering capacity are attributed to protein buffering.
- **Hemoglobin as a Buffer:** The main protein found inside red blood cells, hemoglobin, makes up one-third of the cell's mass. Hemoglobin, which is depleted by oxygen dissociation, buffers hydrogen ions released in the process of converting CO_2 into bicarbonate. By buffering, the pH is kept within the normal range. For CO_2 to re-form in the pulmonary capillaries and then diffuse into the air sacs for exhalation into the atmosphere, the process is reversed there.
- **Phosphate Buffer:** Blood contains two types of phosphates: sodium monohydrogen phosphate ($Na_2HPO_4^{2-}$), a weak base, and sodium dihydrogen phosphate ($Na_2H_2PO_4^-$) a weak acid. Upon encountering a strong acid like HCl, ($Na_2HPO_4^{2-}$), – absorbs a second hydrogen ion from the base, resulting in the formation of the weak acid $Na_2H_2PO_4^-$ and sodium chloride, NaCl. The weak acid, $Na_2HPO_4^{2-}$, turns back into the weak base, producing water, when it comes into contact with a strong base, like sodium hydroxide (NaOH). Although they still exist,

the acids and bases are attached to the ions.

$HCl + Na_2HPO_4 \rightarrow NaH_2PO_4 + NaCl$
(strong acid) + (weak base) → (weak acid) + (salt)
$NaOH + NaH_2PO_4 \rightarrow Na_2HPO_4 + H_2O$
(strong base) + (weak acid) → (weak base) + (water)

- **Bicarbonate-Carbonic Acid Buffer:** Phosphate buffers and the bicarbonate-carbonic acid buffer function similarly. Both sodium and phosphate ions control the levels of bicarbonate and phosphate in the blood. NaCl and carbonic acid (H_2CO_3), a weak acid, are created when sodium bicarbonate ($NaHCO_3$) and a strong acid, like HCl, come into contact. Bicarbonate and water are produced when carbonic acid and a strong base, like NaOH, come into contact.

$NaHCO_3 + HCl \rightarrow H_2CO_3 + NaCl$
(sodium bicarbonate) + (strong acid) → (weak acid) + (salt)
$H_2CO_3 + NaOH \rightarrow HCO_3^- + H_2O$
(weak acid) + (strong base) → (bicarbonate) + (water)

Similar to the phosphate buffer system, a weak acid or weak base in the blood captures free ions to prevent significant changes in pH. The ratio of bicarbonate ions to carbonic acid is 20:1 when the blood pH is normal, making this system most efficient at buffering changes that would acidify the blood.

This is important because many metabolic wastes in the body are acids. Carbonic acid levels are regulated by the lungs expelling CO_2, while carbonic anhydrase in red blood cells helps make the blood less acidic by dissociating the acid.

Bicarbonate levels are maintained by the renal system, where bicarbonate ions are conserved and put back into the blood. Overall, the bicarbonate buffer system is the main buffering system of the interstitial fluid surrounding cells in tissues all over the body.

3.7.6.2. Respiratory Regulation of Acid-Base Balance

The respiratory system controls blood levels of carbonic acid, which helps the body maintain an acid-base balance. The blood's levels of CO_2 and carbonic acid are balanced because CO_2 in the blood easily combines with water to form carbonic acid. Blood pH is lowered when the blood's CO_2 level rises, which happens when a person holds their breath. The extra CO_2 combines with water to create more carbonic acid. A human can exhale more CO_2 by increasing the rate and/or depth of their breathing, which may feel compelled to do after holding their breath. The body loses carbon dioxide, which lowers blood carbonic acid levels and raises pH levels back to normal. As one might expect, this process also operates in the other way. Breathing too deeply and quickly (hyperventilation) removes CO_2 from the blood and lowers the carbonic acid level, making the blood too alkaline. A person can reverse this momentary alkalosis by breathing in again after exhaling into a paper bag. Rebreathing exhaled air will quickly return the blood pH to its normal level.

When blood passes through the pulmonary capillaries in the lungs, chemical reactions take place that control the amounts of carbon dioxide and carbonic acid. By varying the amount of CO_2 exhaled, small breathing adjustments are typically enough to change the blood's pH. Actually, the blood pH would rise by 0.2 if the respiratory rate was doubled for less than a minute to eliminate "extra" CO_2. This is a common scenario when a person exercises intensely for an extended period of time. A person would generate excess CO_2 (and lactic acid if exercising above the aerobic threshold) to maintain the required energy production. The respiration rate increases to remove CO_2 to counteract the increased production of acid. This prevents acidosis from occurring. Through the use of chemoreceptors, which primarily use CO_2 as a signal, the body controls the respiratory rate. The aortic and carotid artery walls contain peripheral blood sensors. If CO_2 levels rise or fall, these sensors send a signal to the brain telling it to immediately adjust the respiratory rate. There are still more sensors within the brain. The respiratory center in the medulla oblongata is impacted by changes in CSF pH and can directly regulate breathing rate to restore the pH to a normal range. Hypercapnia, or abnormally high blood CO_2 levels, can arise from pneumonia or congestive heart failure, among other conditions that affect breathing. Hypercapnia can also arise from reduced breathing (hypoventilation) brought on by medications like morphine, barbiturates, or ethanol, or even from simply holding one's breath. Any cause of hyperventilation that drives off the CO_2, such as salicylate toxicity, elevated room temperatures, fever, or hysteria, results in hypocapnia, or abnormally low blood levels of CO_2.

3.7.6.3. Renal Regulation of Acid-Base Balance

The metabolic part of the buffering system is handled by the kidneys' regulation of the body's acid-base balance. The renal system regulates the blood levels of bicarbonate, while the respiratory system (along with brain breathing centers) regulates the blood levels of carbonic acid by regulating CO_2 exhalation. Blood bicarbonate levels can drop when carbonic anhydrase is inhibited by some diuretics or when diarrhea causes an excessive loss of bicarbonate. Additionally, those with renal damage, such as chronic nephritis, and those with Addison's disease (chronic adrenal insufficiency), in which aldosterone levels are decreased, usually have lower blood bicarbonate levels. Lastly, high levels of ketones, which are common in uncontrolled diabetes mellitus, can cause low blood bicarbonate levels because they bind bicarbonate in the filtrate and prevent it from being conserved.

Although bicarbonate ions, HCO_3^-, which are present in the filtrate, are necessary for the functioning of the bicarbonate buffer system, they are not able to pass through the tubule cells. The preceding diagram illustrates the procedures needed to supply the system with bicarbonate ions, which are summed up as follows:

- **Step 1:** In the apical membranes of cells lining the renal tubule, sodium ions are reabsorbed from the filtrate through an antiport mechanism, exchanging them for H^+.
- **Step 2:** Bicarbonate ions produced by the cells can be transferred to peritubular capillaries.
- **Step 3:** The process is pushed to produce carbonic acid when CO_2 is present, which then splits to produce a hydrogen ion and a bicarbonate ion.
- **Step 4:** The peritubular capillaries allow the bicarbonate ion to enter and exit, returning to the blood. The hydrogen ion is released into the filtrate, from which it can either be eliminated in the urine or reabsorbed as part of freshwater molecules.

Hydrogen ions may also be attracted to salts present in the filtrate, such as phosphates, sulfates, or ammonia. If this happens, there won't be any hydrogen ions available to mix with bicarbonate ions to create CO_2. Bicarbonate ions are not preserved in such circumstances, and this will further lead to an acidotic environment and pH imbalance in the blood. In the renal tubules, the hydrogen ions also face competition from potassium for the exchange of sodium. Greater potassium enters the filtrate when there is more potassium than usual because potassium will be exchanged instead of hydrogen ions. When this happens, less bicarbonate is conserved and fewer hydrogen ions in the filtrate contribute to the conversion of bicarbonate into CO_2. Less potassium causes more hydrogen ions to enter the filtrate and exchange with sodium, resulting in the conservation of more bicarbonate.

The body uses chloride ions to balance out positive ion charges. The body substitutes bicarbonate ions for lost chloride ions in the event of chloride loss. Therefore, when chloride is lost, the kidneys reabsorb bicarbonate more readily (Kristinsson, 2014; Kumar and Bagchi, 2021).

3.8. ANTIOXIDANTS IN FOOD

The roughly one trillion cells that make up the body are in grave danger from viral infections to malnutrition. Free radicals are a class of chemicals that pose a continuous threat. When present in extremely high concentrations, they can harm genetic material and cells. Free radicals are the unavoidable byproducts of the body converting food into energy. In addition, exercise, cigarette smoke, air pollution, and sunlight can all produce free radicals.

There are numerous sizes, forms, and chemical combinations of free radicals. They all have an insatiable hunger for electrons, which they take from any nearby materials that will allow them to be had. The structure or function of the "loser" may be drastically changed by this electron theft. Damage from free radicals can alter the instructions encoded in a DNA strand. It may increase the likelihood that a circulating low-density lipoprotein (LDL, also known as bad cholesterol) molecule will become lodged in the wall of an artery. Alternatively, it can modify a cell's membrane, altering the flow of substances that enter and exit the cell. Oxidative stress, a state where the body has an excessive amount of free radicals over an extended period of time, can harm cells and result in chronic illnesses.

There is protection from free radicals in the system. Because the body is accustomed to this constant assault, it produces a lot of molecules that put out free radicals just like water puts out a fire. Additionally, they remove anti-free radicals from food. These protectors are called "antioxidants."

They function by freely providing free radicals with electrons without becoming themselves substances that scavenge electrons. They also play a role in processes that preserve cell viability and fix DNA. Antioxidants can be derived from hundreds, if not thousands, of different substances. The most well-known ones include the minerals manganese and selenium as well as the vitamins C, E, beta-carotene, and other related carotenoids. Glutathione, coenzyme Q10, lipoic acid, flavonoids, phenols, polyphenols, phytoestrogens, and numerous other substances are joined to them. The majority are found in nature, and their inclusion in food is probably going to stop oxidation or act as a natural barrier against the surroundings.

However, it is misleading to refer to substances as antioxidants. Actually, it's a chemical characteristic—that is, the capacity to donate electrons. Certain substances that function as pro-oxidants—that is, electron grabbers—in one context may function as antioxidants in another. The idea that antioxidants are interchangeable is another common misperception. They're not. Everybody has different biological characteristics and chemical behaviors. They most likely developed as components of complex networks, where each distinct material (or family of substances) had a slightly different function. This implies that no material is able to perform the functions of the entire group.

3.8.1. Types of Antioxidants

Antioxidants are often discussed as a single category, but in reality, they belong to a large family. Antioxidants include beta-carotene, vitamin C, vitamin E, and vitamin A. There are numerous others, each with special advantages.

Other antioxidants include:

- Glutathione
- Coenzyme Q10
- Lipoic acid
- Flavonoids
- Phenols
- Polyphenols
- Phytoestrogens

3.8.2. Antioxidants Benefits

Antioxidants may be able to reduce the risk of developing certain diseases because they can shield from free radical damage. However, the relationship isn't always evident, particularly in the case of supplements. A diet rich in fruits and vegetables is nevertheless beneficial for numerous other reasons. Typically, foods high in antioxidants are:

- High in fiber
- Low in saturated fat and cholesterol
- Full of vitamins and minerals

3.8.2.1. Antioxidants for Cancer Prevention

Antioxidants may be able to stop part of the damage free radicals cause, which has been linked to cancer, according to some research. However, there isn't any proof from science that antioxidant supplements can stop cancer from happening. To find out if antioxidants from food can lower the risk of cancer, more research is needed.

3.8.2.2. Antioxidants for Eyes

Age-related macular degeneration (AMD) is the leading cause of permanent vision loss in adults over 50. It occurs when the macula, located at the back of the eye, starts to deteriorate, leading to a loss of central vision. Antioxidants have been shown to reduce the risk of AMD by up to 25% and can also help individuals with AMD preserve more of their vision. Additionally, vitamins C and E have been found to decrease the likelihood of developing cataracts, which cause cloudiness in the lens of the eye and result in blurred vision. Antioxidants may also slow the progression of cataracts, allowing individuals to maintain better vision for a longer period of time.

3.8.2.3. Antioxidants for Heart Health

The question of whether antioxidants can reduce a person's risk of heart disease is hotly debated. On the one hand, studies have demonstrated a link between increased fruit and vegetable consumption and a decreased risk of heart disease and stroke. According to preliminary studies, antioxidants might be in charge of this advantage. However, additional research on antioxidant supplements hasn't yielded the same results. According to some scientists, this is because supplements contain higher concentrations of antioxidants. Consuming foods high in antioxidants may hold the key, but further investigation is required to determine whether there is a true correlation.

3.8.2.4. Antioxidants for Skin

There is evidence that the antioxidant vitamins C and E have benefits for the skin, but more research is needed to confirm this. According to studies, applying a vitamin C formula topically could be beneficial:

- Improve the appearance of wrinkles
- In addition to broad-spectrum sunscreen, protects skin from UV damage caused by the sun.
- Make dark spots on the skin less visible

As an ingredient in moisturizers, vitamin E appears to help:

- Make skin softer
- Reduce moisture loss
- Protect skin cells from sun damage

Healthy skin is also encouraged by an antioxidant-rich diet that includes lots of fruits, vegetables, nuts, whole grains, and fatty fish.

3.8.3. Foods High in Antioxidants

Antioxidants are chemicals that stop or slow down the cellular damage caused by substances known as free radicals. These highly reactive substances, known as free radicals, can harm cells and cause chronic illnesses like diabetes, cancer, Alzheimer's, Parkinson's, and more. Antioxidants protect the body from the harm that free radicals cause. As a result, they shield cells and aid in illness prevention. Consume a wide variety of plants in your diet, such as fruits, vegetables, nuts, fungi, whole grains, legumes, herbs, spices, and even edible flowers, to enhance your intake of antioxidants overall.

- **Blueberries:** Blueberries are low in calories and high in nutrients. Antioxidants are abundant in wild blueberries, according to a 2017 study.
- **Dark Chocolate:** High-quality dark chocolate contains lots of antioxidants and nutrients. Numerous possible health advantages of dark chocolate have been connected by researchers, including:
  - lower risk of heart disease
  - reduced inflammation
  - less chance of high blood pressure
  - promotion of good cholesterol
- **Artichokes:** Antioxidants and nutrients abound in artichokes. According to research, they might assist in lowering cholesterol and enhancing gut health.
- **Pecans:** Pecans are rich in calories, antioxidants, minerals, and good fat.
- **Strawberries:** Minerals, vitamins, and antioxidants abound in strawberries. The antioxidant properties of anthocyanins give strawberries their red color. Raw strawberries can be eaten on their own as a snack or combined with other foods like salads. While strawberries can be found in some baked goods, they are rarely a healthy option and should not be chosen by those who are trying to reduce weight.
- **Red Cabbage:** In addition to several antioxidants and the vitamins A, C, and K, red cabbage is packed with nutrients. Red cabbage has anthocyanins, just like strawberries and red kale. These antioxidants not only give the vegetable its red hue but also aid in reducing inflammation, preventing cancer, and promoting heart health.
- **Raspberries:** An excellent source of numerous antioxidants is raspberries. They also include

dietary fiber, vitamin C, and manganese.

- **Beans:** Beans are a great source of dietary fiber and protein. Antioxidants are also abundant in some beans, like pinto beans. Kaempferol, a plant flavonoid found in pinto beans, may have anti-inflammatory and anti-cancer effects. Numerous researchers associate kaempferol with the inhibition of particular cancers, such as:
  - breast
  - kidney
  - lung
  - bladder

Despite these encouraging findings, little is known about how kaempferol affects human antioxidant levels. They have mostly used test tubes and animals in their research up to this point. Nonetheless, it is a good idea for people to include beans in their regular diet because they have many potential health benefits.

- **Purple or Red Grapes:** Red and purple grape varieties are rich in antioxidants, selenium, and vitamin C. Anthocyanin and proanthocyanin, two antioxidants found in grapes, may offer some protection against cancer and heart disease. To precisely demonstrate how eating grapes affects heart health and cancer risk, more research is necessary.
- **Broccoli:** Similar to other leafy, dark vegetables, broccoli is a superfood in terms of nutrients. Plants produce phenolics, a type of chemical that helps shield them from oxidative stress, in abundance when they grow broccoli. Phenolics are also essential to human health. These substances may offer protection against illness, inflammation, and allergies due to their high levels of antioxidants and anticancer characteristics.

According to the USDA, one cup (91 g) of chopped broccoli has the following nutrition facts:

Calories 31

Protein 2.6 g

Fat 0.3 g

Carbohydrates 6 g

Fiber 2.4 g

Sugar 1.6 g

Sodium 30 mg

Calcium 43 mg, or 3 percent of the DV

Iron 1mg, or 4 percent of the DV

Magnesium 19 mg, or 5 percent of the DV

Phosphorus 60 mg, 5 percent of the DV

Potassium 288 mg, 6 percent of the DV

Vitamin C 81 mg, or 90% of the DV

Folate 57 mcg, or 14% of the DV

Vitamin A 567 IU, or 11% of the DV

Vitamin K 93 mcg, or 77% of the DV

- **Spinach:** Spinach, a nutritious green leafy vegetable, is packed with essential vitamins, minerals, and antioxidants. With its low calorie content, it is a great option for boosting the nutritional value of salads and main dishes. Spinach contains zeaxanthin and lutein antioxidants that play a key role in promoting eye health by protecting against damage from UV rays and harmful light waves.

- **Walnuts:** Nuts are a great source of unsaturated fats, fiber, and protein for snacking. However, if one had to name just one nut as the healthiest, it would be the walnut, at least in terms of how much nutrition one gets for their money. Walnuts are used in traditional Chinese medicine to treat brain health because of their striking resemblance to the human brain. They also support the health of brain cells and may help with memory.

According to the USDA, one ounce (28 g) of walnuts has the following nutrition facts:

Calories 185

Protein 4.3 g

Fat 18.5 g

Carbohydrates 3.9 g

Fiber 1.9 g

Sugar 0.7 g

Calcium 27.8 mg, or 2 percent of the DV

Iron 0.8 mg, or 4 percent of the DV

- **Beets:** Vegetables called beets are rich in antioxidants that belong to a group of pigments called betalains. Both digestive problems and colon cancer may be prevented by betalains. In addition, beets contain potassium, iron, folate, and dietary fiber. These compounds might be useful in reducing inflammation.
- **Kale:** In addition to being high in antioxidants, kale is also high in vitamins A, C, and K. It is a resilient winter vegetable and well-liked health food in many northern regions. Because red kale contains anthocyanins, it may provide benefits beyond those of green kale. Antioxidants called anthocyanins are found in a wide range of fruits and vegetables. These foods' colors, which range from vivid red to blue, are due to them.
- **Beans:** Although there are many different kinds of edible beans, we'll focus on beans as a whole because most common varieties, such as navy, black, and kidney beans, have comparable nutritional values. Because they are high in fiber, phytochemicals, and protein, beans are a mainstay of plant-based diets like vegetarian and vegan diets. As a matter of fact, beans contain nearly as much protein as meat. Although some people report experiencing digestive discomfort when consuming beans regularly, this usually goes away and is more than made up for by the many health advantages of these high-nutrient nuggets.
- According to the USDA, the following are the nutritional values for 1 cup (266 g) of rinsed and drained canned red kidney beans.

Calories 191

Protein 12.8 g

Fat 1.5 g

Carbohydrates 32.9 g

Fiber 9.5 g

Sodium 329 mg

Calcium 92 mg, or 7 percent of the DV

Iron 2 mg, or 11 percent of the DV

Magnesium 46 mg, or 11 percent of the DV

Phosphorus 186 mg, or 15% of the DV

Potassium 395 mg, or 8% of the DV

Zinc 1 mg, or 9% of the DV

Thiamin 0.1 mg, or 8% of the DV

Niacin 1 mg, or 8% of the DV

Folate 36 mcg, or 9% of the DV

- **Oats:** Oats are currently popular in various forms such as milk, flour, granola, beauty products, and even dog treats. They are packed with antioxidants that can help reduce chronic inflammation associated with heart disease and diabetes. Oats are also beneficial for weight loss due to their high levels of soluble fiber, which can slow digestion and promote feelings of fullness.

According to the USDA, 100 g of cooked oatmeal, or roughly 2/3 cup, has the following nutritional values:

Calories 116

Protein 3.1 g

Fat 1.2 g

Carbohydrates 23.2 g

Fiber 3.0 g

Sugar 9.8 g

Sodium 116 mg

Calcium 91 mg, or 7% of the DV

Iron 0.4 mg, or 3% of the DV

- **Orange Vegetables:** Many orange vegetables are rich in vitamin A and other essential nutrients. They also boast high levels of phytochemicals that are beneficial for preventing heart disease and cancer. Examples of orange vegetables with high antioxidant content include:
  - sweet potatoes
  - carrots
  - acorn squash
  - butternut squash

Kale is a nutrient-dense vegetable packed with vitamins A, C, and K, as well as antioxidants. It is a popular choice for those seeking optimal health and is a resilient winter vegetable commonly found in northern climates. Red kale, in particular, may provide additional health benefits due to its anthocyanin content. By adding antioxidant-rich foods like kale to their diet, people can support heart and eye health, reduce cancer risk, and defend against diseases linked to harmful free radicals.

3.8.4. Antioxidant Supplements

Taking supplements may seem like a convenient way to obtain antioxidants, but nothing compares to the benefits of a diverse, nutritious diet filled with a variety of colorful fruits and vegetables.

Also:

- Dietary supplements are not regulated by the FDA, meaning there is no guarantee of the effectiveness, safety, or accuracy of their ingredients.
- Apart from specific formulas designed for individuals with macular degeneration, antioxidant supplements have not been demonstrated to prevent or treat disease.
- Large amounts of antioxidants may have negative effects.

Although supplements can assist in addressing nutrient deficiencies, they cannot compensate for an inadequate diet. It is important to consult with a doctor before taking any supplements, particularly if an individual is pregnant, nursing, or currently taking medication.

3.8.5. Antioxidant Risks

Although antioxidants offer numerous potential health benefits, it is important to be aware of a few considerations, particularly when consuming them through supplements. Antioxidant supplements have the potential to interact with medications that a person may be taking for other health conditions. For instance, individuals who are on blood thinners should be cautious when taking vitamin E supplements, as they may increase the risk of bleeding. Additionally, smokers who take high doses of beta-carotene may be at an increased risk of developing lung cancer. High doses of vitamin E have also been linked to a higher risk of prostate cancer and stroke. It is crucial to discuss all medications and supplements with the doctor to ensure that they are safe for a person to consume (Kristinsson, 2014; Kumar and Bagchi, 2021)

CASE STUDY

RICKETS DUE TO SEVERE NUTRITIONAL DEFICIENCIES

Vitamin D is a lipid-soluble vitamin that is mainly produced by the body when ultraviolet rays from sunlight make contact with the skin and activate its synthesis, and can also be found in some foods and supplements. It plays a major role in the intestinal absorption of calcium, magnesium, and phosphate, which promotes bone mineralization and growth. Previous studies have also shown that vitamin D can reduce cancer cell growth, help control infections and reduce inflammation.

Vitamin D deficiency is a highly common epidemic and health problem affecting 86.27% of children living in Jeddah, Saudi Arabia. People who are dark-skinned, obese, on a vegan diet, or have limited exposure to sunlight are at a higher risk for developing vitamin D deficiency. Research has shown that vitamin D deficiency has serious clinical consequences, including cardiovascular diseases, cancer, and diabetes. It can also lead to loss of bone density, which contributes to osteoporosis and fractures.

In children, severe vitamin D deficiency can lead to rickets. Rickets is a common childhood skeletal disorder that causes the bones to soften and become weak resulting in bone deformities. The patients can present with symptoms of bone pain, poor growth, and delayed motor skills. This includes fatigue, bone pain, depression, hair loss, muscle weakness, and loss of appetite. Complications of rickets include failure to thrive, abnormally curved spine, bone deformities, dental defects, and seizures.

In this case report, a 19-month-old infant has been described as a case of nutritional rickets caused by severe nutritional deficiencies including vitamin D, zinc, ferritin, magnesium, and sodium.

Case Presentation

A 19-month-old male child was referred to the general pediatric department by his family physician as a highly suspected case of severe rickets. The child was born at term by an emergency cesarean due to multiple previous cesarean deliveries with a birth weight of 3.23 kg and length of 51 cm, within the 90th centile range. He cried actively at delivery with an Apgar score of 9/10 at one and five minutes and suffered no complications. He was exclusively breastfed up to nine months of age, without any vitamin D supplementation. His vaccinations were up to date and he had no known allergies. He had four older siblings and no family history of a similar condition. Upon the first presentation at the age of 17 months, to the family physician, the child's weight was 7.97 kg, his height was 70 cm, and his head circumference was 47 cm, which all fall below the fifth centile.

On examination, it was seen that developmentally, he was delayed and could not walk unsupported. It was observed that there was marked widening of the wrists and bowing of the legs. Harrison sulcus was noted, but the rachitic rosary was not visible. Apart from an umbilical hernia, the remaining examination, including cardiothoracic and neuromuscular examinations, revealed no additional findings. Initial blood tests revealed elevated alkaline phosphatase (>2,200 U/L; normal range (n): 156–369) and low hypophosphatemia (0.64 mmol/L; n: 1.38–2.19), adjusted calcium (2.19 mmol/L; n: 2.25–2.75), and 25-hydroxyvitamin D levels (19.3 nmol/L; n: 50–250); albumin levels were slightly above the normal range (49 g/L; n: 38–47). Wrist X-ray findings were consistent with the clinical findings showing fraying, widening, and cupping of the metaphyseal ends of both the ulna and the radius. Defects in mineralization were evident yielding a picture of active rickets with rachitic changes (Figure 3.5).

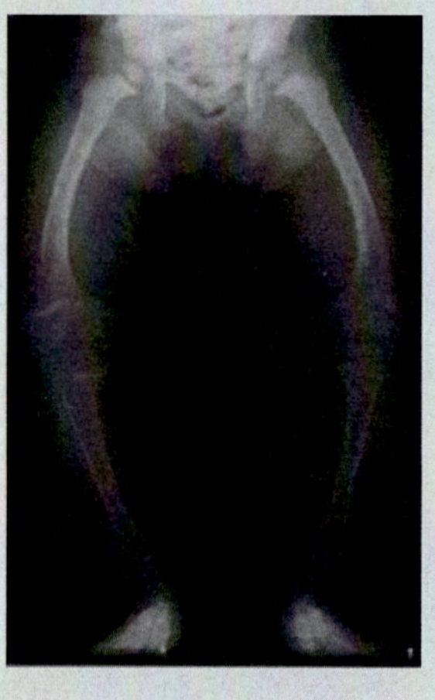

Figure 3.5. X-ray showing rickets.

Source: By The original uploader was Mrich at English Wikipedia. - Image created by Michael L. Richardson, M.D. Sept 28th, 2004de:Bild:Rachitis.jpgOriginally from en.wikipedia; description page is/was here., CC BY-SA 1.0, https://commons.wikimedia.org/w/index.php?curid=59299.

The patient was started on treatment for his multiple nutritional deficiencies including vitamin D and calcium rickets at the age of 19 months. He had no basis for hereditary rickets or other risk factors apart from the diet; being exclusively breastfed till the age of month months, and was then on a regular family diet including fruits, vegetables and meat. He was supplemented with cholecalciferol oral solution 400IU and then 2,000 IU drops for six weeks.

At his follow-up at 20 months of age, a low hemoglobin level (6.7 g/dL; n: 11.0–14.7) was detected. The child had no history of pallor, chest pain, shortness of breath, fatigue, or bleeding. His parents had never noticed signs of jaundice, change in urine or stool color, or change in bowel habits. He had no family history of thalassemia, sickle cell anemia, or glucose-6-phosphate dehydrogenase deficiency although the mother had a history of iron deficiency anemia before her pregnancy. After further testing of ferritin levels, the patient was diagnosed with microcytic hypochromic iron deficiency anemia. Follow-up lab tests showed that the patient had secondary hyperparathyroidism: hypophosphatemia (0.64 mmol/L; n: 1.38 – 2.19), adjusted calcium (2.35 mmol/L; n: 2.25

– 2.75), low 25-hydroxyvitamin D levels (19.3 nmol/L; n: 50 – 250) and high parathyroid hormone (170 pg/mL; n: 14 – 65 pg/mL), which can be another factor that may contribute to anemia in rickets.

Following the lab results at 20 months, he was started on ferrous sulfate 6mg/kg/day for his microcytic hypochromic anemia for months. In addition, he was also started on calcium carbonate 65 mg/kg/day for two weeks. Multivitamin drops of 5 ml orally were prescribed to support his diet. Cholecalciferol oral solution was increased to 4,000IU for four months. Evaluating his socio-motor skills, at 20 months, he showed slight improvement and was able to walk forward but not backward, and he could not run. He was also referred to Endocrinology to follow up on his failure to thrive.

At the latest follow-up at 22 months, he was able to kick a ball, use a spoon and fork, and use six words. He was developmentally up to age, could walk, run, go up and down the stairs (with two feet on the same step), say "baba" and "mama," and other words with two-word sentences, follow simple commands, and express his needs. The patient showed more improvement regarding the symptoms of rickets with a good appetite and a slight increase in weight.

The patient had also been referred to a dietician for diet guidance and at his latest follow-up at 22 months, he was on a regular family diet, which includes two meals (rice, meat, chicken) and was still breastfed six to seven times daily according to the child's needs. The patient was booked for regular check-ups every two months.

Discussion

This report presents a case of severe multiple nutritional deficiencies including hypophosphatemic rickets that were due to being exclusively breastfed and the absence of vitamin D supplementation throughout the first nine months of life. Unfortunately, by the time the patient sought medical assistance, he had already become symptomatic and was developmentally delayed. Rickets is an example of severe vitamin D deficiency, with a peak incidence between three and 18 months of age. This case could have been easily prevented by a systematic supplementation from birth to the age of 12–18 months. The European Society for Paediatric Gastroenterology, Hepatology and Nutrition recommends vitamin D as a supplement for breastfed infants, which should start during the first months of life and continue throughout childhood and adolescence to prevent deficiencies. The optimal daily intake of vitamin D for these groups is 400IU daily. In addition to vitamin D deficiency, the quantity and quality of calcium consumption is important and are highly dependent on the maternal diet, especially in cases of breastfed infants. Therefore, calcium supplements are also recommended in exclusively or partially breastfed infants as it highly depends on the maternal diet such as the quality of the vegetarian or vegan mother's supplementation, which can lead to nutritional deficiencies. It is, therefore, necessary to provide parents with awareness and regular checkups every two to six months to avoid potentially irreversible consequences. A previous case was reported with nutritional rickets due to a lack of vitamin D supplements after birth and presented with clinical signs of rickets, suggesting hypocalcemia as the cause of afebrile seizures. The baby was successfully managed following the intravenous administration of calcium gluconate, oral supplementation with calcium and vitamin D, and dietary changes. The patient was also given iron supplements, as iron-deficiency anemia was also detected.

Iron deficiency anemia has been associated with exclusive breastfeeding as well as rickets and anemia. Consequently, in this case, exclusive breastfeeding could be said to have led to rickets and anemia. Vitamin D deficiency has also been associated with a greater risk of anemia, lower mean hemoglobin, and higher usage of erythrocyte-stimulating agents.

Conclusions

Multiple nutritional deficiencies including Vitamin D, iron, and zinc deficiency are very common, especially in exclusively breastfed infants and can be prevented by early education, regular follow-ups every two to six months for early detection and early supplementation. Therefore, this case report adds the importance of raising awareness of nutritional deficiencies due to exclusive breastfeeding after a certain age, adding solid food and supplements to the diet, and early treatment.

CLASS ACTIVITY

ELECTROLYTE BALANCE AND RED BLOOD CELLS: EFFECTS OF HYPERNATREMIA AND HYPERCHLOREMIA

In this activity, you will investigate the impact of elevated levels of sodium and chloride in the blood, known as hypernatremia and hyperchloremia, on the health of red blood cells. Remember the concept of osmosis and how it affects the cells as you analyze your findings.

Required Materials

- 0.9% Sodium Chloride NaCl solution
- 5% Sodium Chloride NaCl solution
- Sheep's blood (red blood cells)
- Dropper
- Microscope slides
- Coverslips

- Microscope oil
- Compound light microscope
- Absorbent paper
- Lens paper
- Wax pen for labeling
- Gloves

Procedure

1. Take two microscope slides and line them up next to each other on the workbench. Write "0.9%" on one slide and "5% NaCl" on the other.
2. Place one drop of the appropriate NaCl solution onto each slide.
3. Place one drop of Sheep's Blood onto each slide.
4. Gently place a coverslip on top of the blood solution.
5. Under a microscope, look over each slide. To locate the cells, begin at a low magnification and work your way up to the highest magnification of 100x objective with oil.
6. Examine the difference in the shape of red blood cells between the 0.9% and 5% NaCl solutions. Identify and explain the variations in their shapes.
7. Can you identify which solution matches the isotonicity of the red blood cell cytoplasm and provide reasoning for your choice?
8. Please draw the shape of the red blood cells at high magnification (100x) with 0.9% NaCl and 5% NaCl in the space provided below.

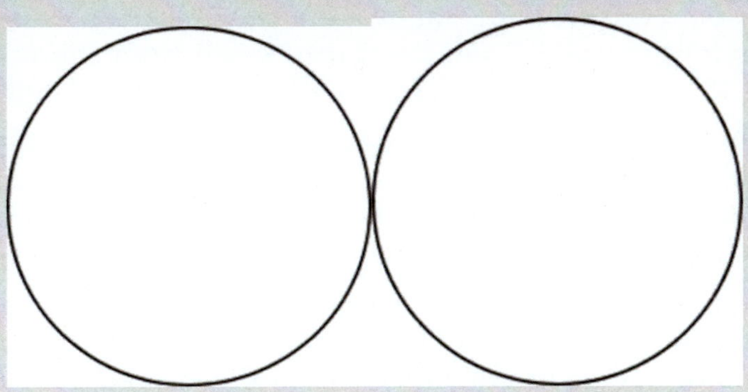

SUMMARY

- Carbohydrates are the perfect nutrient to meet one's body's nutritional needs. They nourish the brain and nervous system, provide energy to all cells (and within proper caloric limits), and help keep the body fit and lean.
- Lipid is a name for a group of organic compounds that are insoluble in water. Dietary fat is one type of lipid that, in nature, is found as a triglyceride in oils and fats. Most foods contain more than one type of triglyceride (polyunsaturated, monounsaturated, or saturated). The three main types of lipids are phospholipids, sterols, and triacylglycerols (also known as triglycerides).
- The proteins in food—such as albumin in egg white, casein in dairy products, and gluten in wheat—are broken down during digestion into constituent amino acids, which, when absorbed, contribute to the body's metabolic pool.
- Energy metabolism is the collection of biochemical processes that take place in living cells to transform nutrients from meals into energy that the body may use for a variety of physiological processes. These nutrients include proteins, lipids, and carbohydrates, which the body's enzymes break down into smaller molecules.
- The basic function of all living things, from bacteria to people, is energy metabolism. The main goal of energy metabolism is to transform food's nutrients into energy that the body may use for a variety of physiological processes. The primary energy unit of the body is ATP (adenosine triphosphate), which fuels vital cellular functions including muscular contraction, nerve conduction, and metabolic synthesis.
- Vitamins are organic compounds found in very small amounts in food and required for normal functioning—indeed, for survival. Humans are able to synthesize certain vitamins to some extent. For example, vitamin D is produced when the skin is exposed to sunlight; niacin can be synthesized from the amino acid tryptophan; and vitamin K and biotin are synthesized by bacteria living in the gut.
- Minerals are important for the body to stay healthy. The human body uses minerals for many different jobs, including keeping their bones, muscles, heart, and brain working properly. Minerals are also important for making enzymes and hormones. There are two kinds of minerals: macrominerals and trace minerals.
- The human body maintains a delicate balance of fluids and electrolytes to help ensure proper functioning and homeostasis. When fluids or electrolytes become imbalanced, individuals are at risk for organ system dysfunction. If an imbalance goes undetected and is left untreated, organ systems cannot function properly and ultimately death will occur.
- Antioxidants are substances that prevent or delay cell damage caused by compounds called free radicals. These free radicals are highly reactive compounds that can damage cells and lead to the development of chronic diseases, such as cancer, diabetes, Alzheimer's disease, Parkinson's disease, and more. Antioxidants counter the damage caused by free radicals.

REVIEW QUESTIONS

1. What do you eat most of: fast-releasing carbohydrates, starches, or fiber?
2. How are carbohydrates absorbed after digestion?
3. Experience the taste sensations of different carbohydrates. What are some foods that satisfy your sweetness sensation?
4. Even though fiber contains calories, albeit less than half of other carbohydrates, why do we generally discount its caloric contribution from our diets?
5. Explain the role of emulsifiers in fat digestion.
6. Name the part of the digestive system where most fat digestion and absorption occurs.
7. Describe the role of bile salts in the digestion of triacylglycerols and phospholipids.
8. Explain how fiber-rich foods affect cholesterol absorption.
9. What are the essential amino acids? Why are they essential?
10. What are the functions of energy metabolism?
11. What are vitamins, and how do they work?
12. List the functions of water in the body.
13. How is the acid-base balance regulated in the body?

REFERENCES

1. Adame, L. (2002). "Leaf absorption of mineral nutrients in carnivorous plants stimulates root nutrient uptake" (PDF). New Phytologist 155: 89–100. doi:10.1046/j.1469-8137.2002.00441.x.
2. Agoro, R.; Taleb, M.; Quesniaux, V.F.J.; Mura, C. Cell iron status influences macrophage polarization. *PLoS ONE* **2018**, *13*, e0196921.
3. Alpert, P. The role of vitamins and minerals on the immune system. *Home Health Care Manag. Pract.* **2017**, *29*, 199–202.
4. Alpert, P. The role of vitamins and minerals on the immune system. *Home Health Care Manag. Pract.* **2017**, *29*, 199–202.
5. Azam, F.; Fenchel, T.; Field, J. G.; Gray, J. S.; Meyer-Reil, L. A.; Thingstad, F. (1983). "The ecological role of water-column microbes in the sea" (PDF). Mar. Ecol. Prog. Ser. 10: 257–263. doi:10.3354/meps010257.
6. Beck, M.A. Trace minerals, immune function, and viral evolution. In *Military Strategies for Sustainment of Nutrition and Immune Function in the Field*; National Academy Press: Washington, DC, USA, 1999; p. 339.
7. Besold, A.N.; Culbertson, E.M.; Culotta, V.C. The Yin and Yang of copper during infection. *J. Biol. Inorg. Chem.* **2016**, *21*, 137–144.
8. Bozonet, S.M.; Carr, A.C. The Role of Physiological Vitamin C Concentrations on Key Functions of Neutrophils Isolated from Healthy Individuals. *Nutrients* **2019**, *11*, 1363.

9. Brand-Miller, J., PhD, et al. "Dietary Glycemic Index: Health Implications." J Am Coll Nutr 28, no. 4, supplement (2009): 446S–49S. http://www.jacn.org/content/28/4_Supplement_1/446S.long.

10. Bscheider, M.; Butcher, E.C. Vitamin D immunoregulation through dendritic cells. *Immunology* **2016**, *148*, 227–236.

11. Bussiere, F.I.; Mazur, A.; Fauquert, J.L.; Labbe, A.; Rayssiguier, Y.; Tridon, A. High magnesium concentration in vitro decreases human leukocyte activation. *Magnes. Res.* **2002**, *15*, 43–48.

12. Carr, A.; Maggini, S. Vitamin C and immune function. *Nutrients* **2017**, *9*, 1211.

13. Clairmont, A.; Tessman, D.; Stock, A.; Nicolai, S.; Stahl, W.; Sies, H. Induction of gap junctional intercellular communication by vitamin D in human skin fibroblasts is dependent on the nuclear Induction of gap junctional intercellular communication by vitamin D in human skin fibroblasts is dependent on the nuclear vitamin D receptor. *Carcinogenesis* **1996**, *17*, 1389–1391.

14. Foster, M.; Samman, S. Zinc and regulation of inflammatory cytokines: Implications for cardiometabolic disease. *Nutrients* **2012**, *4*, 676–694.

15. Gao, H.; Dai, W.; Zhao, L.; Min, J.; Wang, F. The Role of Zinc and Zinc Homeostasis in Macrophage Function. *J. Immunol. Res.* **2018**, *2018*, 6872621.

16. Gniadecki, R.; Gajkowska, B.; Hansen, M. 1,25-dihydroxyvitamin D3 stimulates the assembly of adherens junctions in keratinocytes: Involvement of protein kinase C. *Endocrinology* **1997**, *138*, 2241–2248.

17. Gombart, A.F. The vitamin D–antimicrobial peptide pathway and its role in protection against infection. *Future Microbiol.* **2009**, *4*, 1151.

18. Gombart, A.F.; Borregaard, N.; Koeffler, H.P. Human cathelicidin antimicrobial peptide (CAMP) gene is a direct target of the vitamin D receptor and is strongly up-regulated in myeloid cells by 1,25-dihydroxyvitamin D3. *FASEB J.* **2005**, *19*, 1067–1077.

19. González-Muñoz, M. T.; Rodriguez-Navarro, C.; Martinez-Ruiz, F.; Arias, J. M.; Merroun, M. L.; Rodriguez-Gallego, M. (2010). "Bacterial biomineralization: new insights from Myxococcus-induced mineral precipitation". Geological Society, London, Special Publications 336 (1): 31–50. doi:10.1144/SP336.3.

20. Han, S.N.; Adolfsson, O.; Lee, C.K.; Prolla, T.A.; Ordovas, J.; Meydani, S.N. Vitamin E and gene expression in immune cells. *Ann. N. Y. Acad. Sci.* **2004**, *1031*, 96–101.

21. Haryanto, B.; Suksmasari, T.; Wintergerst, E.; Maggini, S. Multivitamin supplementation supports immune function and ameliorates conditions triggered by reduced air quality. *Vitam. Miner.* **2015**, *4*, 1–15.

22. Hurwitz, B.E.; Klaus, J.R.; Llabre, M.M.; Gonzalez, A.; Lawrence, P.J.; Maher, K.J.; Greeson, J.M.; Baum, M.K.; Shor-Posner, G.; Skyler, J.S.; et al. Suppression of human immunodeficiency virus type 1 viral load with selenium supplementation: A randomized controlled trial. *Arch. Intern. Med.* **2007**, *167*, 148–154.

23. Ibs, K.-H.; Rink, L. Zinc-Altered Immune function. *J. Nutr.* **2003**, *133*, 1452S–1456S.

24. Inoue, M.; Matsui, T.; Nishibu, A.; Nihei, Y.; Iwatsuki, K.; Kaneko, F. Regulatory effects of 1alpha,25-dihydroxyvitamin D3 on inflammatory responses in psoriasis. *Eur. J. Dermatol.* **1998**, *8*, 16–20

25. Jarosz, M.; Olbert, M.; Wyszogrodzka, G.; Młyniec, K.; Librowski, T. Antioxidant and anti-inflammatory effects of zinc. Zinc-dependent NF-κB signaling. *Inflammopharmacology* **2017**, *25*, 11–24.

26. Kirkby, H.; Kirkby, E. A.; Cakmak, I. (1996). "Effect of mineral nutritional status on shoot-root partitioning of photoassimilates and cycling of mineral nutrients" (PDF). Journal of Experimental Biology 47 (S1255): 1255. doi:10.1093/jxb/47.Special_Issue.1255.

27. Kitabayashi, C.; Fukada, T.; Kanamoto, M.; Ohashi, W.; Hojyo, S.; Atsumi, T.; Ueda, N.; Azuma, I.; Hirota, H.; Murakami, M.; et al. Zinc suppresses Th17 development via inhibition of STAT3 activation. *Int Immunol.* **2010**, *22*, 375–386.

28. Kristinsson, Hordur G., ed. *Antioxidants and Functional Components in Aquatic Foods*. Chichester, UK: John Wiley & Sons, Ltd, 2014. http://dx.doi.org/10.1002/9781118855102.

29. Kumar, Abhai, and Debasis Bagchi. *Antioxidants and Functional Foods for Neurodegenerative Disorders*. Edited by Abhai Kumar and Debasis Bagchi. First edition. | Boca Raton : CRC Press, 2021.: CRC Press, 2020. http://dx.doi.org/10.1201/9780429319310.

30. Laires, M.J.; Monteiro, C. Exercise, magnesium and immune function. *Magnes. Res.* **2008**, *21*, 92–96.

31. Lee, G.Y.; Han, S.N. The Role of Vitamin E in Immunity. *Nutrients* **2018**, *10*, 1614.

32. Lin, P.H.; Sermersheim, M.; Li, H.; Lee, P.H.U.; Steinberg, S.M.; Ma, J. Zinc in Wound Healing Modulation. *Nutrients* **2017**, *10*, 16.

33. Lin, Z.; Li, W. The Roles of Vitamin D and Its Analogs in Inflammatory Diseases. *Curr. Top. Med. Chem.* **2016**, *16*, 1242–1261.

34. Macierzanka A, Torcello-Gómez A, Jungnickel C, Maldonado-Valderrama J. Bile salts in digestion and transport of lipids. Advances in Colloid and Interface Science. 2019;274:102045. doi:10.1016/j.cis.2019.102045

35. Matsui, T.; Takahashi, R.; Nakao, Y.; Koizumi, T.; Katakami, Y.; Mihara, K.; Sugiyama, T.; Fujita, T. 1,25-Dihydroxyvitamin D3-regulated expression of genes involved in human T-lymphocyte proliferation and differentiation. *Cancer Res.* **1986**, *46*, 5827–5831.

36. Maywald, M.; Wang, F.; Rink, L. Zinc supplementation plays a crucial role in T helper 9 differentiation in allogeneic immune reactions and non-activated T cells. *J. Trace Elem. Med. Biol.* **2018**, *50*, 482–488.

37. Micronutrient Information Center. Immunity in Depth. Linus Pauling Institute. 2016. Available online: http://lpi.oregonstate.edu/mic/health-disease/immunity (accessed on 10 May 2019).

38. Mihajlovic, M.; Fedecostante, M.; Oost, M.J.; Steenhuis, S.K.P.; Lentjes, E.; Maitimu-Smeele, I.; Janssen, M.J.; Hilbrands, L.B.; Masereeuw, R. Role of Vitamin D in

Maintaining Renal Epithelial Barrier Function in Uremic Conditions. *Int. J. Mol. Sci.* **2017**, *18*, 2531.

39. Palmer, H.G.; Gonzalez-Sancho, J.M.; Espada, J.; Berciano, M.T.; Puig, I.; Baulida, J.; Quintanilla, M.; Cano, A.; de Herreros, A.G.; Lafarga, M.; et al. Vitamin D3 promotes the differentiation of colon carcinoma cells by the induction of E-cadherin and the inhibition of beta-catenin signaling. *J. Cell Biol.* **2001**, *154*, 369–387.
40. Penna, G.; Adorini, L. 1 Alpha,25-dihydroxyvitamin D3 inhibits differentiation, maturation, activation, and survival of dendritic cells leading to impaired alloreactive T cell activation. *J. Immunol.* **2000**, *164*, 2405–2411.
41. Petrovic, J.; Stanic, D.; Dmitrasinovic, G.; Plecas-Solarovic, B.; Ignjatovic, S.; Batinic, B.; Popovic, D.; Pesic, V. Magnesium Supplementation Diminishes Peripheral Blood Lymphocyte DNA Oxidative Damage in Athletes and Sedentary Young Man. *Oxidat. Med. Cell. Longev.* **2016**, *2016*, 2019643.
42. Piemonti, L.; Monti, P.; Sironi, M.; Fraticelli, P.; Leone, B.E.; Dal Cin, E.; Allavena, P.; Di Carlo, V. Vitamin D3 affects differentiation, maturation, and function of human monocyte-derived dendritic cells. *J. Immunol.* **2000**, *164*, 4443–4451.
43. Reichel, H.; Koeffler, H.P.; Tobler, A.; Norman, A.W. 1 alpha,25-Dihydroxyvitamin D3 inhibits gamma-interferon synthesis by normal human peripheral blood lymphocytes. *Proc. Natl. Acad. Sci. USA* **1987**, *84*, 3385–3389.
44. Rigby, W.F.; Denome, S.; Fanger, M.W. Regulation of lymphokine production and human T lymphocyte activation by 1,25-dihydroxyvitamin D3. Specific inhibition at the level of messenger RNA. *J. Clin. Investig.* **1987**, *79*, 1659–1664.
45. Ross, A.C. Vitamin A and retinoic acid in T cell-related immunity. *Am. J. Clin. Nutr.* **2012**, *96*, 1166s–1172s.
46. Sacks FM, Yu JHY, Appel LJ, Creager MA, Kris-Etherton PM, Robinson JG, Stone NJ, Van Horn LV. Dietary fats and cardiovascular disease: A presidential advisory from the American Heart Association. Circulation. 2017;136:e1-e23. doi:10.1161/CIR.0000000000000510.
47. Saeed, F.; Nadeem, M.; Ahmed, R.; Nadeem, M.; Arshad, M.; Ullah, A. Studying the impact of nutritional immunology underlying the modulation of immune responses by nutritional compounds—A review. *Food Agric. Immunol.* **2016**, *27*, 205–229.
48. Sakakeeny, L.; Roubenoff, R.; Obin, M.; Fontes, J.D.; Benjamin, E.J.; Bujanover, Y.; Jacques, P.F.; Selhub, J. Plasma Pyridoxal-5-Phosphate Is Inversely Associated with Systemic Markers of Inflammation in a Population of U.S. Adults. *J. Nutr.* **2012**, *142*, 1280–1285.
49. Schrauzer, GN (2002). "Lithium: Occurrence, dietary intakes, nutritional essentiality". Journal of the American College of Nutrition 21 (1): 14–21. doi:10.1080/07315724.2002.10719188. PMID 11838882.
50. Shankar, A.H.; Prasad, A.S. Zinc and immune function: The biological basis of altered resistance to infection. *Am. J. Clin. Nutr.* **1998**, *68*, 447s–463s.
51. Sheikh, A.; Shamsuzzaman, S.; Ahmad, S.M.; Nasrin, D.; Nahar, S.; Alam, M.M.; Al Tarique, A.; Begum, Y.A.; Qadri, S.S.; Chowdhury, M.I.; et al. Zinc influences

innate immune responses in children with enterotoxigenic Escherichia coli-induced diarrhea. *J. Nutr.* **2010**, *140*, 1049–1056.

52. Sigmundsdottir, H.; Pan, J.; Debes, G.F.; Alt, C.; Habtezion, A.; Soler, D.; Butcher, E.C. DCs metabolize sunlight-induced vitamin D3 to 'program' T cell attraction to the epidermal chemokine CCL27. *Nat. Immunol.* **2007**, *8*, 285–293.
53. Sly, L.M.; Lopez, M.; Nauseef, W.M.; Reiner, N.E. 1alpha,25-Dihydroxyvitamin D3-induced monocyte antimycobacterial activity is regulated by phosphatidylinositol 3-kinase and mediated by the NADPH-dependent phagocyte oxidase. *J. Biol. Chem.* **2001**, *276*, 35482–35493.
54. Soliman GA. Dietary cholesterol and the lack of evidence in cardiovascular disease. Nutrients. 2018;10(6):780. doi:10.3390/nu10060780.
55. Tanaka, H.; Hruska, K.A.; Seino, Y.; Malone, J.D.; Nishii, Y.; Teitelbaum, S.L. Disassociation of the macrophage-maturational effects of vitamin D from respiratory burst priming. *J. Biol. Chem.* **1991**, *266*, 10888–10892.
56. Topilski, I.; Flaishon, L.; Naveh, Y.; Harmelin, A.; Levo, Y.; Shachar, I. The anti-inflammatory effects of 1,25-dihydroxyvitamin D3 on Th2 cells in vivo are due in part to the control of integrin-mediated T lymphocyte homing. *Eur. J. Immunol.* **2004**, *34*, 1068–1076.
57. Ueland, P.M.; McCann, A.; Midttun, O.; Ulvik, A. Inflammation, vitamin B6 and related pathways. *Mol. Asp. Med.* **2017**, *53*, 10–27.
58. Wang, T.T.; Nestel, F.P.; Bourdeau, V.; Nagai, Y.; Wang, Q.; Liao, J.; Tavera-Mendoza, L.; Lin, R.; Hanrahan, J.W.; Mader, S.; et al. Cutting edge: 1,25-dihydroxyvitamin D3 is a direct inducer of antimicrobial peptide gene expression. *J. Immunol.* **2004**, *173*, 2909–2912.
59. Warren, L. A.; Kauffman, M. E. (2003). "Microbial geoengineers". Science 299 (5609): 1027–9. doi:10.1126/science.1072076. JSTOR 3833546. PMID 12586932.
60. Weber, G.; Heilborn, J.D.; Chamorro Jimenez, C.I.; Hammarsjö, A.; Törmä, H.; Ståhle, M. Vitamin D induces the antimicrobial protein hCAP18 in human skin. *J. Investig. Dermatol.* **2005**, *124*, 1080–1082
61. Wessels, I.; Rink, L. Micronutrients in autoimmune diseases: Possible therapeutic benefits of zinc and vitamin D. *J. Nutr. Biochem.* **2019**, 77, 108240.
62. Wintergerst, E.; Maggini, S.; Hornig, D. Contribution of selected vitamins and trace elements to immune function. *Nutr. Metab.* **2007**, *51*, 301–323.
63. Wintergerst, E.; Maggini, S.; Hornig, D. Immune-enhancing role of vitamin C and zinc and effect on clinical conditions. *Ann. Nutr. Metab.* **2006**, *50*, 85–94.
64. Wishart, K. Increased micronutrient requirements during physiologically demanding situations: Review of the current evidence. *Vitamin. Miner.* **2017**, *6*, 1–16.
65. World Health Organization; Food and Agricultural Organization of the United Nations. Part 2. Evaluating the public health significance of micronutrient malnutrition. In *Guidelines on Food Fortification with Micronutrients*; WHO: Geneva, Switzerland, 2006.

66. Wu, D.; Lewis, E.D.; Pae, M.; Meydani, S.N. Nutritional modulation of immune function: Analysis of evidence, mechanisms, and clinical relevance. *Front. Immunol.* **2019**, *9*, 3160.

67. Wu, D.; Meydani, S.N. Age-associated changes in immune function: Impact of vitamin E intervention and the underlying mechanisms. *Endocr. Metab. Immune. Disord. Drug Targets* **2014**, *14*, 283–289.

68. Yin, Z.; Pintea, V.; Lin, Y.; Hammock, B.D.; Watsky, M.A. Vitamin D enhances corneal epithelial barrier function. *Investig. Ophthalmol. Vis. Sci.* **2011**, *52*, 7359–7364.

69. Yoshii, K.; Hosomi, K.; Sawane, K.; Kunisawa, J. Metabolism of Dietary and Microbial Vitamin B Family in the Regulation of Host Immunity. *Front. Nutr.* **2019**, *6*, 48.

70. Yuan G, Al-Shali KZ, Hegele RA. Hypertriglyceridemia: Its etiology, effects and treatment. Canadian Medical Association Journal. 2007;176(8):1113-1120. doi:10.1503%2Fcmaj.060963

71. Zarse, Kim; Terao, Takeshi; Tian, Jing; Iwata, Noboru; Ishii, Nobuyoshi; Ristow, Michael (2011). "Low-dose lithium uptake promotes longevity in humans and metazoans". European Journal of Nutrition 50 (5): 387–9. doi:10.1007/s00394-011-0171-x. PMC 3151375. PMID 21301855.

72. Zhang, Y.; Leung, D.Y.M.; Richers, B.N.; Liu, Y.; Remigio, L.K.; Riches, D.W.; Goleva, E. Vitamin D inhibits monocyte/macrophage proinflammatory cytokine production by targeting MAPK phosphatase-1. *J. Immunol.* **2012**, *188*, 2127–2135.

CHAPTER 4

Nutrition for Fitness, Athletics and Sports

LEARNING OBJECTIVES

After studying this chapter, you will be able to:

- Know the importance of exercise for physical fitness
- Identify requirements for moderate exercise
- Discuss nutritional needs for exercise
- Understand exercise and energy
- Focus on nutritional needs for athletic performance
- Explain ergogenic aids.

KEY TERMS FROM THIS CHAPTER

Body composition	Body displacement
Calories	Casein
Circulation	Electrolytes
Energy demands	Ergogenic aids
Exercise	Flexibility
Fluids	Macronutrients
Micronutrients	Non-communicable chronic diseases
Obesity	Oxygen

INTRODUCTION

Young athletes' sport performance is greatly influenced by nutrition, which also promotes ideal growth and development. The right ratios of macronutrients, micronutrients, and fluids are necessary to supply energy for development and activity. Young athletes must understand what, when, and how to eat and drink before, during, and after exercise to maximize performance.

4.1. EXERCISE FOR PHYSICAL FITNESS

Exercise should be a consistent, regular part of daily life, just as eating three meals a day is a consistent part of life. Activities suitable for lifetime participation must be prioritized in school physical education programs. This will guarantee the majority of students' physical health and the best possible performance. A few of them might turn into athletes.

4.1.1. Enjoy Your Activities and Exercise

Energy conservation is the human body's guiding principle. A person must learn to enjoy spending energy if they are to maintain a healthy balance as adults.

Exercise that they voluntarily engage in is a great way to use energy because it keeps us in shape and presents well. It strengthens muscles, promotes joint flexibility, and helps with posture. It enhances circulation, which is referred to as the "pink glow of health," and fortifies the heart, one of the most important muscles. It naturally reduces stress and increases the enjoyment of life. Plus, there's a lower chance of developing chronic illnesses.

4.1.2. How to Select an Exercise Program

Each person has a different preference for an exercise regimen. The word choice is crucial because there are countless methods to achieve a person's fitness objective. The guidelines for selecting it are as follows:

1. Choose an exercise a person likes so that they will enjoy doing it.
2. Choose one that can be done in all seasons.
3. Choose one that can be done preferably near residence.
4. Choose one for which a person has facilities.
5. Choose one that can be done regularly.
6. Choose an exercise that involves the large muscles in the body.

4.1.3. Physical Fitness Movement

Among people, interest in and knowledge of physical fitness are growing. It seems to be a positive trend rather than a passing fad. One becomes more dedicated to regular exercise the longer they do it. More people who are not athletically inclined are now able to participate thanks to walking and gentle workouts. Many of these individuals are middle-aged adults who manage their health issues with a modest exercise regimen. They feel more in charge of their lives as a result.

4.1.4. Beginning an Exercise Program

Starting an exercise program with a well-thought-out plan is crucial. Long-term participation in an exercise program requires a strong sense of motivation. A few fundamental pointers before getting started are:

1. Record height, weight, waist, hip circumference, and chest.
2. Launch the program incrementally. For those who have been inactive, particularly those who are elderly, as little as three to five minutes may be adequate. Extend the duration by three to four weeks to 20 to 30 minutes.
3. Intense exercise should be preceded by a 3 to 5-minute warm-up that consists of gentle, stretching exercises. Reduce the intensity gradually after a vigorous workout to give the body time to cool down.
4. The best types of exercise to improve physical fitness are aerobic ones.

Those who are overweight or have a chronic illness should see their doctor before beginning, and they should also have their blood pressure, heart health, and muscle and joint function checked. Setting goals and choosing an exercise regimen may involve consultation with the doctor (James & Schofield, 1990).

4.2. REQUIREMENTS FOR MODERATE EXERCISE

4.2.1. Fluids and Oxygen

4.2.1.1. Fluids

With increased exercise comes a greater need for water. It is commonly known that replacing lost fluids during exercise is necessary. Every cell function occurs in an aqueous medium. Water travels through the bloodstream to and from the cells carrying nutrients and waste products. Therefore, adequate blood volume is necessary to guarantee heat loss through perspiration and the skin. Sweating and breathing cause the body to lose water. Blood volume is decreased as a result of the removal of water for sweat, which could jeopardize cardiac health. Significant fluid loss may impair blood flow and sweating, which would disrupt the body's ability to regulate temperature. A 4–5% water loss causes a 20–30% reduction in work capacity; a 10% loss could cause circulatory collapse. Exercise duration, intensity, outside temperature, and humidity all affect how much fluid is lost during exercise. When a person doesn't exercise, their body produces 500–700 ml of sweat per day on average. In humid conditions and with prolonged exercise, it may rise to 8 to 12 liters per day.

4.2.1.2. Electrolytes

Although sweat contains potassium, magnesium, chloride, and electrolytes, electrolyte losses do not impair performance. A diluted salt solution (1/2 teaspoon salt per liter) can be used as a rehydration drink during training in the hot months to counteract excessive sweat losses. The body temperature rises during exercise as heat is released during the production of energy. Sweating is an attempt to regulate body temperature. To prevent dehydration, one must replenish the water lost through perspiration. During the hot summer months, a lot more water is required than during the cool months.

4.2.1.3. Oxygen

The act of breathing guarantees the body's continuous oxygen supply. Exercise causes the body to require more oxygen because it releases more energy. Aerobic capacity is the body's capacity to produce the necessary amount of oxygen. The fitness of the heart, blood vessels, lungs, and other tissues involved in oxygen intake and transportation, as well as the body's composition, determine an individual's aerobic capacity. Aerobic capacity is a measure of body fitness. The body's ability to supply and use oxygen in response to increased demands during exercise is known as aerobic capacity. It is expressed as the quantity of oxygen used per kilogram of body weight, and it varies depending on body size. As a result, it serves as a proxy for the respiratory system's health.

4.2.1.4. Body Composition

A greater portion of oxygen is used by the body's muscle mass, also known as lean body mass or the active metabolic tissues. The percentage of body fat and lean mass determines the aerobic capacity. These two aspects of body weight comprise body composition.

4.3. NUTRITIONAL NEEDS FOR EXERCISE

For young athletes to grow properly and compete at their best in sports, they must eat a healthy diet. Young athletes must be taught which foods are high in energy when to eat them, how to fuel up during competition, and what to eat to recover from exercise. To provide enough energy for growth and activity, a well-balanced diet with the right proportions of macronutrients (protein, carbs, and fat) and micronutrients (vitamins and minerals) is necessary. To maintain hydration and enhance growth and athletic performance, fluids are also necessary.

4.3.1. Macronutrients

Carbohydrates, protein, and fats are examples of macronutrients that provide the fuel for exercise and sports participation.

- **Carbohydrates:** Since they supply the glucose needed for energy, carbohydrates are the most crucial fuel source for athletes. An energy-dense gram of carbohydrates is about four kilocalories. In the liver and muscles, glucose is stored as glycogen. When working muscle needs energy, muscle glycogen is the most accessible source and releases energy faster than other energy sources. For individuals aged four to 18, 45–65% of their total caloric intake should come from carbohydrates. Carbohydrate-rich foods include fruits, vegetables, whole grains, milk, and yogurt.
- **Protein:** Proteins help to grow and repair skin, hair, nails, and muscles. Proteins don't serve as the main energy source during brief or moderate exercise. However, through liver gluconeogenesis, proteins aid in blood glucose maintenance as exercise duration increases. Four kilocalories of energy are found in one gram of protein. For children aged four to 18, protein should make up between 10–30% of their total energy intake. Lean meat and poultry, fish, eggs, dairy products, beans, and nuts—including peanuts—are all excellent sources of protein.
- **Fats:** Fat is required for insulation, the provision of essential fatty acids, the absorption of fat-soluble vitamins (A, D, E, and K), and the protection of critical organs. Furthermore, fat gives one a sated feeling. Although it is more difficult to use, it is a calorie-dense source of energy (one gram contains nine kilocalories). For children aged four to 18, fats should make up 25–35% of their total energy intake (Otten, Hellwig, Meyers, 2006). The maximum percentage of calories from saturated fats should be 10% of the total calories consumed. Lean meat and poultry, fish, nuts, seeds, dairy products, and olive and canola oils are good sources of fat. Reduce intake of fat from chips, candies, fried foods, and baked goods.

4.3.2. Micronutrients

While many vitamins and minerals are necessary for optimal health, athletes should pay special attention to making sure they get enough calcium, vitamin D, and iron. For healthy bones, normal enzyme activity, and muscle contraction, calcium is essential. For children ages four to eight, the recommended daily intake of calcium is 1,000 mg, and for those ages nine to 18, it is 1,300 mg. Many foods and drinks, such as milk, yogurt, cheese, broccoli, spinach, and fortified grain products, contain calcium. In addition to being involved in calcium absorption and regulation, vitamin D is essential for healthy bones. 600 IU per day is the recommended amount for children aged four to 18. Vitamin D normal values also differ based on race and geographic location. Athletes who train indoors, such as dancers, gymnasts, and figure skaters, or who reside in northern latitudes, are more likely to be vitamin D deficient. Sunlight exposure and fortified foods, like milk, are sources of vitamin D. Yogurt and other dairy products that aren't milk aren't vitamin D-rich.

The transportation of oxygen to bodily tissues depends on iron. More iron is needed during adolescence to support lean muscle mass and blood volume increases in addition to growth. Iron-deficiency anemia and the depletion of iron stores can be prevented in boys and girls aged nine to 13 by consuming eight milligrams per day. More iron is needed by adolescents between the ages of 14 and 18—up to 11 mg for males and 15 mg for females per day. Due to diets low in meat, fish, and poultry, or increased iron losses in sweat, urine, feces, or menstrual blood, athletes frequently experience iron deficiency. As a result, long-distance runners, vegetarians, and athletes—especially female athletes—should undergo routine iron status screenings. Foods high in iron include lean meat, eggs, fortified whole grains, and leafy green vegetables.

4.3.3. Fluids

For athletes, fluids—especially water—are crucial nutrients. What, how much, and when an athlete drinks can all have an impact on their performance. Fluids replace lost sweat during exercise and aid in controlling body temperature. The temperature and humidity of the surrounding air can have an impact on how much an athlete perspires and how much fluid intake is necessary. An individual will sweat more in hotter and more humid conditions, so drinking more fluids will be necessary to stay hydrated. Dehydrated athletes run the risk of experiencing heat exhaustion or heat stroke, which can impair performance.

Drinking fluids before, during, and after exercise or other activities is necessary for proper hydration. Age and body size are just two of the many variables that determine how much fluid is needed. Athletes should drink 400 mL to 600 mL of cold water two to three hours before their activity. Athletes should drink 150 mL to 300 mL of fluid every 15 to 20 minutes when participating in sports. Water is plenty for events that last under an hour. Sports drinks with 6% carbohydrates and 20–30 mEq/L of sodium chloride are advised for events longer than 60 minutes and/or held in hot, muggy weather in order to replenish energy reserves and electrolyte and fluid losses. Athletes should consume enough fluids to replace fluids lost through sweat after exercise. Usually, this calls for drinking 1.5 L of fluid for every kilogram of body weight lost. After exercise, consuming sodium-containing liquids and snacks promotes thirst and fluid retention, which aids in rehydration. Sports drinks with added carbohydrates should be avoided by people who are not athletes because they can cause them to consume too many calories and raise their risk of being overweight, obese, and developing dental caries (Table 4.1).

Table 4.1. The Approximate Requirements for Carbohydrates

Exercise intensity	Minutes of activity per day	Recommended carbohydrate intake (g/kg)
Low	<60	3-5
Moderate	60	5-7
High	60-180	6-10
Very high	>180	8-12

Source: Amawi A, AlKasasbeh W, Jaradat M, Almasri A, Alobaidi S, Hammad AA, Bishtawi T, Fataftah B, Turk N, Saoud HA, Jarrar A and Ghazzawi H (2024) Athletes' nutritional demands: a narrative review of nutritional requirements. Front. Nutr. 10:1331854. doi: 10.3389/fnut.2023.1331854.

4.3.4. Nutritional Recommendations for Athletes

Every athlete knows that nutrition is crucial, and at the highest levels, when training and talent are comparatively equal, it can mean the difference between winning and losing.

The American College of Sports Medicine (ACSM) released information on suggested nutrition for athletes that they gathered from a variety of sources. They claim that "optimal nutrition enhances physical activity, athletic performance, and recovery from exercise." Athletes are therefore encouraged to closely monitor how much food they eat each day to ensure that their bodies are getting the nutrients necessary for muscle growth, maintenance, and repair. There are numerous suggestions for what foods athletes should eat, but we've focused on the ACSM and the U.S. official recommendations from the U.S. Anti-Doping Agency (USADA). Ultimately, the most crucial advice is for an athlete to monitor their daily energy expenditure and ensure that they are replenishing this energy with a diet that is balanced in terms of fat, protein, and carbs. Eating 0.68 grams of carbohydrates per pound of body weight within the first 30 minutes of training and then every two hours for four hours is the recommended replenishment for maintaining blood sugar levels in the body. It is important to note that studies have shown that athletes who rest one or two days in between training bouts do not necessarily need to follow this post-training carbohydrate recommendation as long as they are meeting their daily requirements, which are explained below. An athlete who has finished their training for the day should have a meal or snack with about 122 g of carbohydrates. Examples of such foods include an oatmeal bar and a multigrain bagel with peanut butter.

Athletes are advised by the ACSM and numerous other publications to never train on an empty stomach. It is advised that an athlete consume 200–300 g of carbohydrates three to four hours before engaging in physical activity. To help slow down the breakdown of carbohydrates and help regulate energy levels by supplying carbohydrates to muscles at a steady rate throughout the training, the USADA advises adding small amounts of protein. But they also state that it depends on the athlete and how big of a meal they can eat before working out. To determine what pre-training meal they are most comfortable with, athletes should try a variety of options and eat that meal at a time that is consistent with when they eat the pre-training meal before a game. Athletes competing in events longer than an hour can significantly improve their endurance performance by eating between

30 and 60 grams of carbohydrates per hour, according to research. This is equivalent to consuming a food bar, banana, or other item with the same amount of carbohydrates every hour. The American College of Sports Medicine (ACSM) goes one step further and states that eating small snacks or bite-sized portions of carbohydrates (10–20 grams each time) every 15–20 minutes is far more advantageous to the body's energy storage than waiting and consuming 60–120 grams every two hours. Athletes can consume 6–12 ounces of a sports drink with a ratio of 6–8% carbohydrates to water, according to the USADA. The chance of experiencing cramps and upset stomach increases with a ratio higher than 10%. Similarly, studies are now demonstrating that a maximal acute anabolic effect can be achieved with just 0.2 grams of high-quality protein consumed pre- and post-training, or about 4 ounces of chicken breast. The USADA and almost all other organizations advise that an athlete's daily calorie intake should consist of roughly 50–70% carbohydrates, 10–35% protein, and 20–30% fat. An athlete must therefore schedule their meals based on these percentages in order to ensure that they consume an adequate amount of each type of calorie throughout the day. An athlete's dietary requirements for each nutrient are broken down by body weight in the following comprehensive recommendation from ACSM:

1. **Carbohydrates**

1 hour of training per day	2.5 grams per pound of body wt per hour of training
More than 4 hours of training per day	No more than 6 grams per pound of body wt

2. **Proteins**

Endurance training	54–0.64 grams per pound of body wt
Strength training (to gain muscle mass) body	0.72–0.81 grams per pound of wt
Strength training (to maintain wt) body	0.54–0.64 grams per pound of wt

3. **Fats**

   20% of daily calories

   A person can consume a wide variety of fats, so they do not recommend a specific number of grams per pound of fat. Most athletes already eat foods that contain fat, so they only need to be careful to choose foods that contain the right amount of fat. Most athletes do not need to try to consume more fat. All researchers agree that, for the athlete seeking nutritional guidance, meeting daily requirements for fat, protein, and carbohydrates is crucial to optimizing the benefits of training, even though there is some degree of flexibility in the optimal timing of nutrients.

4.3.5. Recovery Foods

Consuming recovery foods within 30 minutes of exercise and again within 1 to 2 hours of exercise will help replenish muscles with glycogen and facilitate a full recovery. Carbs and protein should be included in these diets. Sports drinks with fruit and cheese, yogurt with fruit, and graham crackers with peanut butter and juice are a few examples.

4.3.6. Meal Planning

The most difficult thing to coordinate is scheduling meals around sporting events. Meal timing is crucial and must be customized for each person. It's critical for athletes to identify foods they enjoy eating to optimize performance. On the day of the competition, they shouldn't try any new foods or routines.

Meals should be consumed at least three hours before an event in order to facilitate adequate digestion and reduce the risk of gastrointestinal distress while exercising, according to general recommendations. Carbs, protein, and fat should all be present in meals. Fiber needs to be restricted. Eat something low in fat before working out to prevent delayed gastric emptying, sluggish feelings in athletes, and poor performance. A liquid meal or snack one to two hours before exercise, followed by a full breakfast post-event, will help ensure that there is enough energy for optimal performance during early morning practices or events. Liquid meals or pre-game snacks should be consumed one to two hours before an event to allow for digestion before beginning physical activity. Fresh fruit, dried fruit, cereal with milk, juice, or fruit smoothies can all be considered snacks. Sports drinks, fruit, or granola bars can be consumed during an event to help sustain and replenish energy levels.

4.3.7. Reaching the Finish Line

Proper nutrition is crucial for young athletes to support growth and maximize performance in sports. A balanced diet should consist of 45–65% carbohydrates, 10–30% protein, and 25–35% fat. It is also important to stay hydrated by consuming fluids before, during, and after workouts to prevent dehydration. Timing of meals is key for optimal performance, with main meals eaten at least 3 hours before exercise and snacks consumed 1–2 hours beforehand. After exercise, recovery foods should be eaten within 30 minutes and again within 1–2 hours to aid muscle repair and promote proper recovery.

4.4. EXERCISE AND ENERGY

The body goes through a lot of physiological and nutritional demands when exercising. Muscle contraction raises the need for oxygen, hydrogen, and other essential nutrients. The human body needs energy constantly to carry out its numerous tasks. Exercise increases energy demands; if these demands are not met, the exercise will come to an end.

4.4.1. Factorial Estimation of Total Energy Expenditure and Physical Activity Level

The heterogeneity of adult populations with varying geographic, cultural, and economic backgrounds with respect to body size, body composition, and habitual physical activity precludes the universal application of energy requirements based on TEE measured with DLW (or HRM) in lifestyle groups. Therefore, factorial calculations were used to estimate TEE to account for the variations in physical activity. These calculations combined the time spent on habitual activities with the energy cost associated with those activities. Examples of these calculations are presented in Table 4.2. The physical activity ratio (PAR), which is a multiple of BMR per minute, was used to calculate the energy cost of activities and express the 24-hour energy requirement as a multiple of BMR per 24 hours. This allowed for variations in body composition and size. When the physical activity limit or PAL is known or derived using BMR estimated from gender- and age-specific predictive equations based on the population's average body weight, it can be used in conjunction with the population's BMR to estimate TEE and, consequently, the mean energy requirement for that population.

To streamline computations, the prior expert consultation categorized adult population group PAL as light, moderate, or heavy based on their employment status or other activities. The PAL was then multiplied by the corresponding BMR to determine the necessary amounts. During the current consultation, it was determined that the 24-hour PAL should not solely be determined by the physical demands of an occupational job. This is because some individuals with light occupations engage in intense physical activity during their free time, while those with heavy jobs spend most of their days sitting still. The decision was made to base the factorial estimates of energy requirements on the energy costs related to lifestyles that incorporate both voluntary and occupational physical activity.

This consultation determined that the average energy cost of activities should be consistent for both men and women when expressed as a multiple of BMR or PAR.

However, there is a difference in energy expenditure between genders when the PAR value is converted into energy units, as men typically have higher BMR for their body weight compared to women due to their heavier weight. As a result, most activities listed in Table 4.2 in relation to BMR can be applied to both men and women. The only exceptions are vigorous activities that require a level of effort proportional to muscle mass and strength, which is often higher in men (such as lifting heavy objects, chopping wood, or using a sledgehammer).

4.4.2. Estimation of Basal Metabolic Rate

BMR makes up a significant portion of total energy expenditure in adults, typically ranging from 45 -70%. It is mainly influenced by factors like gender, body size, body composition, and age.

Accurate measurement of BMR can be achieved through direct or indirect calorimetry under specific conditions, such as being awake, lying down, several hours after a meal, and after a period of rest without vigorous physical activity. BMR measurements are typically conducted in laboratory settings and on small, representative groups of individuals. When estimating TEE using the factorial approach, population-level BMR estimates are necessary. To simplify this process, predictive equations based on easily obtainable measurements like body weight and height are commonly used to calculate a group's average BMR.

Table 4.2. Factorial Calculations of Total Energy Expenditure for a Population Group

Main Daily Activities	Time Allocation Hours	Energy Cost[a] PAR	Time × Energy Cost	Mean PAL[b] Multiple of 24-Hour BMR
Sedentary or Light Activity Lifestyle				
Sleeping	8	1	8.0	
Personal care (dressing, showering)	1	2.3	2.3	
Eating	1	1.5	1.5	
Cooking	1	2.1	2.1	
Sitting (office work, selling produce, tending shop)	8	1.5	12.0	
General household work	1	2.8	2.8	
Driving car to/from work	1	2.0	2.0	
Walking at varying paces without a load	1	3.2	3.2	
Light leisure activities (watching TV, chatting)	2	1.4	2.8	
Total	**24**		**36.7**	**36.7/24 = 1.53**
Active or Moderately Active Lifestyle				
Sleeping	8	1	8.0	

Personal care (dressing, showering)	1	2.3	2.3	
Eating	1	1.5	1.5	
Standing, carrying light loads (waiting on tables, arranging merchandise)[c]	8	2.2	17.6	
Commuting to/from work on the bus	1	1.2	1.2	
Walking at varying paces without a load	1	3.2	3.2	
Low intensity aerobic exercise	1	4.2	4.2	
Light leisure activities (watching TV, chatting)	3	1.4	4.2	
Total	**24**		**42.2**	**42.2/24 = 1.76**
Vigorous or Vigorously Active Lifestyle				
Sleeping	8	1	8.0	
Personal care (dressing, bathing)	1	2.3	2.3	
Eating	1	1.4	1.4	
Cooking	1	2.1	2.1	
Non-mechanized agricultural work (planting, weeding, gathering)	6	4.1	24.6	
Collecting water/wood	1	4.4	4.4	
Non-mechanized domestic chores (sweeping, washing clothes and dishes by hand)	1	2.3	2.3	
Walking at varying paces without a load	1	3.2	3.2	
Miscellaneous light leisure activities	4	1.4	5.6	
Total	**24**		**53.9**	**53.9/24 = 2.25**

Source: https://www.fao.org/4/y5686e/y5686e07.htm.

[a] Energy costs of activities, expressed as multiples of basal metabolic rate.

[b] PAL = physical activity level, or energy requirement expressed as a multiple of 24-hour BMR.

[c] Composite of the energy cost of standing, walking slowly and serving meals or carrying a light load.

- **Examples**
  - **Sedentary or Light Activity:** TEE = 1.53 × 5.40 = 8.26 MJ (1 975 kcal) or 150 kJ (36 kcal)/kg/d would be the result if this PAL came from a female population aged 30 to 50, with a mean weight of 55 kg and a mean BMR of 5.40 MJ/day (1 290 kcal/day).
  - **Active or Moderately Active:** TEE = 1.76 × 5.60 = 9.86 MJ (2 355 kcal)/kg/d if this PAL came from a female population of 20–25 years old with a mean weight of 57 kg and a mean BMR of 5.60 MJ/day (1 338 kcal/day).
  - **Vigorous or Vigorously Active:** A male population aged 20 to 25 with a mean weight of 70 kg and a mean BMR of 7.30 MJ/day (1 745 kcal/day) would be the source of this PAL. TEE would be 2.25 × 7.30 = 16.42 MJ (3 925 kcal), or 235 kJ (56 kcal)/kg/d.

Equations primarily from Western European and North American studies were used in the 1985 FAO/WHO/UNU expert consultation report (Schofield, 1985). There have been concerns raised regarding the generalizability of the adult equations because nearly half of the data came from studies done on Italian men with relatively high BMR values in the late 1930s and early 1940s. There have also been doubts about the majority of studies' use of closed-circuit indirect calorimetry because this method may overestimate energy and oxygen consumption. Predictive equations from a database with more diverse geographic and ethnic representation were assessed for the current consultation. The published BMR measurements of adults from various parts of the world that were not included in the databases used to create the predictive equations were compared to the predictive accuracy of the new equations and those from 1985. This consultation found that while the new equations had some benefits—such as slight decreases in prediction error and the overestimation bias among men—they were not reliable enough to support their adoption at this time. For the time being, it was determined to keep the equations that Schofield (1985) proposed (Table 4.3). Additionally, a more in-depth examination of the available data will be pursued, or a prospective study with wide global geographic and ethnic representation will be supported.

Table 4.3. Equations for Estimating BMR from Body Weight*

Age Years	No.	BMR: MJ/day	See	BMR: kcal/day	see
Males					
< 3	162	0.249 kg – 0.127	0.292	59.512 kg – 30.4	70
3–10	338	0.095 kg + 2.110	0.280	22.706 kg + 504.3	67
10–18	734	0.074 kg + 2.754	0.441	17.686 kg + 658.2	105

18–30	2,879	0.063 kg + 2.896	0.641	15.057 kg + 692.2	153
30–60	646	0.048 kg + 3.653	0.700	11.472 kg + 873.1	167
³ 60	50	0.049 kg + 2.459	0.686	11.711 kg + 587.7	164
Females					
< 3	137	0.244 kg – 0.130	0.246	58.317 kg – 31.1	59
3–10	413	0.085 kg + 2.033	0.292	20.315 kg + 485.9	70
10–18	575	0.056 kg + 2.898	0.466	13.384 kg + 692.6	111
18–30	829	0.062 kg + 2.036	0.497	14.818 kg + 486.6	119
30–60	372	0.034 kg + 3.538	0.465	8.126 kg + 845.6	111
³ 60	38	0.038 kg + 2.755	0.451	9.082 kg + 658.5	108

Source: Schofield, 1985. https://www.fao.org/4/y5686e/y5686e07.htm.

** Weight is expressed in kg. Predictive equations for children and adolescents are presented for the sake of completeness.*

4.4.3. Physical Activity Level

A healthy, well-nourished adult's average PAL plays a significant role in determining how much energy they need overall. Since adult energy needs are not influenced by growth, PAL can be calculated or estimated using the average 24-hour BMR and TEE (i.e., PAL = TEE/BMR). The real energy needs are obtained by multiplying the PAL by the BMR. For instance, the mean energy requirement of a male with a PAL of 1.75 and a mean BMR of 7.10 MJ/day (1 697 kcal/day) would be 1.75 × 7.10 = 12.42 MJ/day (2 970 kcal/day). Table 4.2 provides additional examples of these computations at the bottom of each panel.

Numerous studies have computed PAL using TEE measurements and estimates or measurements of BMR. Although some research has been conducted in developing nations, where many people lead lifestyles linked to levels of physical activity that differ from those in industrialized nations, the majority of the data on the TEE of adults currently available comes from studies conducted in those societies (Coward, 1998). A modal value for PAL of 1.60 (range 1.55 to 1.65) was found for both men and women in a meta-analysis of studies involving 411 individuals aged 18 to 64 (Black et al., 1996). The majority of the subjects came from wealthy, developed nations. While everyone was in good health, 13% of the females and 9% of the males had a BMI of 30 or higher and were overweight or obese. The typical subpopulations were housewives, students, professionals or white-collar workers, unemployed people, and retirees; only three people were identified as manual workers specifically. Therefore, the study participants were defined by the authors of the meta-analysis as Westerners who lead primarily sedentary lifestyles.

A somewhat lower PAL range of 1.50 to 1.55 was recommended by an expert panel of the International Obesity Task Force (IOTF) as being indicative of sedentary people.

About 1.40 to 2.40 are the PAL values that free-living adult populations can maintain over an extended length of time. During this consultation, it was determined that regular physical activity, either during work or leisure time, at a level and for a duration that will lower the risk of obesity and several chronic non-communicable diseases that are typically co-morbidities with it, as well as the risk of becoming overweight.

This is consistent with PAL values greater than 1.75. However, there was no definition of a minimum maintenance energy requirement, which reiterated the expert consultation's previous position that any figure selected would reflect a value judgment on what activity levels above the minimum for survival could be appropriately included in the term maintenance.

4.4.3.1. Classification of Physical Activity Levels

Energy needs are strongly influenced by regular physical activity. Similar to the 1981 FAO/WHO/UNU expert consultation, this one divided the intensity of a population's habitual physical activity into three categories. Nevertheless, a range of PAL values rather than a mean PAL value was established for each category, which is different from the 1981 consultation. Moreover, for the reasons outlined, the same PAL values were utilized to place men and women in separate PAL categories.

Table 4.4 displays categories that correspond to varying activity levels linked to a population's lifestyle. These categories show the most common types of physical activity that the majority of people in the population engage in over time. It can be defined as one month or more, even though there is no physiological basis for determining its length. Since there are groups of people with light or sedentary occupations who regularly engage in vigorous discretionary activities, the term "lifestyle" was preferred over "occupational work," as it better reflected the types of people who belong in the active or vigorously active categories. It's important to keep in mind that certain populations experience periodic shifts in their way of life. Examples of these include those associated with the agricultural cycle in traditional rural societies or with the seasons, where hot or mild summers alternate with cold winters. These populations' energy needs will fluctuate according to the energy demands of their cyclical lifestyles.

Table 4.4. Classification of Lifestyles in Relation to the Intensity of Habitual Physical Activity, or PAL

Category	PAL Value
Sedentary or light activity lifestyle	1.40–1.69
Active or moderately active lifestyle	1.70–1.99
Vigorous or vigorously active lifestyle	2.00–2.40*

* PAL values > 2.40 are difficult to maintain over a long period of time.

4.4.3.2. Examples of Lifestyles with Different Levels of Energy Demands

- **Sedentary or Light Activity Lifestyles:** These individuals work in jobs that don't require a lot of physical exertion, aren't required to walk far, typically travel by car, don't regularly exercise or play sports, and spend the majority of their free time sitting or standing with little movement of the body (e.g., speaking, writing, listening to the radio, viewing television, and using computers). Male urban office workers are one example; they sporadically partake in physically demanding activities during or after work. Another example would be rural women who spend most of their time selling produce at home or in the market, taking care of their children in or around their homes, and doing light household chores. These women live in villages with piped water, electricity, and close access to paved roads.

- **Active or Moderately Active Lifestyles:** These individuals work in jobs that don't require a lot of energy, but they do require more energy than those associated with sedentary lifestyles. Alternatively, they could be sedentary workers who make it a point to engage in moderate-to-intense physical activity for a portion of their daily schedule, either required or voluntary. For example, a person's average PAL can increase from 1.55 (the sedentary category) to 1.75 (the moderately active category) by engaging in one hour of moderate to vigorous exercise per day, either continuously or in several bouts throughout the day. Examples of this type of exercise include cycling, aerobic dancing, jogging/running, and various sports activities. Additional instances of moderately active lifestyles are linked to professions like masons and construction workers, or to rural women in less developed traditional villages who walk great distances to retrieve fuelwood and water.
- **Vigorous or Vigorously Active Lifestyles:** These individuals frequently spend several hours doing physically demanding work or leisure activities. Examples include women in non-sedentary jobs who spend an average of two hours dancing or swimming, or non-mechanized agricultural laborers who walk long distances over difficult terrain while carrying heavy loads, and who work with a machete, hoe, or axe for several hours each day.
- **Extremes of Low and High PALs:** While extremely low energy expenditure is necessary for survival, it is incompatible with long-term health, freedom of movement, and earning a living. Such levels have been documented in resting adults confined to a whole body calorimeter, adolescents with cerebral palsy or myelodysplasia, and elderly mental patients. In times of crisis, the mean PAL of 1.21—which is comparable to the baseline energy requirement of 1.27 as determined in the 1985 report—is recommended for the temporary survival of completely sedentary dependent individuals. According to the current consultation, this kind of value is too low and shouldn't be applied to emergency relief efforts because people don't become totally immobile during times of crisis and because the stresses placed on them can raise their energy needs. For short-term

relief interventions, the consultation therefore recommends food supplies to satisfy a PAL of 1.40, which represents the lower limit of the sedentary lifestyle range shown in Table 4.4.

Conversely, research has demonstrated that for three weeks of competitive cycling or sled hauling across the Arctic, PAL values can reach as high as 4.5 to 4.7. These kinds of energy expenditures, however, are not long-term viable (James & Schofield, 1990).

4.5. ATHLETIC PERFORMANCE

Athletes who participate in rigorous training regimens regularly ought to be cognizant of their daily dietary requirements. Sustaining an intense training program and maximizing immune system functions depends on eating a balanced diet that supplies enough energy and nutrients.

4.5.1. Energy Requirements

Athletes need to eat enough calories to meet their energy requirements. The body will use fat and lean tissue as fuel if energy requirements are not satisfied. One will lose stamina and strength as a result of this. Moreover, there will be a compromise in musculoskeletal, endocrine, and immune function. Low-calorie intake over time may cause deficiencies in vital vitamin and mineral intake as well as a slower resting metabolic rate. Weight class athletes who use drastic methods to drop weight quickly before a competition run the risk of suffering from the negative effects of low energy intake. Examples of these athletes include boxers, kickboxers, and mixed martial artists. Such energy limitations may impair athletic performance and result in muscle loss.

4.5.2. Carbohydrate Requirements

Sugars, both simple and complex, are considered carbohydrates. In order to power exercise, carbohydrates keep blood sugar levels stable. They also restore glycogen, the form that muscles use to store carbohydrates. Athletes should consume between 6 and 10 g of carbohydrates per kilogram of body weight each day.

4.5.3. Protein Requirements

Muscle tissue is composed of proteins. It also serves a plethora of other purposes throughout the human body. It is recommended that endurance athletes consume 1.2–1.4 grams of protein per kilogram of body weight per day. Protein intake should be slightly higher for ultra-endurance athletes who train nonstop for several hours or days in a row; however, consuming more than 2 grams of protein per kilogram of body weight is not advised. Protein intake for strength athletes should be between 1.2–1.7 g/kg of body weight. Generally speaking, obtaining this amount without the use of supplements is simple with a regular diet. Good sources of protein, like soy, whey, or casein, work just as well for maintaining, repairing, and synthesizing muscle proteins.

4.5.4. Fat Requirements

A sufficient intake of fat is required for many metabolic processes that support optimum health. For example, fat is necessary for the proper absorption of vitamins A, D, and E. An athlete's daily fat intake should be between 20–35% of total calories. 10% of total fat intake should come from monounsaturated fat, 10% from polyunsaturated fat, and no more than 10% from saturated fat, according to current dietary guidelines. Studies have not demonstrated any advantages associated with a diet high in fat (>70% of total energy).

4.5.5. Vitamin and Mineral Requirements

Micronutrients perform a number of tasks that promote health. They have a role in the synthesis of blood, the immune system, bone health maintenance, energy production, and the avoidance of oxidative damage. Additionally, they support the process of muscle and tissue healing after an injury or period of exercise. In general, athletes who eat a balanced diet don't need to take extra micronutrient supplements.

A multivitamin supplement, however, might be suitable if the athlete is on a diet or stays away from particular foods or food groups. If a medical practitioner diagnoses a deficiency, supplements of specific nutrients, like iron, might be necessary. Although taking large amounts of antioxidant supplements (such as vitamins C, E, and B-carotene) is becoming common among athletes, there isn't much proof that these supplements improve performance. Athletes should exercise caution when megadosing on these vitamins as larger dosages may have detrimental effects. Low intakes of iron, calcium, vitamin D, riboflavin, zinc, and vitamin B12 may be a concern for athletes who follow a vegetarian diet. It is advised to speak with a sports dietitian to make sure a person is getting enough of these nutrients.

4.5.6. Timing of Food and Fluid Intakes

4.5.6.1. Before Exercise

Athletes should be accustomed to eating a pre-exercise meal or snack that is low in fat and high in fiber to facilitate rapid digestion and reduce the risk of gastrointestinal distress. The meal ought to have a moderate amount of protein and lots of carbs. 200–300 grams of carbohydrates consumed three to four hours before an event are typically the amount of carbohydrates that have been shown to improve performance.

Drinking fluids at least four hours before an exercise session is recommended. Water or an energy drink are suitable. For best results, the athlete should aim for 5–7 mL/kg (2–3 mL/lb) of fluid per body weight. Since there is no proof that hyperhydration with water and glycerol mixtures improves performance, this approach should be avoided.

4.5.6.2. During Exercise

If an athlete exercises for more than an hour, sports drinks with 6–8% carbohydrates are good for them. It is recommended for athletes to consume 0.7 grams of carbohydrates/kg of body weight (or 30–60 grams per hour) during extended workouts or events. Studies have indicated that this practice results in an increased capacity for endurance. When athletes exercise first thing in the morning after fasting through the night and their liver glycogen levels are low, sports drinks provide a substantial advantage.

Supplementing with carbohydrates during this kind of exercise is also beneficial for athletes who are cutting calories in order to lose weight and for those who have not eaten a pre-exercise meal. Sports drinks have been shown to provide the biggest performance gains when consumed at intervals of 15–20 minutes to stay hydrated. Simple sugar mixtures in liquid form, such as glucose, fructose, or others, work just as well. Fructose by itself, however, is less efficient and can result in diarrhea. To help maintain a fluid and electrolyte balance, drink an electrolyte-containing sports beverage. Drinks high in potassium and sodium will help replenish electrolytes lost through perspiration. Consuming sodium will also promote the need to stay hydrated. Enough fluid replacement during exercise is crucial because sweat losses of more than 2% of body weight have a negative impact on athletic performance.

4.5.6.3. After Exercise

The duration, intensity, and timing of the subsequent exercise session all influence the post-exercise meal. Carbs should be ingested within 30 minutes of exercise if multiple training sessions are anticipated in a single day. This will ensure that the muscle glycogen stores are fully restored. It is advised to consume 1.0–1.5 grams of carbohydrates/kg of body weight every two hours for a maximum of six hours. Combining simple sugars with exercise enhances post-exercise glycogen synthesis more than fructose alone. Consuming some protein after working out will support the growth and repair of muscle tissue. To recover from exercise-induced dehydration, consume 16–24 oz (450–675 mL) of fluid for each pound (0.5 kg) lost through perspiration. Restoring electrolyte balance can also be aided by eating foods high in salt (James & Schofield, 1990).

4.6. ERGOGENIC AIDS

Each and every sports competitor wants to win. Of course, they want to use anything that can help them with their activity. These compounds are referred to as ergogenic aids, as they produce work. With the exception of steroids, most of these are useless. The use of steroids carries a risk. These are synthetic forms of hormones used in sex. They serve two purposes:

a. Tissue growth (anabolic); and
b. Masculinization (androgenic).

These are employed to improve the size, power, and functionality of muscles. Athletes frequently consume these at megadoses of 10 to 20 times the body's typical production. Steroid use is prohibited and athletes who use steroids are not allowed to compete in the Olympics.

Steroid use is discouraged since it negatively impacts an athlete's health. The harmful physiological effects include sterility, liver damage, heart damage, normal skeletal development being stunted, and many more. Unwanted personality changes, such as excessive aggression and mood swings from depression to violent rage, occur in steroid users. The development of a drug addiction can seriously impair one's ability to lead a normal life. Because of these long-term negative effects, athletes should avoid using hormones as performance enhancers. The adults who mentor them—their parents, teachers, trainers, and others—must make sure that the children are inspired to set long-term objectives for a healthy competitive environment and lifetime health rather than giving in to the allure of temporary fame, which eventually leads to ill health, an unhappy attitude, and the loss of a normal, respectable human existence. Enhancing one's physical, social, and ethical standards is the goal of sports. All athletes and sportsmen must abide by this fundamental idea.

4.6.1. Ergogenic Foods for Sports Performance

The following is a list of foods that are ergogenic (beneficial to performance).

4.6.1.1. Watermelon Juice

Watermelon consumption before exercise reduces recovery time and improves performance. Citrulline, an amino acid that can be converted into the conditionally necessary amino acid arginine for humans, is abundant in watermelon. The nitrogenous substrate utilized in the production of nitric oxide is arginine. It helps the immune system and cardiovascular systems by speeding up the removal of lactic acid. It also improves physical performance by enabling athletes to engage in more rigorous training. A quicker recovery after every workout is the ultimate outcome. 500 mL of pure watermelon juice is the dosage.

4.6.1.2. Beetroot Juice

Leafy greens and beetroot are excellent providers of nitrates. Dietary nitrate leaves the stomach and small intestine quickly absorbed. Nitrate supplementation in the form of beetroot juice (~0.5 L/d) for six days has been shown to increase tolerance to high-intensity work rates and decrease pulmonary oxygen uptake (VO_2) during submaximal exercise, indicating that nitrate can be a powerful ergogenic aid.

4.6.1.3. Green Tea

According to a study evaluating the benefits of green tea extract (GTE), taking a 0.5% GTE supplement over a 10-week period improved exercise performance and endurance by up to 8–24%. Researchers at Kao Corp.'s Biological Sciences Laboratories the 8–24% increase in swimming time-to-exhaustion was correlated with higher rates of fat oxidation and lower respiratory quotients, according to Tochigi, Japan. The findings support the theory that stimulating fatty acid utilization is a viable tactic for enhancing endurance capacity and show that GTE is advantageous for doing so. The equivalent of four cups of green tea per day produced the results.

4.6.1.4. Coconut Water

Saat et al. conducted a study of eight fit male participants who worked out in the heat at 60% of their maximum oxygen consumption (VO2max) until 2.78 +/– 0.06% of their body weight (BW) was gone. Following their workout, the participants rested in a thermoneutral setting for two hours and consumed a combination of plain water, coconut water, and a carbohydrate electrolyte beverage, which combined accounted for 120% of their fluid loss. Comparing coconut water to a carbohydrate electrolyte beverage and plain water, it was discovered that the former was much sweeter, produced less nausea and fullness without an upset stomach, and was also simpler to take in larger quantities. After exercise, drinking fresh young coconut water can help rehydrate the entire body.

4.6.1.5. Peppermint

Natural peppermint has anti-inflammatory, antispasmodic, analgesic, antioxidant, and vasoconstrictor properties. One 500 ml bottle of mineral water containing 0.05 ml of peppermint essential oil was consumed for 10 days by 12 healthy male students participating in the study.

The young male students who completed a standard treadmill exercise test showed positive effects from peppermint essential oil on blood pressure, respiratory rate, gas analysis, performance, and spirometry parameters. The bronchial smooth muscle relaxation, the rise in brain oxygen concentration and ventilation, and the drop in blood lactate levels were the likely causes.

4.6.1.6. Protein

According to the International Society of Sports Nutrition (ISSN), an individual who exercises should consume between 1.4 to 2.0 grams of protein per kilogram of body weight per day.

The precise amount required will vary depending on the type and intensity of the exercise, the quality of the protein consumed, and the individual's energy and carbohydrate intake. It has been shown that in healthy, active individuals, concerns about protein intake within this range being unhealthy are unfounded. Ideally, one should try to meet their needs for protein by consuming whole foods (e.g., eggs, poultry, fish, meat, and dairy products; also includes sources of vegetables like soy and legumes). High-quality dietary protein can also be conveniently and safely consumed by taking supplements. Timing one's protein consumption to coincide with their workout can speed up their recovery and increase their gains in fat-free mass.

4.6.1.7. Fat: Medium Chain Triglycerides (MCT)

The impact of consistent medium-chain triglyceride (MCT) consumption on swimming endurance. The primary metabolic effects of muscle adaptation to extended MCT administration during endurance training were increased activity of energy-producing and macronutrient-utilization-related enzymes.Because MCTs cross the mitochondrial membrane more quickly than long-chain fatty acids and don't require the presence of carnitine, they provide roughly 10% fewer calories and are absorbed by the body and metabolized as fuel more quickly. Less fat is stored and more is converted into fuel for instant use as a result of this increased metabolic conversion.

MCTs have gained in popularity with athletes seeking to increase energy levels and enhance endurance during high-intensity exercise, as well as serving as an alternative energy source (i.e., nutritional ketosis) for athletes on high-protein, low-carbohydrate diets. Furthermore, during the post-exercise recovery phase, MCTs can be swiftly mobilized to promote muscle growth and inhibit catabolism.

The MCT dosage (i.e., extra virgin coconut oil that is organic) is roughly 1/4 teaspoon, several times a day. MCTs should only be taken as tolerated; otherwise, start low and avoid nausea and gastric discomfort.

4.6.1.8. Carbohydrates

During extended endurance exercise, skeletal muscle tissue oxidizes two main fuel sources: fats and carbohydrates. The length and intensity of the exercise determine how much each fuel source contributes; as exercise intensity rises, carbohydrates contribute more. Carbohydrate availability affects both endurance capacity and performance. Thus, increasing the availability of carbohydrates during extended exercise by consuming carbohydrates is a crucial performance-enhancing tactic.

Consuming carbohydrates during extended periods of moderate-to-high-intensity exercise (>2 hours) can greatly enhance endurance performance. This is probably connected to the preservation of glycogen in skeletal muscle, the avoidance of liver glycogen depletion, and the eventual development of hypoglycemia.

Even though endogenous carbohydrate stores are unlikely to be limiting, small amounts of carbohydrate ingestion during exercise may also improve the performance of shorter (45–60 min) more intense (>75% peak oxygen uptake; VO($_2$peak)) exercise bouts. The central nervous system is most likely the mechanism through which the ingestion of carbohydrates during brief, higher-intensity exercise sessions produces ergogenic effects.

Athletes in team sports or intermittent sports also benefit from consuming carbohydrates while exercising. These athletes should feed themselves carbohydrates similarly to endurance athletes, but they should adjust their exogenous carbohydrate intake according to the intensity and duration of the activity. An investigation by Lunn et al. scientists investigated the use of chocolate milk as a post-exercise beverage. The consumption of fat-free chocolate milk was investigated for its impact on muscle glycogen, kinetic and cellular markers of protein turnover, and performance during recovery from endurance exercise.

When comparing milk to a beverage that only contains carbohydrates, there may be distinct advantages to milk when it comes to post-exercise muscle protein fractional synthetic rate (FSR), leucine kinetics, signaling molecules of skeletal muscle protein turnover, and performance measures. A growing body of research supports the consumption of chocolate milk and low-fat milk by both regular athletes and individuals after strength or endurance training.

Numerous studies indicate that commercial sports drinks may not be as effective as fat-free or chocolate milk in aiding in the recovery process following strength and endurance training. In fact, these milk types may be even more effective. The extra nutrients and vitamins that milk contains are not found in commercial sports drinks, which is an added bonus. There are lactose-free options available for people who are intolerant to lactose. Additionally, raw cacao and local honey combined with organic grass-fed milk could make a healthier alternative to "conventional" milk.

4.6.1.9. Bananas

In terms of sports drinks and carbohydrates, bananas have been proven to be a beneficial choice. Nieman et al. conducted a study comparing the impact of consuming bananas versus a 6% carbohydrate drink on cycling performance, inflammation, oxidative stress, and innate immune function after a 75-km ride. The research showed that eating bananas before and during extended and strenuous exercise is an effective method for enhancing performance, with the benefits of fuel utilization and cost efficiency.

4.6.1.10. Raisins

Glycemic indexes of dietary carbohydrates vary, and they may have different effects on endurance and metabolism. An investigation by Kern et al. investigated possible variations in cycling performance and metabolism following the ingestion of raisins versus a high-glycemic commercial sports gel. Eight cyclists with endurance training finished two trials in a random order. 45 minutes before working out on a cycle ergometer at 70% VO2max, the subjects were given 1 g of carbohydrates per kilogram of body weight in the form of either sports gel or raisins.

Following 45 minutes of moderate exercise, participants undertook a 15-minute performance test. Blood samples were taken before and after the workout to measure levels of glucose, insulin, lactate, free fatty acids, triglycerides, and beta-hydroxybutyrate in the serum. Results showed no significant differences in performance between the raisin and gel groups. The study suggested that raisins may be a more affordable choice for pre-workout carbohydrate intake during brief exercise sessions compared to sports gels (James & Schofield, 1990).

CASE STUDY

TAILORING NUTRITION AND TRAINING TO A FOOTBALLER'S MENSTRUAL CYCLE

The menstrual cycle can affect an athlete's dietary patterns and their performance capacity. Sports Nutritionist, exercise physiologist and personal trainer, Simone do Carmo shares her experiences through this case study.

"Ultimately, what matters as a practitioner is the person in front of you, and understanding a woman's unique experiences around her menstrual cycle."

In a perfect world, we could notice the changes in our symptoms and fluctuations in eating habits, cravings and weight, and relate that to the menstrual cycle. However, more often we feel guilty about eating a chocolate bar or lack of willpower to stick to our usual eating patterns.

The same can happen with training. Fluctuations in energy levels and hormonal changes can strongly affect motivation levels, increase mental and physical fatigue, and impact physical performance.

Recent research has shown that performance can be reduced during the early follicular phase (when estrogen and progesterone are low) compared to the other phases (1). When working with a client, it is important to consider the literature and the unique experiences of the person in front of you.

Commonly, athletes use oral contraceptive pills to manipulate their menstrual cycle (i.e., control when to have their withdrawal bleeds), especially around important competitions, and to diminish symptoms that negatively impact their performance. This produces a different hormonal profile. A meta-analysis has shown that oral contraceptive users might experience a slight decline in performance compared to non-users with a natural menstrual cycle (2). However, variability in response to an oral contraceptive was high. This further supports the argument for a personalized approach.

Introducing Anna

Anna, a 22-year-old university student and club-level footballer, trained in the gym twice a week alongside her club twice a week. She usually played one match per week during the university sports season.

Her menstrual cycle was a major concern. Anna didn't use an oral contraceptive. She experienced debilitating PMS symptoms, specifically bloating, painful cramping and a change in bowel function, with a spike in hunger levels and cravings a few days before her period. She also experienced large fluctuations in energy levels and noted that it affected her performance on the pitch. Her weight also fluctuated between 64 kg and 67 kg during the month. After taking an in-depth health history, the medical team agreed that a better understanding of her menstrual cycle, and how this affected her eating habits and training, would be the best thing to focus on. This would be done alongside (i) optimizing her current diet, (ii) learning how to fuel properly for training and refuel to support recovery, and (iii) working together in the gym to increase her overall strength levels.

Initial Intervention Strategies Considering the Menstrual Cycle

We started simply by tracking Anna's menstrual cycle with an app on her smartphone. This allowed Anna to record her symptoms such as energy levels, sleep, digestive issues, emotional triggers and how heavy her monthly bleed was.

Anna's existing diet was good, mainly plant-based, with a regular intake of eggs and cheese during the week. However, she didn't know where to begin with regard to fueling herself properly for training or how to use nutrition to optimize her recovery. We focused on two metrics: energy levels during training and muscle soreness, which she rated as 5/10 for both during this first session (1 being extremely poor and 10 being excellent).

Look at Anna's overall diet and implemented five changes:

1. On average, Anna was close to meeting her overall energy needs but was very inconsistent on a day-to-day basis, including under-fueling on training days. We developed a plan to bring more consistency to her eating and make sure she was getting more fuel on training days than non-training days.
2. An increase in her daily protein intake from ~1.2 g/kg body weight/day to ~1.6 g/kg body weight/day, with an even distribution throughout the day to support recovery, promote a positive protein balance, and support strength-related adaptations (3).
3. Increased omega-3 and iodine intake, which I identified as low in her three-day food diary analysis. Anna was happy to eat fish and seafood – she didn't eat these foods regularly only because they hadn't been a staple at home growing up, and her flatmates didn't cook fish or seafood. We agreed on at least two portions of fish per week (both oily for more omega-3s and white for more iodine) from her local fishmonger, more chia seeds, flaxseeds and walnuts (for omega-3s), and to incorporate seaweed salt (iodine).
4. Although the dietary analysis hadn't indicated that Anna's iron intake was low, we optimized it by (i) focusing on specific plant-based foods (such as legumes, nuts, seeds and leafy greens), (ii) reducing her consumption of coffee with or right after her meal, and (iii) increasing her consumption of vitamin C-rich foods (such as citrus fruits, greens, okra and parsley) with iron-rich foods to support absorption (4).
5. Anna's intake of B vitamins (including B12) was within healthy ranges. Again, we focused on optimizing these by tweaking her diet and bringing in more nuts/

seeds, leafy greens and grains. She regularly included eggs, cheese, nutritional yeast and fortified plant-based milks in her diet, so I wasn't overly concerned with her vitamin B12 levels at this point. Adding fish to her diet would also contribute towards more iron and B vitamins, especially B12.

It is recommended that Anna get her vitamin D levels checked before considering supplementation, as living in Scotland and going into winter increased her risk of deficiency.

After six weeks of implementing these recommendations, we turned our attention to more sport-specific nutritional strategies: pre-, during and post-exercise strategies.

- **Pre-Exercise (Roughly One Hour Before a Training Session, Usually Late Afternoon):** we focused on good-quality and digestible carbohydrates with a bit of protein to provide amino acids to her system. This helped regulate Anna's hunger levels especially during her football sessions.
- **During Exercise**: Anna would drink only water during her football training sessions (which could last up to two hours), so we also introduced a DIY sports drink made with diluted cranberry juice at a 6 percent solution with a pinch of sea salt. Gym sessions lasted for 1hr, so water sufficed.

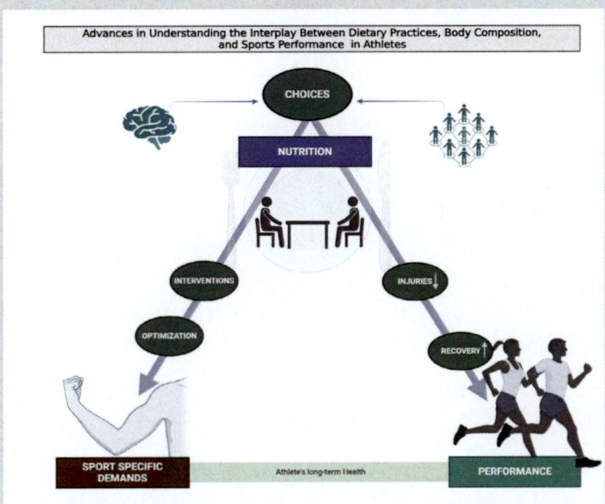

Figure 4.1. *Graphical abstract of this comprehensive exploration of the complex interplay between dietary practices, body composition, and sports performance.*

Source: Martín-Rodríguez A, Belinchón-deMiguel P, Rubio-Zarapuz A, Tornero-Aguilera JF, Martínez-Guardado I, Villanueva-Tobaldo CV, Clemente-Suárez VJ. Advances in understanding the interplay between dietary practices, body composition, and sports performance in athletes. Nutrients. 2024; 16(4):571. https://doi.org/10.3390/nu16040571.

- **Post-Exercise**: sometimes Anna would have her main meal about an hour after training. So I wasn't too concerned if she didn't eat something after her training session, considering she had a snack before training and could finish any remaining DIY sports drink. But I recommended a few options for when she knew she would only eat her main meal two hours or more after her training session – even if it meant eating less for her main meal if she wasn't that hungry. Anna could have these as soon as possible after training to provide roughly 20 g of protein and carbohydrates to start the recovery process (Figure 4.1).

Lesson Learned

The more time we spent together, the better I became at anticipating how she felt, and I adapted her training accordingly. Had I planned a more intense workout on a particular day, but she felt low in energy, The doctor modified the workout and kept notes of these modifications to support her profile.

Over five months, we built a personalized profile of Anna's menstrual cycle. Her vitamin D levels were insufficient, so we agreed to incorporate a vitamin D3 supplement at 1,000 IU per day for the rest of the winter and then re-test. Her metrics also improved from 5/10 to 7/10 (energy levels) and 8/10 (muscle soreness).

Anna's nutrition and training were adjusted accordingly:

1. Anna craved more chocolate and sweetness about a week before her period (the luteal phase when both estrogen and progesterone are high and then drop as menstruation starts). To support her during these days, we discussed healthier options than a standard chocolate bar, such as:

- Homemade DIY chocolate milk (also a good post-exercise snack) made with plant-based milk, unflavored plant-based protein blend powder, cacao, and sweetened with good-quality honey. A banana could also be added to thicken the milk and provide more sweetness.
- A few Brazil nuts coated in dark chocolate.
- Homemade 3-ingredient chocolate energy bites made with blended dates, cocoa and walnuts.
- A chocolate mousse – a blend of avocado, banana and cacao.

2. As Anna had elevated hunger levels, we increased her protein to 1.8 g/kg body weight/day during this week to increase satiety (5).
3. We realized that Anna was increasing her coffee intake during this week (up to four or five cups a day) to curb cravings and hunger, and give her an energy 'boost.' We discussed how this reliance on caffeine could disrupt her sleep and exacerbate some of her digestive upsets and bowel-function changes since research shows that caffeine can increase colonic motility (6, 7). She agreed to reduce her intake to one cup of coffee in the morning during this phase and to focus more on non-caffeinated teas like rooibos and water infused with lemon and mint or berries, to ensure proper hydration.
4. As Anna's energy levels and mood were also low a few days before and during her period, we talked about viewing training more as 'movement' than 'planned exercise.' Sometimes, she would understandably have to cancel one or both gym

sessions this week because of her symptoms. We agreed that she could choose what she wanted to do in the gym instead: keep the weights light and focus on technique or do something more restorative like a Pilates session as I'm a qualified Pilates teacher. This week would essentially be a de-load week and by giving Anna a choice helped promote adherence and supported her, especially from a mental health perspective. As for her football training, we discussed that it would be best to have a chat with her team captain and coach to see how they could support her and other teammates who experience such debilitating symptoms.

5. After menstruation (early follicular phase characterized by low progesterone levels and slowly increasing estrogen levels) until ovulation (peak in estrogen), Anna's energy levels and mood would improve. She performed well in this phase, and we focused on pushing the intensity of her training sessions. She was already eating enough carbohydrates during her training sessions and throughout the day. Still, we prioritized the DIY sports drink for all the sessions, even if they were only for an hour, to support the intensity.
6. During ovulation (peak in estrogen), Anna noticed that she was more anxious and aroused than usual, so we agreed to incorporate slow and deep breathing exercises to help modulate these symptoms and promote parasympathetic activity (8).
7. After ovulation, when Anna entered the luteal phase (when progesterone rises), her mood and motivation usually dropped, but there was no noticeable difference in energy levels. We started to gradually decrease the intensity of her gym sessions during this phase, removed the DIY sports drinks from the gym sessions, and included more recovery and restorative strategies, such as a monthly massage or joining her friend at a local pool and sauna. These strategies could also help her prepare mentally for the end of the luteal phase and menstruation when she experiences all the negative symptoms Alaunyte et al (2015).

CLASS ACTIVITY

NUTRITION CROSSWORD PUZZLE

- **Project Skill**: utilizing prior learning about nutrients from previous activities to solve a crossword puzzle.
- Life skills:
  - Recognizing self-worth
  - Communicating
  - Working with others

- Making decisions
- Solving problems

- **Preparation**: Ensure that there are enough copies of the Nutrition Crossword Puzzle for every group. Please remind everyone to bring pencils for this exercise.
- **Introduction**
  - Inform everyone that in this exercise, they will apply the knowledge they have gained from earlier tasks to solve a crossword puzzle utilizing nutritional concepts.
- **Experience**
  - Members of your group should divide into smaller groups of three to four individuals.

Make an effort to have at least one person with sufficient skills in each small group to assist the others if there are some individuals who struggle with reading and writing.

- Each group should receive a copy of the Nutrition Crossword Puzzle.

Describe how the numbered clues listed for the Across and Down words match the numbers in the crossword puzzle squares.

Read clue number one to the participants and find out what they believe the word is, if you think that would be helpful. If this is their first time solving a crossword puzzle, walk them through filling in the squares once they have determined the correct word. Otherwise, don't provide them with any more answers.

- **Process**
  - Instruct the groups to work on the puzzle until it is finished.

As you go between the groups, assist anyone who might be struggling.

- **Apply**

After everyone has finished the puzzle, pose the following debate topics:

- What did you learn from this activity?
- Did the activity help you to remember more about the basic nutrients?
- Were there any words you had difficulty thinking of?
- Were some words easy for you to think of?
- Why do you think some words are easy to think of and other words are harder to think of?
- How can you use what you learned from doing this crossword puzzle?

- **Across**
  - A deficiency disease caused by a lack of iodine.
  - An enlargement of the _____ gland is caused by iodine deficiency.
  - Iodized salt is a good source of ____.

- Sorghum is a good source of carbohydrates in which country?
- Vitamin A helps keep the ____to stay healthy.
- The nutrient needed to control body temperature.
- Most meats and beans are a good source of this nutrient.
- A good source of carbohydrates.

- **Handout: Crossword Puzzle**

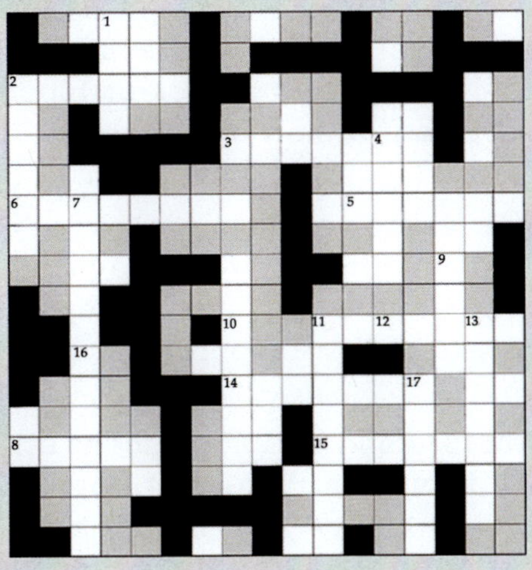

- **Down**
  1. The body needs _____ to provide energy for the body and carry certain vitamins through the body.
  2. Dark _____ vegetables are a good source of Vitamin A.
  3. _____ helps the body to make red blood cells to carry oxygen from the lungs to cells throughout the body.
  4. A good source of protein is the _____ nut.
  5. _____ are good sources of protein and fats.
  6. _____ worms are a good source of protein and iron.
  7. Carbohydrates are the body's most important source of _____.
  8. A deficiency disease caused by a lack of iron in the blood.
  9. A lack of this nutrient leads to the disease called pellagra

SUMMARY

- Just as eating three meals regularly is a consistent part of daily life, so should exercise be a consistent, regular part of daily life. In physical education in schools, activities that are appropriate for life-long participation need to be emphasized.
- It is important to begin an exercise program with a proper plan. Strong motivation is necessary to pursue an exercise program on a long-term basis.
- More water is needed with the increase in exercise. The need to replace fluid losses during exercise is well recorded. All cellular activities take place in an aqueous medium. Water carries nutrients and waste products via the blood stream to and from the cells. Thus sufficient blood volume is a must to ensure loss of heat produced through skin and sweat.
- Breathing ensures a constant supply of oxygen to the body. The need for oxygen increases with exercise, as more oxygen is needed to release extra body energy. The ability of the body to provide the oxygen needed is known as aerobic capacity. The aerobic capacity is dependent on the fitness of tissues involved in oxygen intake and transport — lungs, heart and blood vessels and the body composition.
- Proper nutrition is vital for child and adolescent athletes to attain proper growth and perform optimally in sports. Young athletes need to learn what foods are good for energy, when to eat certain foods, how to eat during an event, and when and what to eat to replenish after activity.
- Macronutrients, such as carbohydrates, protein and fats, provide the fuel for physical activity and sports participation. Although there are many vitamins and minerals required for good health, particular attention should be devoted to ensuring that athletes consume proper amounts of calcium, vitamin D and iron.
- Many physiological and nutritional demands occur within the body during exercise. As muscles contract, the demand for oxygen, hydrogen and other key nutrients increases. The human body requires a continuous supply of energy to perform its many functions. As energy demands increase with exercise, additional energy must be supplied or the exercise will end.
- Athletes regularly engaging in strenuous exercise programs should be aware of their daily nutritional needs. Maintaining a healthy diet that provides adequate energy and nutrients is vital to support intense training as well as to optimize immune system functions.
- Winning is the aim of all competitors in sports. Naturally, they would like to use any substance that can aid their activity. These substances are known as ergogenic (work producing) aids. Most of these are worthless, except the steroids. The consumption of steroids is dangerous. These steroids are synthetic sex hormones.

REVIEW QUESTIONS

1. What are the exercises to be physically fit?
2. What are the nutritional needs during exercise?
3. Why is nutrition good for exercise?
4. Why is energy important for exercise?
5. What are the energy requirements?
6. How does exercise help energy balance?
7. What are the nutritional needs of an athlete?
8. Why is nutrition important for athletes?
9. How should you select an exercise program?
10. What are legal nutritional ergogenic aids?

REFERENCES

1. Alaunyte, I., et al. (2015). Iron and the female athlete: A review of dietary treatment methods for improving iron status and exercise performance. *Journal of the International Society of Sports Nutrition, 12*(38). https://doi.org/10.1186/s12970-015-0099-2.
2. Cermak, N. M., & van Loon, L. J. (2013). The use of carbohydrates during exercise as an ergogenic aid. *Sports Medicine, 43*(11), 1139–1155. https://doi.org/10.1007/s40279-013-0079-0.
3. Clark, I., & Landolt, H. P. (2017). Coffee, caffeine, and sleep: A systematic review of epidemiological studies and randomized controlled trials. *Sleep Medicine Reviews, 31*, 70–78. https://doi.org/10.1016/j.smrv.2016.01.006.
4. Elliott-Sale, K. J., et al. (2020). The effects of oral contraceptives on exercise performance in women: A systematic review and meta-analysis. *Sports Medicine, 50*(10), 1785–1812. https://doi.org/10.1007/s40279-020-01317-5.
5. Hoch, A. Z., Goossen, K., & Kretschmer, T. (2008). Nutritional requirements of the child and teenage athlete. *Physical Medicine and Rehabilitation Clinics of North America, 19*(2), 373–398. https://doi.org/10.1016/j.pmr.2007.12.001.
6. Itiondo-DeHond, A., et al. (2021). Effects of coffee and its components on the gastrointestinal tract and the brain-gut axis. *Nutrients, 13*(1), 88. https://doi.org/10.3390/nu13010088.
7. Jäger, R., et al. (2017). International Society of Sports Nutrition Position Stand: Protein and exercise. *Journal of the International Society of Sports Nutrition, 14*(20). https://doi.org/10.1186/s12970-017-0177-8.
8. James, W. P. T., & Schofield, E. C. (1990). *Human Energy Requirements: A Manual for Planners and Nutritionists*. Oxford Medical Publications under arrangement with FAO. https://openknowledge.fao.org/server/api/core/bitstreams/14ec1032-cf6d-4c2a-b629-82c033573fae/content (accessed on 11 September 2024).

9. Kern, M., Heslin, C. J., & Rezende, R. S. (2007). Metabolic and performance effects of raisins versus sports gel as pre-exercise feedings in cyclists. *Journal of Strength and Conditioning Research, 21*(4), 1204–1207. https://doi.org/10.1519/R-21226.1.
10. Lunn, W. R., Pasiakos, S. M., Colletto, M. R., Karfonta, K. E., Carbone, J. W., Anderson, J. M., & Rodriguez, N. R. (2012). Chocolate milk and endurance exercise recovery: Protein balance, glycogen, and performance. *Medicine & Science in Sports & Exercise, 44*(4), 682–691. https://doi.org/10.1249/MSS.0b013e3182364162.
11. McNulty, K. L., et al. (2020). The effects of menstrual cycle phase on exercise performance in eumenorrheic women: A systematic review and meta-analysis. *Sports Medicine, 50*(10), 1813–1827. https://doi.org/10.1007/s40279-020-01319-3.
12. Meamarbashi, A., & Rajabi, A. (2013). The effects of peppermint on exercise performance. *Journal of the International Society of Sports Nutrition, 10*(15). https://doi.org/10.1186/1550-2783-10-15.
13. Morell, P., & Fiszman, S. (2017). Revisiting the role of protein-induced satiation and satiety. *Food Hydrocolloids, 68*, 199–210. https://doi.org/10.1016/j.foodhyd.2016.08.003.
14. Nieman, D. C., Gillitt, N. D., Henson, D. A., Sha, W., Shanely, R. A., Knab, A. M., Cialdella-Kam, L., & Jin, F. (2012). Bananas as an energy source during exercise: A metabolomics approach. *PLOS ONE, 7*(5), e37479. https://doi.org/10.1371/journal.pone.0037479.
15. Otten, J. J., Hellwig, J. P., & Meyers, L. D. (2006). *Dietary Reference Intakes: The Essential Guide to Nutrient Requirements*. National Academies Press. https://nap.edu/openbook.php?record_id=11537 (accessed on 11 September 2024).
16. Zaccaro, A., Piarulli, A., Laurino, M., Garbella, E., Menicucci, D., Neri, B., & Gemignani, A. (2018). How breath-control can change your life: A systematic review on psycho-physiological correlates of slow breathing. *Frontiers in Human Neuroscience, 12*, 353. https://doi.org/10.3389/fnhum.2018.00353.

CHAPTER 5

Disorders of Nutrition

LEARNING OBJECTIVES

After studying this chapter, you will be able to:

- Identify nutrient deficiencies
- Focus on nutrient toxicities
- Understand diet and chronic disease
- Know food-drug interactions
- Discuss food allergies and intolerances
- Identify toxins in foods
- Explain foodborne illnesses
- Describe botanicals and functional foods.

KEY TERMS FROM THIS CHAPTER

Anemia	Anorexia nervosa	Cancer
Chronic illnesses	Civilization	Diarrhea
Digestion	Edema	Eggs
Gastrointestinal tract	Hygiene	Iron
Malnutrition	Marasmus	Milk
Undernutrition	Urbanization	Intestinal malabsorption

INTRODUCTION

One health issue common to all developing nations, including India, is malnutrition. The most common type of malnutrition is undernutrition, or inadequate diet. Undernutrition has a wide range of frequently linked causes. One of the primary causes of undernutrition is low purchasing power brought on by poverty, as impoverished families are unable to purchase enough food for themselves. Some communities may not have year-round access to wholesome foods like milk, particularly those in rural areas. Malnutrition may therefore occasionally result from a family's inability to purchase wholesome food, even though they are able to do so.

Another factor contributing to the prevalence of malnutrition is ignorance of the connection between diet and health. For instance, even though the family may be the ones producing the milk or fishing the fish themselves, there are situations when protein-rich foods like milk, eggs, and fish are withheld from the child because they are unhealthy. Malnutrition is also a result of growing urbanization. Many families move to cities from rural areas in search of better opportunities for employment. For these families, the quality of life is frequently worse than it was in the villages. Among the issues faced are contaminated water supplies, unhygienic surroundings, overcrowding, and poor hygiene. Furthermore, insufficient food lowers immunity to infections and puts family members at risk for frequent bouts of diarrhea and other illnesses. This makes the issue worse because the little food that is eaten is lost rather than absorbed.

In impoverished communities, where both parents must work to support their family, the children, especially the young ones, often lack proper care and regular meals. Older siblings, who are sometimes just a few years older than their younger siblings, are left to take care of them. Without adult supervision, these children may not receive adequate nutrition, leading to the development of deficiency diseases. These diseases occur when the body does not receive enough essential nutrients found in various foods. Common deficiency diseases include Protein-Energy Malnutrition (PEM), lack of vitamin A, anemia from insufficient iron intake or absorption, and deficiency of vitamin B complex.

5.1. NUTRIENT DEFICIENCIES

Chronic undernutrition, which affects over 925 million people globally, is the most significant nutrition-related disease, despite the so-called diseases of civilization, such as diabetes, cancer, heart disease, and stroke. The main symptoms of undernutrition, which occurs when there is not enough food to meet energy needs, include weight loss, failure to thrive, and the wasting of muscle and fat in the body.

A person experiencing chronic, ongoing hunger will likely have low birth weight, inadequate growth and development in children, impaired mental function, and increased susceptibility to disease, to name just a few of the numerous negative effects on those living in poverty in both developed and developing nations. Though sub-Saharan Africa experiences the worst levels of hunger, Asia is home to the greatest number of people who suffer from chronic hunger. About 20,000 people—mostly children—died every day from preventable diseases related to undernutrition at the beginning of the 21st century. Many of these children die as a result of their mothers' inadequate nutrition and the opportunities that poverty imposes on them.

A very small proportion of hunger-related deaths are the result of starvation brought on by severe food shortages. In the 1990s, for instance, complex social and political problems, the devastation of war, and natural disasters like droughts and floods were more frequently the cause of global famine (epidemic failure of the food supply). A protracted shortage or excess of total energy or particular nutrients, such as protein, essential fatty acids, vitamins, or minerals, can lead to malnutrition, which is the impaired function that follows. This disorder may be brought on by prolonged vomiting (as in bulimia nervosa) or difficulty swallowing, poor digestion and intestinal malabsorption, or chronic illnesses that cause loss of appetite (e.g., AIDS, cancer, etc.). Inadequate food availability, poor dietary decisions, and obsessive use of dietary supplements can also lead to malnutrition.

5.1.1. Protein-Energy Malnutrition

Protein-energy malnutrition (PEM), the most prevalent type of malnutrition in the world, is the main symptom of chronic undernutrition. PEM, also referred to as protein-calorie malnutrition, is a spectrum in which individuals—far too frequently children—consume insufficient amounts of protein, energy, or both. Kwashiorkor, which is defined as a severe protein deficiency, is at one end of the continuum, and marasmus, which is an extreme food deprivation with glaringly low amounts of energy and protein, is at the other.

A baby with marasmus has lost most or all of its subcutaneous fat and is severely underweight. The child's body appears to be "skin and bones," and it is extremely weak and prone to infections. The cause is a diet extremely low in calories from all sources (including protein), frequently due to poverty, which starts early weaning the child onto bottled formula made with tainted water and diluted. A vicious cycle of gastroenteritis and gastrointestinal tract deterioration, brought on by poor hygiene and ongoing depletion, hinders the body's ability to absorb nutrients from the limited food that is available and weakens its defenses against infection. If left untreated, marasmus can cause heart failure or starvation to death.

Kwashiorkor, a Ghanaian term for the illness that the first child contracts when the other child is born, usually appears when a child switches from breast milk, which is high in protein, to a diet high in carbohydrates and low in protein. In addition to being weak and growing slowly, children with this disease—which is characterized by an enlarged abdomen from edema, or fluid retention—are also more vulnerable to infectious infections, which can cause deadly diarrhea. Apathy, hair discoloration, and dry, peeling skin with open sores are some additional signs and symptoms of kwashiorkor. In addition to mild muscle and body fat wasting, weight loss may be concealed by the presence of intestinal parasites, enlarged fatty livers, and edema.

In addition, prolonged intravenous glucose therapy (IVG) can cause kwashiorkor and marasmus in hospitalized patients, such as those recuperating from surgery, or in patients suffering from diseases that impair appetite or nutritional absorption. Individuals who suffer from eating disorders, cancer, AIDS, and other ailments that impair appetite or nutrient absorption may experience a loss of muscle and organ mass in addition to fat reserves. There are three parts to PEM treatment. (1) It is imperative to address life-threatening conditions, such as infections and imbalances in fluid and electrolytes. (2) It is imperative to promptly and safely restore nutritional status, as a starving child may experience rapid weight gain in as little as two weeks. (3) Afterwards, ensuring long-term nutritional rehabilitation becomes the main focus of the treatment. The degree of malnutrition, the promptness of treatment, and the sufficiency of continued support all affect how quickly and successfully a patient recovers. Starvation may cause reduced brain growth and intellectual functioning, which may not fully recover, especially in the first year of life.

5.1.2. Carbohydrates

Carbohydrates, including simple sugars, complex carbohydrates like starches, and dietary fiber, an indigestible plant carbohydrate, are not strictly necessary for a healthy diet in most cases. A few types of cells, including brain cells, need glucose, a simple carbohydrate, as fuel. If dietary carbohydrates are insufficient, the breakdown of amino acids—which come from both dietary and body protein—as well as the fat-derived chemical glycerol is what drives the synthesis of glucose. Long-term insufficiency of carbohydrates leads to an increase in the synthesis of organic compounds called ketones (a state called ketosis), which gives breath a unique sweet odor. 50 to 100 grams of carbohydrates per day can help avoid ketosis and other undesirable effects of a very low carbohydrate diet; however, it is recommended and typical of human diets to obtain at least half of the daily energy intake from carbohydrates, which equates to at least 250 grams of carbohydrates (1,000 calories in a 2,000-calorie diet). A balanced diet rich in carbohydrates, such as fruits, vegetables, legumes, and whole-grain cereals, also ensures a sufficient intake of dietary fiber.

5.1.3. Essential Fatty Acids

There is also a minimum requirement for fat, but it only applies to the fatty acids linoleic acid (also known as an omega-6 fatty acid) and alpha-linolenic acid (also known as an omega-3 fatty acid). It does not apply to total fat. Deficits in these two fatty acids have been observed in infants given low-fat formulas, young children fed low-fat diets or nonfat milk, patients with medical conditions affecting fat absorption, and hospitalized patients fed exclusively with intravenous fluids containing no fat for weeks. Hair loss, dry skin, and slowed wound healing are signs of deficiency. A person can meet their daily requirements for essential fatty acids, which are a few grams, by eating about one tablespoon of polyunsaturated plant oils. Omega-3 fatty acids are also abundant in fatty fish. Even people who eat a low-fat diet usually get enough fat in their diet to meet their needs.

5.1.4. Vitamins

A deficiency in one of the eight B-complex vitamins may impact the functioning of other vitamins as they work in tandem with various enzyme systems and metabolic pathways.

5.1.4.1. Vitamin A

The primary cause of avoidable blindness in children is vitamin A deficiency, which is a serious issue in developing nations, particularly in Africa and Southeast Asia. Each year, hundreds of thousands of children in the poorest nations lose their sight as a result of vitamin A deficiency. Immune function can be compromised by even a mild deficiency, which lowers resistance to illness. An early indicator of vitamin A deficiency is night blindness, which is followed by abnormal dry eyes and, in the end, xerophthalmia, or corneal scarring. Additional symptoms include reduced growth and development, dry skin, and hardening of epithelial cells in other parts of the body (such as mucous membranes). Giving kids a single, high-dose vitamin A shot every six months is helping to lower the incidence of vitamin A deficiency in many endemic areas. The use of so-called "golden rice," a genetically modified variety of rice that contains beta-carotene, a precursor to vitamin A, has the potential to significantly lower the incidence of vitamin A deficiency, though its application is debatable.

5.1.4.2. Vitamin D

When ultraviolet light from the sun reacts with a precursor compound in the skin, the body produces vitamin D, also referred to as vitamin D hormone. As a result, in the absence of sufficient dietary sources of vitamin D, a vitamin D deficiency may develop when sunlight exposure is restricted. Rickets is a disease marked by inadequate mineralization of bone, growth retardation, and skeletal abnormalities such as bowed legs that is brought on by a lack of vitamin D in children. Osteomalacia, the adult form of rickets, causes weak bones and muscles in addition to weakness. Insufficient vitamin D levels could potentially be a factor in the bone loss associated with osteoporosis. A vitamin D deficiency can occur in people who get little or no sun exposure (including women who cover their bodies entirely out of religious conviction), are elderly or housebound, and have dark skin, especially if they live in a northern latitude. Since very few foods naturally contain vitamin D, fortification of milk and other foods (e.g., foods high in vitamin C such as breads, cereals, and margarine) have protected populations where sun exposure is insufficient. Since elderly people produce and activate vitamin D less effectively, even when exposed to sunlight, supplemental vitamin D may also help prevent bone fractures in this population.

5.1.4.3. Vitamin E

Vitamin E deficiency is uncommon in humans, but it can occur in premature infants and individuals with issues absorbing or metabolizing fat. In premature infants, it can cause red blood cell fragility, while in those with prolonged deficiency, it can lead to neuromuscular dysfunction affecting the spinal cord and retina, resulting in symptoms such as loss of reflexes, impaired balance, muscle weakness, and visual disturbances. While the specific metabolic role of vitamin E is not fully understood, it plays a crucial role in the body's antioxidant system, protecting cells and membranes from damage caused by free radicals. The need for vitamin E increases with higher consumption of polyunsaturated fatty acids, and individuals who smoke or are exposed to air pollution may require additional vitamin E to shield their lungs from oxidative damage.

5.1.4.4. Vitamin K

Vitamin K is involved in bone metabolism and is required for the liver to produce prothrombin and other blood-clotting factors. Bacteria in the colon produce a form of the vitamin that is somewhat useful. Even in the absence of injury, a deficiency in vitamin K results in impaired blood clotting and internal bleeding. Because the placenta does not effectively transport vitamin K, newborns in developed nations are usually given the vitamin intramuscularly or orally within six hours of birth to prevent hemorrhagic disease of the newborn. Adults rarely experience a vitamin K deficiency, except for liver disease, syndromes with impaired fat absorption, and conditions when taking certain anticoagulant medications that interfere with vitamin K metabolism. Patients whose intestinal bacteria have been destroyed by antibiotics may experience bleeding as a result of a vitamin K deficiency.

5.1.4.5. Thiamin

Beriberi is a disease that has been endemic in populations where white rice has been the main source of nutrition due to a prolonged deficiency of thiamin (vitamin B_1). In regions where white rice or flour makes up the majority of the diet and thiamin lost during milling is not replenished through enrichment, thiamin deficiency is still observed. The dry form of beriberi is characterized by weakness, sore calf muscles, loss of appetite, confusion, and other mental symptoms, as well as tingling, paralysis, and poor coordination. Wet beriberi is characterized by edema and raises the risk of heart failure and an enlarged heart. Populations that consume a lot of raw fish and have intestinal microbes that carry the enzyme thiaminase may also be deficient in thiamin. In the developed world, Wernicke-Korsakoff syndrome, which causes rapid eye movements, loss of muscle coordination, mental confusion, and memory loss, is primarily associated with chronic alcoholism and poor diet. Thiamin deficiency can cause this condition.

5.1.4.6. Riboflavin

Ariboflavinosis, or a deficiency in riboflavin (vitamin B_2), is unlikely to occur in the absence of concurrent deficiencies in other nutrients. Symptoms include cracks in the skin at the corners of the mouth, lip fissures, and an inflamed, magenta-colored tongue after several months of riboflavin deficiency. Vitamin B_{12} is given to jaundiced infants receiving light therapy because riboflavin is easily oxidized by UV light. The main dietary sources of riboflavin, milk, milk products, and cereals, are packed to keep light exposure from reaching them.

5.1.4.7. Niacin

About two months after niacin is removed from the diet, pellagra symptoms appear. The "three Ds" of pellagra are dementia, dermatitis, and diarrhea. If the illness is not treated, it can lead to death. Pellagra, which primarily affects people who eat corn, was widespread in the early 1900s in parts of the southern United States and continues to occur in some regions of Africa, China, and India. The majority of the niacin in corn and other cereal grains is bound, making it difficult for the body to absorb. For centuries, Native American communities have soaked corn in lime water to release bound niacin, which helps prevent pellagra. Additionally, tryptophan, an amino acid that can partially be converted to niacin, is lacking in corn compared to other cereals. Even in cases where niacin intake is insufficient, tryptophan-containing high-quality protein can serve as a preventive measure against niacin deficiency.

5.1.4.8. Vitamin B_6

Vitamin B_6, including pyridoxine and related compounds, plays a crucial role in the body's protein metabolism, neurotransmitter synthesis, and other important functions. Symptoms of deficiency can include dermatitis, small and pale red blood cells, weakened immune system, depression, confusion, and seizures. While severe vitamin B_6 deficiency is rare, many people, especially the elderly and those with alcoholism or liver diseases like cirrhosis and hepatitis, may be at risk of inadequate levels. Several medications, such as isoniazid for tuberculosis, can also disrupt vitamin B_6 metabolism.

5.1.4.9. Folic Acid

Two B vitamins, vitamin B_{12} and folic acid, also known as folate, have numerous closely related roles, most notably playing a part in DNA synthesis. Because of this, individuals with deficiencies in either vitamin experience many of the same symptoms, such as fatigue and weakness from megaloblastic anemia, a condition in which the red blood cells are large and immature due to a lack of DNA for cell division. A lack of folic acid also impairs the production of white blood cells and platelets and disrupts cell division throughout the gastrointestinal tract, leading to chronic diarrhea. Neural tube defects in the fetus may result from inadequate vitamin intake during the early stages of pregnancy. Therefore, in addition to consuming foods high in folic acid, such as fresh fruits and vegetables (especially leafy greens) and legumes, women who are capable of becoming pregnant are advised to take 400 micrograms (μg) of folic acid daily from supplements, fortified foods (such as fortified cereals), or both. Methotrexate, a cancer medication, disrupts the metabolism of folic acid, leading to adverse effects like diarrhea and hair loss. Heavy alcohol consumption can also cause folic acid deficiency by interfering with the vitamin's ability to be absorbed.

5.1.4.10. Vitamin B_{12}

Similar to folic acid, a deficiency in vitamin B_{12} (cobalamin) causes megaloblastic anemia, or large, immature red blood cells, by interfering with normal DNA synthesis. Furthermore, vitamin B_{12} preserves the myelin sheath that shields nerve fibers; as a result, an untreated vitamin B_{12} deficiency can cause nerve degeneration and ultimately paralysis. Excessive folic acid intake (more than 1,000 μg daily) may mask or even make an underlying vitamin B_{12} deficiency worse.

A vitamin B_{12} deficiency may manifest as weakness, exhaustion, pain, dyspnea, tingling or numbness in the hands or feet, altered mental state, and visual issues. The only trustworthy sources of vitamin B_{12} are animal-based diets.

Vegans, who abstain from eating anything derived from animals, run the risk of developing a vitamin B_{12} deficiency and need to supplement or eat foods fortified with the vitamin. A vitamin deficiency is unlikely in individuals who consume animal products regularly unless there is an issue with absorption. Vitamin B_{12} needs to bind to intrinsic factor, a material secreted by the stomach, to be absorbed. Vitamin absorption will be restricted if intrinsic factor is lacking (caused by an autoimmune disease called pernicious anemia) or if the stomach is not producing enough hydrochloric acid. The most common treatment for pernicious anemia, which primarily affects the elderly, is injections or high oral doses (1,000 µg) of vitamin B_{12}.

5.1.4.11. Pantothenic Acid

Under normal circumstances, a deficiency of pantothenic acid is unlikely due to its widespread distribution in food. Deficiency has only been observed in subjects administered a pantothenic acid antagonist or in those fed semisynthetic diets lacking in the vitamin.

Fatigue, irritability, insomnia, stomach discomfort, and neurological symptoms like tingling in the hands are signs of deficiency. When World War II prisoners of war in Asia showed signs of "burning feet" syndrome, which is characterized by tingling and numbness in the toes along with other neurological symptoms, it was suspected that they were deficient in the vitamin pantothenic acid.

5.1.4.12. Biotin

Biotin deficiency is uncommon, possibly due to bacteria in the colon synthesizing the vitamin. However, the significance of this source is not fully understood. Some cases of biotin deficiency have been linked to consuming large amounts of raw egg white, which contains a glycoprotein that inhibits biotin absorption. Infants with a rare genetic defect may require biotin supplements if they cannot absorb the vitamin from food. Chronic use of specific anticonvulsant medications can also hinder biotin absorption. Symptoms of biotin deficiency include skin issues, hair loss, and neurological problems over time.

5.1.4.13. Vitamin C

Ascorbic acid, another name for vitamin C, is a water-soluble antioxidant that also works as an enzyme cofactor in a number of different enzyme systems, including those that synthesize neurotransmitters and connective tissue components. Scurvy is a deficiency of vitamin C that manifests as joint pain, bleeding gums, under-the-skin pinpoint hemorrhages (petechiae), and poor wound healing. Scurvy is uncommon in developed nations, but it can occur in certain cases in those on restricted diets, especially those with few fruits and vegetables, or in infants given boiled cow's milk without any vitamin C supplement. Vitamin C can prevent scurvy with relatively small doses (10 mg/day); however, recommended intakes are closer to 100 mg/day in order to provide adequate antioxidant protection. An individual's requirement for vitamin C may increase due to illness conditions, drugs, environmental toxins, and other stressors. For instance, smokers might need an extra 35 mg of the vitamin each day to keep their vitamin C levels at par with those of nonsmokers (Suskind et al., 2009; Kaur et al., 2019).

5.1.5. Minerals

5.1.5.1. Iron

Of all nutritional deficiencies, iron deficiency is the most prevalent, affecting a large portion of the global populace in some form. The most vulnerable are young children and women who are not yet menopausal. Hemoglobin, the red blood pigment that transports oxygen from the lungs to other tissues, is primarily formed by iron.

Bleeding can deplete the body's iron reserves because each milliliter of blood contains 0.5 mg of iron (as a component of hemoglobin). Microcytic hypochromic anemia is a disorder that develops when iron stores are reduced. It is typified by small red blood cells with lower hemoglobin levels than usual. Severe iron deficiency anemia is characterized by low resistance to cold temperatures, weakness, apathy, pale skin, and difficulty breathing on exertion. Iron deficiency in childhood can impact growth and development, behavior, and learning capacity. Maternal mortality and pregnancy complications are at higher risk in cases of severe anemia. The times when iron deficiency anemia is most common are late infancy and early childhood, when the body's iron stores are depleted and milk, which is low in iron, is a staple food; the adolescent growth spurt; and the childbearing years, when women's blood loss during menstruation and the extra iron requirements of pregnancy cause the condition. Adults who experience intestinal blood loss and consequent iron deficiency anemia may also have ulcers, hemorrhoids, tumors, or long-term use of certain medications, like aspirin. The combination of low dietary iron intake and blood loss from infections like hookworm worsens iron deficiency in both adults and children in developing nations.

5.1.5.2. Iodine

The most common cause of avoidable brain damage, which affects an estimated 50 million people worldwide, is iodine deficiency disorders. Severe iodine deficiency during pregnancy may hinder fetal development and cause miscarriage and stillbirth in addition to cretinism, an irreversible mental retardation characterized by short stature and developmental abnormalities. Some more widespread effects of long-term iodine deficiency include milder impairments in the nervous system and muscles. Iodine from the ocean is a consistent source, but iodine in food varies and is mostly determined by the amount in the soil away from coasts. When there is a persistent lack of iodine in the body, the thyroid gland grows larger in an effort to retain more iodide (the form in which iodine functions in the body) in the blood for the synthesis of thyroid hormones. Eventually, this results in the formation of a goiter, which is a visible lump at the front of the neck. Certain foods, including those in the cabbage family, millet, cassava, sweet potatoes, and some beans, contain substances called goitrogens that interfere with the synthesis of thyroid hormones. These substances are destroyed during cooking, so eating goitrogenic foods can be a major factor in the lives of people who also have iodine deficiency. World iodine status has improved dramatically since 1993, when a strategy of universal iodization of salt was adopted. Millions of people are still at risk, though, who live in iodine-deficient regions, mainly in Central Africa, Southeast and Central Asia, and even central and eastern Europe.

5.1.5.3. Zinc

Zinc is a vital component of enzymes and plays a crucial role in protein structure and gene regulation. The first cases of zinc deficiency in humans were reported in Egypt and Iran in the 1960s, where individuals, particularly children and adolescent boys, showed symptoms such as stunted growth and undeveloped genitalia due to lack of zinc in their diet. This deficiency was linked to a regional diet low in meat and high in legumes, unleavened breads, and whole-grain foods that inhibit zinc absorption. Other factors contributing to zinc deficiency included the practice of consuming clay, which interferes with mineral absorption. Severe zinc deficiency has been observed in patients with inadequate zinc intake from intravenous solutions and in individuals with the zinc-responsive syndrome known as acrodermatitis enteropathica. Symptoms of zinc deficiency can include skin lesions, diarrhea, increased susceptibility to infections, night blindness, reduced taste and smell sensation, poor appetite, hair loss, slow wound healing, low sperm count, and impotence. Zinc is found in abundance in protein-rich foods like red meat and shellfish, and low zinc levels may be seen in individuals with protein-energy malnutrition. Even in developed countries, vulnerable populations such as young children, pregnant women, the elderly, strict vegetarians, individuals with alcoholism, and those with malabsorption issues should be mindful of their zinc intake to prevent deficiency.

5.1.5.4. Calcium

The skeleton acts as a storehouse for calcium required in the blood and other tissues, with the majority of calcium in the body being found in the teeth and bones. Adequate calcium intake is essential for bone growth and calcification during childhood and adolescence. Low calcium consumption in childhood, particularly during the adolescent growth spurt, may put a person at risk for osteoporosis, a condition marked by decreased bone mass, in later life. Because of their weakened state brought on by a loss of density, bones are more susceptible to fractures, especially those involving the hip, which can be fatal or severely incapacitate an individual. In industrial societies, postmenopausal women are especially prone to osteoporosis. Osteoporosis is not a calcium-deficiency disease per se, but it is strongly influenced by heredity; the risk of the condition can be reduced by making sure one consumes enough calcium throughout life and by regularly lifting weights. In the early postmenopausal years, adequate calcium intake does seem to slow bone loss, albeit not as much as bone-conserving medications.

5.1.5.5. Fluoride

In addition to preventing tooth decay, fluoride helps to mineralize teeth and bones. The natural fluoride content of waterways and the incidence of dental caries are inversely correlated, according to epidemiological research conducted in the US in the 1930s and 1940s. For children older than six months, prescription fluoride supplements are advised in areas where the fluoride content of the drinking water is low. Periodically, dentists may also apply fluoride rinses or gels to their patients' teeth. Both children and adults, who still benefit from fluoride intake, can get their fill from fluoridated toothpastes.

5.1.5.6. Sodium

Food typically contains sufficient amounts of sodium, even in the absence of added table salt (sodium chloride). Furthermore, even in people who follow low-sodium diets, sodium deficiency is uncommon due to the body's highly developed mechanisms for conserving sodium. Depletion of sodium can happen when vomiting, diarrhea, or prolonged heavy sweating occurs, as well as when kidney disease is present. Hyponatremia, or low blood sodium, can cause weakness, nausea, dizziness, cramping in the muscles, and finally shock and coma. Salty food and beverages (also referred to as sports drinks) can help restore sodium balance after extended periods of high-intensity exercise in the heat. Additionally, consuming one liter of water with two milliliters, or one-third teaspoon, of table salt should be adequate. The conditions under which the body loses chloride are similar to those under which sodium is lost. The condition known as metabolic alkalosis (excess alkalinity in bodily fluids) is brought on by severe chloride depletion.

5.1.5.7. Potassium

Potassium is abundant in many foods and is usually not lacking in the average diet. However, certain diuretics used for treating high blood pressure can lead to potassium depletion. In addition, potassium can also be lost through persistent vomiting, diarrhea, or excessive use of laxatives. Symptoms of low potassium levels include weakness, decreased appetite, muscle cramps, and confusion. Severe hypokalemia can even cause abnormal heart rhythms. Consuming potassium-rich foods like bananas or oranges and taking potassium chloride supplements under the guidance of a healthcare professional can help replenish lost potassium levels.

5.1.6. Water Deficiency (Dehydration)

The body is composed mostly of water, making up more than half of body weight. In order to replenish lost fluids, adults typically need to drink 2 to 4 liters of water a day in cooler climates, depending on their level of activity, and up to 8 to 16 liters in very hot climates.

Dehydration can occur if water intake does not meet the body's thirst signals, if the thirst mechanism is impaired (such as during intense exercise), or if there is excessive fluid loss due to conditions like diarrhea or vomiting. By the time thirst is felt, dehydration has already begun, defined as a fluid loss of at least 1–2% of body weight.

Symptoms can worsen rapidly, including dry mouth, sunken eyes, poor skin elasticity, cold extremities, weak and fast heartbeat, rapid and shallow breathing, confusion, fatigue, and even coma. Severe dehydration, with a fluid loss of over 10% of body weight, can be deadly. The elderly, sick individuals, and air travelers are particularly at risk of dehydration. Children with poor nutrition who develop gastroenteritis can also become dangerously dehydrated from diarrhea or vomiting. Treatment involves giving fluids either through an IV or orally, containing glucose and salts (Suskind et al., 2009; Kaur et al., 2019).

5.2. NUTRIENT TOXICITIES

Every nutrient has a safe or desirable range of requirements, beyond which there is a chance of unfavorable outcomes. If consumed in extremely high concentrations, any nutrient—including water—can be toxic. Acute toxicity, or poisoning, can result from nutrient overdosing, and iron poisoning can even be fatal. A habitual excess of most nutrients can have long-term harmful effects on health (chronic toxicity). Long-term overindulgence in the calorie-producing nutrients (protein, fat, and carbohydrates) and alcohol raises the risk of obesity and certain chronic illnesses; additionally, using isolated amino acids can cause toxicities and imbalances. But for the majority of people, there is little chance of injury from consuming too much of certain vitamins or minerals through diet. The Tolerable Upper Intake Level (UL), which was established by the U.S. Institute of Medicine in 1997 for a specific set of nutrients, serves as a model for other nations. The upper limit of the recommended daily intake (UL) is the maximum amount of nutrients that, for the majority of the general population, are unlikely to have any negative health effects. It is not intended for use by those who are under medical supervision. The majority of ULs for newborns, kids, and teenagers are significantly lower than the "safe intakes" for adults that are covered below.

5.2.1. Vitamins

Water-soluble vitamins, like vitamin B_{12}, are easily eliminated in the urine if taken in excess, but fat-soluble vitamins, especially vitamins A and D, have the potential to be more toxic because they can be stored in the liver and fatty tissue. However, if ingested as supplements or in food that has been fortified, water-soluble vitamins may be poisonous.

Acute vitamin A poisoning symptoms include headache, nausea, vomiting, dizziness, blurred vision, abdominal pain, and lack of muscle coordination. Adults typically need to take a dose of at least 15,000 µg (50,000 IU) to experience symptoms. Chronic hypervitaminosis A can have a variety of effects, such as liver damage and bone loss, and is typically brought on by a daily intake of 30,000 µg (100,000 IU) for months or years. A swelling of the fontanelles, or soft spots, in young infants may be an indication of vitamin A toxicity brought on by elevated intracranial pressure. Large doses of vitamin A taken by a pregnant woman can also cause abnormalities in the fetus's development, especially if taken during the first trimester; the exact threshold at which these defects can occur is unknown, but a daily intake of less than 3,000 µg (10,000 IU) of the vitamin seems to be safe. While the majority of vitamins that are found naturally in food do not have any negative effects, some animals may have toxic levels of vitamin

A in their livers. For instance, it's said that eating polar bear liver poisoned early Arctic explorers. Increased beta-carotene intake—from supplements, foods high in beta-carotene, or carrots—may give the skin a yellowish cast after a few weeks, but it doesn't have the same harmful effects as preformed vitamin A.

Increased vitamin D intake can cause several detrimental effects, most notably soft tissue calcification and damage to the kidneys and heart. While most people are not concerned, young children are particularly susceptible to vitamin D toxicity. The safe daily intake of 50 µg (2,000 IU) may be exceeded by people who take numerous supplements or who consume large amounts of fortified milk or fish. Due to its antioxidant properties, vitamin E has become widely used in supplements at high dosages (many 100 milligrams daily) to prevent heart disease and other chronic illnesses. These dosages, which are many times greater than what is typically found in food, seem safe for the majority of individuals, but there is no proof that they can prevent disease or slow down aging. Over 1,000 mg per day should not be taken as this could disrupt blood clotting and result in hemorrhagic effects.

High dosages of niacin (nicotinic acid), which is administered for its cholesterol-lowering properties, can cause burning, tingling, and itching sensations in addition to skin reddening. This is the first sign of niacin excess, also referred to as a "niacin flush," and it is the reason for the safe daily intake of 35 mg of niacin. Several grams of niacin per day have also been linked to liver toxicity and other negative side effects. Huge amounts of vitamin B6 have been consumed with the intention of curing ailments like premenstrual syndrome and carpal tunnel syndrome. One major side effect of this supplementation has been severe sensory neuropathy in the extremities, which has included walking immobility. It is thought to be safe to take up to 100 mg per day; however, only 1 to 2 mg are needed for optimal health.

Since chemist and Nobel laureate Linus Pauling proposed in 1970 that vitamin C could fend off colds, vitamin C supplements have become widely used. Numerous studies have not found a significant effect on incidence, but some have found a modest benefit of vitamin C in reducing the duration and severity of common-cold episodes. Diarrhea and other gastrointestinal symptoms are the most frequent adverse effect of a high vitamin C intake. This is probably because the unabsorbed vitamin passes through the intestines. The prevention of these gastrointestinal symptoms forms the basis of the recommended daily dosage of 2,000 mg. While there have been studies on the potential negative effects of high vitamin C intake, none of them have been proven in healthy individuals.

5.2.2. Minerals

A desirable range is typically found for the minerals' dietary intake. A high intake of one mineral may negatively impact the absorption or utilization of another due to interactions. While consuming fortified foods or supplements raises the risk of toxicity, excessive intake from food alone is unlikely. For some populations, there are additional risks associated with exposure to potentially toxic levels of minerals in the workplace or the environment. The widespread use of calcium supplements, particularly by women wishing to avoid osteoporosis and children who don't drink milk, has sparked worries about potential negative effects of excessive calcium intake. Kidney stones, or nephrolithiasis, have been a big worry; most of them are made of a calcium oxalate compound. For many years, individuals who were at risk of kidney stones were advised to follow a low-calcium diet; however, the effectiveness of this advice was questionable, and the approach was supported by a fair amount of research. A recent study, however, has shown conclusively that a diet relatively low in animal protein and sodium combined with normal levels of calcium (1,200 mg daily) is far more effective than the conventional low-calcium diet in preventing the formation of recurrent stones. In fact, because dietary calcium aids in the intestinal oxalate-binding process, it might offer protection against kidney stones. Although constipation is a common side effect of high calcium intake, adults and children at least one year old can safely consume up to 2,500 mg of calcium per day.

It is possible to experience nausea, diarrhea, and cramping in the abdomen when taking medications that contain magnesium salts, like laxatives and antacids. A person who has impaired kidney function is more vulnerable to magnesium toxicity. Foods by themselves are unlikely to contain too much magnesium. High-dose iron supplements can have gastrointestinal side effects, including constipation, even though they are frequently used to treat iron deficiency anemia. Up to 45 mg of iron per day carries a minimal risk of gastrointestinal distress. One of the main risks of iron supplement poisoning for young children is acute toxicity, which can result in death. Iron can accumulate to dangerous levels in people who have repeated blood transfusions or the genetic disease hereditary hemochromatosis, which is characterized by the overabsorption of iron. This can cause serious organ damage, especially to the liver and heart. Men and postmenopausal women are advised to refrain from taking iron supplements and consuming high amounts of iron-fortified foods. There have been cases of dietary iron toxicity documented in South Africa and Zimbabwe among consumers of traditional beers with exceptionally high iron contents.

It has been documented that gastrointestinal symptoms like nausea and vomiting are brought on by excess zinc. Chronic high zinc consumption can lower levels of high-density lipoprotein cholesterol (the "good" cholesterol), affect immune function, and interfere with the body's ability to use copper. Food alone is unlikely to surpass the recommended daily intake of 40 mg of zinc, but zinc lozenges and supplements, which are popular despite the lack of evidence regarding their safety or effectiveness, may. When taken in excess, selenium is toxic. Chronic selenium toxicity, or seleninosis, causes a variety of symptoms, including skin rashes, brittleness, hair and nail loss, and an odor similar to garlic in the breath. Acute toxicity and fatalities from ingesting grams of the mineral have also been documented. Whether consumed as selenomethionine, the primary form found in food, or in the inorganic forms typically found in supplements, too much selenium can be toxic. It is unlikely that a daily intake of 400 µg from all sources will result in selenium toxicity.

Overconsumption of iodine from food, iodized salt, or pharmaceutical products can lead to impaired thyroid function, goiter, and other negative health effects. While most people are unlikely to consume dangerous levels of iodine, individuals with certain health conditions, like autoimmune thyroid disease, may be more sensitive to excessive intake. Excessive fluoride consumption during tooth development can result in mottled enamel, but this is mostly a cosmetic concern. In adults, too much fluoride can lead to issues such as increased bone density, joint pain, and even skeletal fluorosis in severe cases.

However, instances of skeletal fluorosis are rare, even in areas where natural fluoride levels in water are significantly higher than recommended levels.

Excessive consumption of phosphorus (as phosphate) can negatively impact the metabolism of calcium and cause problems with the absorption of trace elements like iron, copper, and zinc.

It is unlikely that safe levels will be exceeded, even with the consumption of phosphate additives in cola beverages and a variety of foods. Again, not likely to result from diet, manganese toxicity from dust inhaled at work is a well-known occupational hazard that causes damage to the central nervous system and mimics symptoms of Parkinson's disease. In a similar vein, excessive dietary consumption is unlikely to cause copper toxicity outside of people with inherited or acquired disorders of copper metabolism.

5.2.3. Alcohol

The immediate consequences of consuming a significant amount of alcohol are widely understood. Cognitive decline typically begins at a blood alcohol concentration of around 0.05%. Unconsciousness is common at 0.40%, and a level of 0.50% can be deadly. Accidents and acts of violence, frequently linked to alcohol, are leading causes of death among young individuals. Pregnant women who consume alcohol run the risk of causing physical and cognitive harm to their babies, resulting in fetal alcohol syndrome. Additionally, alcohol can dangerously interact with various medications like tranquilizers, antidepressants, and pain relievers.

While it is true that people who drink in moderation or moderation have lower rates of heart disease and generally live longer than people who do not drink at all or who consume large amounts of alcohol, increasing chronic alcohol consumption also carries significant risks, including those of liver disease, pancreatitis, suicide, hemorrhagic stroke, mouth, esophageal, liver, and colorectal cancers, as well as possibly breast cancer. Nutritionally deficient alcoholism can arise from nutrient-dense food being replaced, as well as from gastrointestinal disorders and systemic metabolic changes. Alcoholism is characterized by thiamin deficiency, which calls for immediate treatment. This deficiency is seen in Wernicke-Korsakoff syndrome, a neurological condition (Suskind et al., 2009; Kaur et al., 2019).

5.3. DIET AND CHRONIC DISEASE

Chronic diseases are defined as conditions that worsen over time and do not go away on their own. This makes the relationship between diet and these diseases complex, as many diseases take years to develop and it can be very challenging to pinpoint a specific dietary cause. To get around this challenge, some prospective epidemiologic studies track their subjects over an extended period of time. Even so, it is challenging to establish a causal relationship because of the extremely intricate nature of the diet and the complex etiology of chronic illnesses. Furthermore, rather than acting as individual agents, many substances found in food appear to act synergistically, and single-agent studies may overlook these interactive effects.

The idea of "risk factors" has been a common topic for many years, starting with the influential Framingham Heart Study that began in 1948. This study, published in the early 1960s, identified cigarette smoking, high blood cholesterol, and high blood pressure as predictors of heart disease mortality. Subsequent research has confirmed and expanded upon these findings, showing strong associations between certain conditions or behaviors and specific diseases. While not everyone with a risk factor will develop a disease, the presence of known risk factors increases the likelihood of disease development, especially when multiple risk factors are present. Some risk factors, like diet, exercise, and substance use, can be modified, although making such changes can be challenging even when facing potential health issues. On the other hand, factors like genetics, age, and sex are not modifiable. There are varying degrees to which certain risk factors can be altered, including exposure to radiation, biological agents, and chemical pollutants that may contribute to genetic mutations linked to increased disease risk, particularly cancer.

5.3.1. Cardiovascular Disease

The most common cause of death in developed nations is cardiovascular disease, a broad term that includes conditions affecting the heart and blood vessels. The most prevalent and lethal type of cardiovascular disease is coronary heart disease (CHD), also referred to as coronary artery disease or ischemic heart disease. When the arteries that supply the heart with oxygen and nutrients narrow and obstruct, coronary heart disease (CHD) results.

Atherosclerosis, a disorder where fibrous plaques (deposits of lipid and other material) accumulate on the inner walls of arteries, causes them to stiffen and become less responsive to changes in blood pressure, is typically the cause of this narrowing.

A myocardial infarction, also known as a heart attack, may transpire if there is a disruption in blood flow within the coronary arteries encircling the heart. Peripheral arterial disease is the result of narrowing in the arteries of the legs or the abdominal aorta, as well as major branches of the aorta, which can restrict blood flow to the brain causing a hemorrhage or blood clot. The majority of heart attacks and strokes are caused by blood clots, which form more easily in areas where small plaques have already partially blocked the arteries, rather than by the plaque completely blocking the arteries.

Atherosclerosis typically begins in childhood with the development of fatty streaks, which can eventually lead to plaque formation in the arteries. This process is an inflammatory response aimed at repairing damage to the arterial walls. Factors such as smoking, hypertension, diabetes, high LDL cholesterol levels, and certain infections can contribute to vessel injury and inflammation.

Middle-aged men with a family history of heart disease, as well as those with inherited conditions like familial hypercholesterolemia, are at higher risk for premature coronary heart disease (CHD). Modifying risk factors like high LDL cholesterol, low HDL cholesterol, hypertension, and diabetes through diet and weight loss can help prevent CHD.

However, the impact of diet on these risk factors is not as clear as the risk factors themselves. Combining dietary changes with smoking cessation, exercise, and drug therapy (such as statins, bile acid sequestrants, niacin, aspirin, anticoagulants, and antihypertensive medication) can be beneficial in reducing the risk of CHD. While endogenous estrogen may protect against CHD in premenopausal women, the value of hormone therapy in reducing CHD risk in postmenopausal women is now being questioned by recent studies.

5.3.1.1. Blood Lipoproteins

Due to their nonpolar and insoluble nature in water, lipids such as cholesterol, triglycerides, and phospholipids require binding to proteins to form lipoproteins for transport in the bloodstream. Among these lipoproteins, low-density lipoproteins are key in carrying cholesterol from the liver to various cells in the body, including those in the arteries, where it can lead to plaque formation. Extensive research indicates a strong relationship between high levels of LDL cholesterol and the development of coronary heart disease, with LDL being the primary target for treatment. Additionally, clinical studies have shown that lowering LDL levels can reduce the risk of heart attacks and strokes in individuals with existing coronary heart disease.

Conversely, it is believed that high-density lipoproteins move extra cholesterol to the liver for elimination, which helps to stop plaque from forming. The goal of intervention efforts is to raise HDL cholesterol levels because there is an inverse relationship between HDL cholesterol and the risk of CHD. Although less so than LDL and HDL, very-low-density lipoprotein (VLDL), another type of blood lipoprotein, is also a risk factor for CHD. Being the main blood triglyceride (fat) carrier, VLDL is notably elevated in individuals with metabolic syndrome, diabetes, and obesity. While HDL cholesterol is frequently referred to as "good" cholesterol and LDL cholesterol as "bad" cholesterol, the lipoprotein form—rather than the cholesterol it contains—is what is linked to the risk of coronary heart disease. Although a complete lipoprotein evaluation is more informative, total cholesterol levels—which have a strong correlation with LDL cholesterol levels—are usually utilized for initial screening. A blood lipid profile that is considered desirable has total cholesterol below 200 mg/dl, at least 40 mg/dl of HDL cholesterol, less than 150 mg/dl of fasting triglycerides, and less than 100, 130, or 160 mg/dl of LDL cholesterol, depending on the risk of a heart attack.

5.3.1.2. Dietary Fat

It is commonly understood that reducing fat intake can lower blood cholesterol levels and protect against heart disease. On the other hand, consuming high levels of fat can sometimes lead to obesity, which can increase the risk of heart disease. However, the relationship between fats and heart health is complex, as different fatty acids have varying effects on cholesterol-carrying lipoproteins. Additionally, when certain fats are removed from the diet, they may be replaced by other components that pose a risk. High-carbohydrate diets, for instance, may actually raise the risk of cardiovascular disease in some individuals, particularly those with metabolic syndrome or type 2 diabetes. Genetic factors also influence how an individual responds to different dietary changes.

Typically, saturated fatty acids, mainly found in animal products, tend to raise LDL and total cholesterol levels. However, certain saturated fatty acids such as lauric, myristic, and palmitic acids, which can come from both plant and animal sources, are the most cholesterol-raising. On the other hand, stearic acid, obtained from animal fat and cocoa butter, is considered neutral as it does not significantly impact blood cholesterol levels. Replacing saturated fatty acids with unsaturated fatty acids, such as monounsaturated or polyunsaturated fats, usually results in lower LDL and total blood cholesterol levels, primarily due to the reduction in saturated fat intake. However, polyunsaturated fats may reduce HDL cholesterol levels while monounsaturated fats tend to maintain them. Oleic acid, the primary monounsaturated fatty acid found in both animals and plants, can be sourced from olive, canola, and high-oleic safflower oils, as well as avocados, nuts, and seeds. The traditional Mediterranean diet, which includes high consumption of olive oil, has been associated with low rates of coronary heart disease mortality. Olive oil consumption is also linked to reduced amyloid-beta plaque formation and decreased brain inflammation, both of which are indicative of Alzheimer's disease.

Omega-3 and omega-6 fatty acids are the two forms of polyunsaturated fatty acids that can be found in food. Vegetable oils like sunflower, safflower, and corn oils are the main source of linoleic acid, which is the main omega-6 fatty acid in most diets. The idea that omega-3 fatty acids may be protective against coronary heart disease (CHD) was first sparked by the low rates of cardiovascular disease among Eskimo populations who ate traditional diets rich in these fatty acids. The main way that omega-3 fatty acids change lipid levels is by lowering blood triglycerides. By lowering blood pressure, decreasing blood clotting, preventing irregular heart rhythms, and acting as anti-inflammatory agents, omega-3 fatty acids may also protect the heart and blood vessels. The shorter-chain member of the same family, alpha-linolenic acid, is the source of the long-chain omega-3 fatty acids docosahexaenoic acid (DHA) and eicosapentaenoic acid (EPA). High in EPA and DHA are fatty fish, including mackerel, tuna, salmon, herring, and sardines. Alpha-linolenic acid, which the body can transform into long-chain omega-3 fatty acids, is abundant in flaxseed. Egg yolks, walnuts, hazelnuts, almonds, canola oil, soybean oil, and dark green leafy vegetables like spinach are additional sources of omega-3 fatty acids. A diet rich in polyunsaturated fats should be combined with increased intakes of the antioxidant vitamin E because it may accelerate atherosclerosis by increasing LDL lipid oxidation. Supplementing with fish oil is not recommended without medical supervision due to potential side effects, including hemorrhage.

The safety of trans unsaturated fatty acids, as opposed to naturally occurring cis unsaturated fatty acids, has come under scrutiny because trans fats in the diet can lower HDL cholesterol and raise LDL cholesterol like that of saturated fats. While trans-fatty acids are created during the hydrogenation process—which hardens and stabilizes unsaturated oils—they are also naturally present in some animal fats, including milk, butter, and beef. Trans-fatty acids are primarily found in foods like deep-fried goods, baked goods, snack foods, and certain margarines.

5.3.1.3. Dietary Cholesterol

Cholesterol found in food and cholesterol present in the blood are often mixed up, but they are separate. Dietary cholesterol is found in animal-based foods, with higher amounts in egg yolks and organ meats. While consuming cholesterol from the diet can raise LDL cholesterol levels, it is not as significant as the effect of saturated fats. In cases where dietary cholesterol is already high, consuming more may not further raise blood cholesterol due to feedback control mechanisms. Responses to dietary cholesterol can vary greatly among individuals. It is recommended for healthy individuals to consume less than 300 mg of cholesterol daily, but since the body produces its own cholesterol, none is necessary in the diet.

5.3.1.4. Other Dietary Factors

Consuming soluble fiber, an indigestible plant material that is part of dietary fiber, lowers blood cholesterol levels, particularly LDL and total cholesterol, and has been associated with a lower risk of dying from cardiovascular disease. Whole oats, barley, legumes, certain vegetables, and fruits—especially apples, plums, apricots, blueberries, strawberries, and citrus fruits—are good sources of soluble fiber. Psyllium and other fiber supplements may also be advised. It is unknown how soluble fiber lowers cholesterol, but it is most likely due to its capacity to bind with bile acids and cholesterol in the stomach and draw them out of the bloodstream. There could be additional factors at play, like the fermentation of fiber by bacteria in the colon, which produces substances that prevent the synthesis of cholesterol.

Drinking up to two drinks a day for men and one drink a day for women is linked to a decrease in the risk of coronary heart disease. This is mainly due to alcohol's ability to increase HDL cholesterol levels and potentially prevent blood clot formation. The consumption of alcohol may help to explain the lower rates of heart disease in France compared to the United States, despite similar CHD risk profiles. Wine, in particular, contains antioxidants like resveratrol from grape skins, which can help to prevent LDL oxidation, although the positive effects of these substances are likely not as significant as those of alcohol.

Dietary sodium (salt) intake is strongly correlated with mortality from heart disease and stroke, but only in overweight people who may be more sensitive to dietary sodium. Beyond its impact on blood pressure, which in turn affects the risk of stroke, sodium consumption also seems to directly affect the risk of stroke.

Conversely, diets high in potassium have been associated with a lower risk of stroke. Foods containing soy have been linked to lower levels of total and LDL cholesterol in the blood, as well as other vascular benefits that lower the risk of CHD. Soy foods such as tofu, tempeh, miso, soy flour, soy milk, and soy nuts contain compounds called isoflavones, which are similar to estrogen and are thought to have positive effects on the cardiovascular system. Vitamin C, Vitamin E, and beta-carotene (a plant-based precursor to vitamin A) are examples of antioxidant substances that can be obtained through diet supplements or found in food. Antioxidants in the diet may reduce the risk of CHD, however, this idea has not yet been validated by clinical trials.

5.3.1.5. Other Factors

The risk of heart disease increases with blood pressure levels that are equal to or higher than the "prehypertension" threshold of 120/80 millimeters of mercury (mm Hg). Atherosclerosis and hypertension (140/90 mm Hg and above) feed off each other. Hypertension damages artery walls, which promotes the development of plaque; once plaque has formed and arteries lose their elasticity, hypertension worsens. The incidence of congestive heart failure, stroke, and CHD declines with treatment of hypertension.

Diabetes is a significant risk factor for heart disease and also calls for aggressive intervention because it is frequently associated with hypertension, high blood triglyceride levels, and obesity. Moreover, there is an exceptionally high death rate among diabetics who experience a heart attack, either right away or over the course of the following years. Vascular complications can be reduced by strictly controlling blood glucose levels. Additionally, obesity plays a significant role in cardiovascular disease, mostly through its impact on other concurrent risk factors. Obese people frequently have abnormal blood lipoprotein levels, including higher triglyceride and lower HDL cholesterol levels, as well as diabetes, hypertension, and an abnormal glucose tolerance. Compared to fat accumulation around the hips (pear shape), fat accumulation around the waist (the so-called apple shape) increases the risk of developing premature heart disease. A high-risk waist circumference is defined as one that is larger than 102 cm (40 inches) for men or 88 cm (35 inches) for women. Regular exercise is believed to lower the risk of coronary heart disease (CHD) in addition to aiding with weight control. It does this by slowing the advancement of atherosclerosis, boosting blood flow to the heart muscle, raising HDL cholesterol, lowering VLDL levels, enhancing glucose tolerance, and lowering blood pressure. It is advised to engage in moderate aerobic activity, like brisk walking, for at least 30 minutes most days.

Insulin resistance, low HDL cholesterol, elevated blood triglycerides, high blood pressure, and abdominal obesity are characteristics of a recently identified cluster of CHD risk factors known as metabolic syndrome. American endocrinologist Gerald Reaven initially identified this condition as Syndrome X in 1988. Individuals who are susceptible to it should avoid high-carbohydrate diets. A diet higher in unsaturated fat and lower in carbohydrates and saturated fat is beneficial for those with metabolic syndrome, as is regular exercise and weight loss. The hereditary hemochromatosis gene causes a person to absorb iron excessively, which can lead to dangerously high levels that can harm the liver, heart, and other organs. Of those of European descent, about 1 in 9 are carriers (i.e., possess one of the two potential genes) for the illness and are more likely to develop heart disease. Nevertheless, research investigating the potential impact of dietary iron on the risk of heart disease in individuals without the hemochromatosis gene has yielded conflicting results.

Elevated blood levels of the amino acid homocysteine can cause atherosclerosis and damage to the arteries. Blood homocysteine levels can be raised by insufficient consumption of vitamin B_6, vitamin B_{12}, or folic acid, with folic acid deficiency being the most frequent cause. Even though increased homocysteine isn't a known risk factor for CHD yet, it's still a good idea to make sure a person is getting enough folic acid in their diet.

5.3.1.6. Dietary Recommendations

Even though childhood is the prime time for plaque formation, babies and young children under the age of two shouldn't have any dietary restrictions on fat or cholesterol. Dietary recommendations to lower the risk of coronary heart disease (CHD) after the age of two typically center on limiting the consumption of total fat, saturated and trans fatty acids, and dietary cholesterol in addition to physical activity and weight management. Due to the widespread occurrence of atherosclerosis, these diets are thought to be beneficial for both the general population and those who have high LDL cholesterol or other cardiovascular disease risk factors. Adults following a preventive diet could consume 20–35% of their calories from fat, 10% or less of their calories from saturated and trans fats, and no more than 300 mg of cholesterol per day. Even more stringent guidelines apply to a therapeutic diet, which should be overseen by a registered dietitian or other certified nutrition specialist. Reducing consumption of fatty spreads, organ meats, fatty meats, egg yolks, full-fat dairy products, baked goods, and fried foods are some realistic recommendations. Another is to remove the skin from poultry and carefully read food labels to minimize hidden fats in processed foods. Along with nonfat or low-fat dairy products, fish, legumes, poultry, and lean meats, a diet high in oats and other whole grains, vegetables, and fruits is likely to improve overall health and cardiovascular health.

5.3.2. Hypertension

In developed nations, hypertension, also known as high blood pressure, is one of the most prevalent medical conditions. It is a significant risk factor for various illnesses, including aneurysms, kidney disease, congestive heart failure, coronary heart disease, and stroke. The majority of people with high blood pressure have essential hypertension, also known as primary hypertension, for which there is no known cause. The disease is partly hereditary, but it can also be influenced by factors that can be changed, like being overweight, being inactive, drinking a lot of alcohol, and eating a diet high in salt. African Americans have some of the highest rates of hypertension worldwide, for unclear reasons. The standard definition of hypertension is a blood pressure of 140/90 mm Hg or higher, i.e., equivalent to the force applied by a mercury column 140 mm high during the heart's systole and 90 mm high during its diastole; hypertension may cause an increase in either or both of these blood pressure measurements. People who have hypertension may go years without showing any symptoms before having a heart attack or stroke that proves fatal. Complication risk can be greatly reduced by controlling and preventing hypertension. It's critical to diagnose hypertension early on to initiate lifestyle modification as soon as feasible. Compared to lean individuals, those who are overweight, particularly those with excess abdominal fat, are far more likely to develop hypertension. Even 4.5 kg (10 pounds) of weight loss can have a significant impact on lowering high blood pressure. Increasing physical activity appears to lower blood pressure on its own, in addition to its obvious benefits for weight control. Extensive research on the relationship between blood pressure and salt intake in various communities worldwide has unequivocally demonstrated that dietary intake of sodium (salt) is positively correlated with blood pressure. Low-sodium primitive societies have very little hypertension, and they do not experience the normal aging-related rise in blood pressure that industrialized societies experience. On the other hand, hypertension is prevalent and stroke is a major cause of death in nations with extraordinarily high salt consumption. Additionally, research conducted through experiments suggests that consuming less sodium can lower blood pressure.

It seems that some people have a genetic sensitivity to salt. Most people eat more salt than they actually need, even though restricting salt may only help people who are sensitive to it reduce their blood pressure. The general public is generally advised by dietary recommendations to consume no more than 2,400 mg of sodium per day, or slightly more than one teaspoon of salt. This level can be attained by limiting the amount of salt used in cooking, abstaining from adding salt at the table and cutting back on processed foods, which frequently contain sodium that is hidden, and so-called fast foods. Particularly high in sodium are luncheon meats, breakfast cereals, and canned vegetables. Hypertension is linked to heavy alcohol consumption, defined as consuming more than two drinks per day. People who eat no animal products, such as milk or eggs, and who are vegetarians or vegans, typically have lower blood pressure than people who eat meat. Reducing blood pressure will also improve cardiovascular health, so the diet that is advised will focus on fruits, vegetables, low-fat dairy products, whole grains, chicken, fish, and nuts, with very little red meat and sugary foods and drinks. Lowering salt consumption ought to improve the diet's efficacy even more. Many medications, some of which have negative effects on nutrition, are used to treat hypertension. For instance, thiazide diuretics cause the body to lose more potassium, which usually means consuming more of it. Potassium is abundant in foods like potatoes, citrus fruits, bananas, and vegetables. It is not advisable to use potassium-based salt substitutes without first consulting a physician.

5.3.3. Cancer

Cancer is a leading cause of death for adults 45 years of age and older, second only to cardiovascular disease in many parts of the world. In addition to differences in affected cell type and location within the body, the different types of cancer also differ in terms of treatment options, disease progression, and potential contributing or causative factors. Research on identical twins shows that environmental factors still play a major role in determining the risk of most cancers, even in people who share the same genetic makeup. Studies of immigrant populations provide additional evidence that heredity plays a limited role in most cancers; in these populations, cancer rates tend to follow a group's adopted country more closely with each generation that passes. For instance, people who migrate from rural Asia to the United States eventually experience higher rates of breast and colorectal cancer in order to keep up with the country's higher cancer rates. Conversely, the risk of stomach cancer declines over time following Japanese immigration to the US. While specific dietary components that influence the changing disease rates have been harder to identify, nutrition is one of the important environmental and lifestyle factors examined in migration studies. According to estimates from several cancer organizations worldwide, a suitable diet could prevent between 30–40% of cancer cases.

It is likely that the majority of chemicals known as carcinogens enter the body through the digestive tract through food and drink. Ingesting large amounts of certain pesticides, food additives, and foodborne toxins can be carcinogenic, but the foods themselves are the main cause of cancer. Certain food habits or ingredients may encourage cancer, while others may prevent it. The multistage process of cancer development (carcinogenesis) can be influenced by substances in the diet or other environmental factors at any point: initiation, when a cell's DNA, or genetic material, is altered; promotion, when cells with altered DNA multiply; and progression, when cancer cells spread to nearby tissue and distant locations (metastasis).

Research attempting to link different types of fat or total fat to different types of cancer has yielded inconsistent results. Consuming a lot of fat in food may increase the risk of cancer, but this may be partially because fat contains additional calories. Numerous cancers, including colorectal, prostate, uterine, pancreatic, and breast cancers, are linked to obesity. The higher levels of estrogen, insulin, and other hormones in the blood that go hand in hand with increased body fat could be one explanation for this effect. Additionally, numerous studies have demonstrated that regular exercise lowers the risk of colon and breast cancers. Restricting energy intake is the most effective way to lower the risk of cancer in laboratory animals; most experimentally induced tumors and many spontaneous tumors grow more slowly when an animal is chronically underfed. Excessive alcohol intake has also been linked to the emergence of multiple cancers, particularly in the mouth, throat, liver, and esophagus (where it works in concert with tobacco), as well as potentially in the breast, colon, and rectum. The beneficial effects of moderate alcohol consumption on cardiovascular disease serve as a reminder of how intricate and possibly perplexing the relationship is between nutrition and health.

Additionally, foods contain compounds that provide some cancer-prevention benefits. For instance, eating fresh fruits and vegetables—and more especially, vitamins C and E—with nitrate-containing foods, like ham, bacon, sausages, frankfurters, and luncheon meats, can help prevent stomach cancer by inhibiting the production of nitrosamine. A substantial correlation has been observed by several 100 studies between diets rich in fruits and vegetables and a decreased risk of different types of cancer. However, it has proven more challenging to pinpoint the precise protective elements in these diets. The body's antioxidant systems work with vitamins C, and E, beta-carotene (a plant-based precursor to vitamin A), and selenium, a trace mineral, to help stop reactive molecules called free radicals from damaging DNA. Sulforaphane and other compounds known as isothiocyanates are found in certain vegetables, particularly the cruciferous vegetables (broccoli, cauliflower, Brussels sprouts, kale, and other members of the cabbage family). These compounds have been shown in animal studies to protect against cancer by inducing enzymes that detoxify carcinogens. Plant foods also contain dietary fiber, which may have a protective effect on the body because it binds to and dilutes potential carcinogens while also accelerating their passage through the gut. Phytochemicals, or biologically active plant substances, are abundant in fruits and vegetables and are currently being studied for their possible anticarcinogenic properties. Both black and green tea contain antioxidants called polyphenols, which may be able to slow the spread of cancer, according to research on animals. Regular tea drinking has been linked to a lower risk of several cancers, including stomach cancer, though the data is mixed, particularly in China and Japan where green tea is the preferred variety.

A diet that emphasizes plant-based foods like fruits, vegetables (especially cruciferous ones), whole grains, beans, and nuts, limits animal fat intake, balances energy intake and physical activity to maintain a healthy body weight, and consumes alcohol in moderation or not at all is the most likely to lower the risk of cancer. It is also possible to lower the intake of carcinogenic compounds by cutting fat and burning off portions of meat before eating.

5.3.3.1. Colorectal Cancer

Eating meat, especially processed and red meat, is linked to a slight increase in the risk of colorectal cancer. It is unclear, though, whether this effect is caused by a particular type of meat, the possibility that a diet high in meat may deplete other nutrients, like fiber, or the production of carcinogens like polycyclic aromatic hydrocarbons and heterocyclic amines, which are created during high-temperature broiling and grilling, especially of fatty muscle meats. There is also evidence connecting a higher risk of colorectal cancer with high alcohol consumption and low calcium and folic acid intakes. While numerous studies have indicated that eating foods high in fiber may also protect against colorectal cancer, efforts to distinguish the protective effects of dietary fiber from the non-fiber components of fruits and vegetables have yielded conflicting results. In both men and premenopausal women, obesity is a significant risk factor for colorectal cancer, and there is a strong correlation between a reduced risk of colon cancer and mild to moderate physical activity.

5.3.3.2. Prostate Cancer

An increasing amount of research suggests that eating a diet high in fruits and vegetables, especially cruciferous ones, and low in fat and animal products can prevent prostate cancer. Pectin, a fiber that can be found in fruits and vegetables and has been shown to have anticancer properties, may help to explain some of this protection. Eating tomatoes and tomato-based products has been linked to a lower risk of prostate cancer because they are high in the carotenoid antioxidant lycopene. Although there is no proof that soy directly prevents prostate cancer, prostate cancer rates are low in nations like Japan where soy products are frequently consumed. Research is being done to determine whether vitamin E and the mineral selenium can prevent prostate cancer.

5.3.3.3. Breast Cancer

It's unclear if diet and breast cancer are related. Based on research on animals and global correlations between fat intake and breast cancer rates, high-fat diets have been linked to the disease. Large prospective studies have not, however, supported this link, although a high-fat diet may not be recommended for other reasons. In a similar vein, while eating a lot of fruits and vegetables is healthy, it doesn't seem to protect against breast cancer. Although drinking alcohol is linked to breast cancer, only heavy drinking seems to be associated with an increased risk. Because obesity is linked to an increased risk of postmenopausal breast cancer, regular exercise may help prevent breast cancer for the rest of one's life. The two main known factors that affect a person's risk of breast cancer are genetics and estrogen levels throughout a lifetime.

In the developed world, there has been a surge in interest recently in soy foods and products as a breast cancer preventive. While Japanese women have a lifetime exposure to high levels of dietary soy and have low rates of breast cancer, their situation is not always similar to that of Western women who supplement with soy isoflavones (compounds that resemble estrogen) in their middle age. It seems that isoflavones and estrogen compete (e.g., in premenopausal women), and so mitigate its impact; in low-estrogen settings (e.g., for example, they have minimal estrogenic effects on postmenopausal women. The effects of dietary soy once cancer has begun are unknown (estrogen itself is a cancer promoter), and there is currently no conclusive evidence that soy in the diet offers protection against breast cancer or any other cancer. Consumption of soy foods like tofu is advised, as research on the health benefits of soy is still ongoing. However, it is not recommended to consume isolated soy constituents like isoflavones, as these may pose unknown risks.

5.3.4. Diabetes Mellitus and Metabolic Disorders

Diabetes mellitus is a collection of metabolic disorders pertaining to the metabolism of carbohydrates. It is typified by elevated blood glucose levels, or hyperglycemia, and is typically brought on by either insufficient insulin production (type 1 diabetes) or an inefficient cell response to insulin (type 2 diabetes). Insulin, which the pancreas secretes, is necessary for transferring blood glucose, or sugar, into cells. Diabetes is a major cause of adult blindness and a significant risk factor for cardiovascular disease. Kidney failure, nerve damage, and lower limb amputation from compromised circulation are additional long-term complications.

Although type 1 diabetes (also referred to as juvenile-onset or insulin-dependent diabetes) can strike at any age, it typically starts in late childhood when the pancreas is unable to secrete enough insulin. Although there is a strong genetic component to type 1 diabetes, the majority of cases are caused by an autoimmune disease that may be brought on by a virus, foreign protein, or environmental toxin. Diabetes is not caused by sugar or carbohydrates in the diet, despite the fact that elevated blood sugar is a key symptom of the condition. Insulin injections are used to treat type 1 diabetes, along with tiny, frequently spaced meals and snacks that help to distribute glucose intake throughout the day and reduce blood glucose swings.

Type 2 diabetes, previously known as adult-onset or non-insulin-dependent diabetes, is the more common form of diabetes, accounting for 90–95% of cases. In this condition, cells become resistant to insulin, preventing them from taking in glucose, and leading to elevated blood sugar levels. While type 2 diabetes typically develops in middle age, it is increasingly being diagnosed in children, particularly those who are overweight. Genetic factors may play a role in the development of type 2 diabetes, particularly in individuals with excess body fat, especially around the abdomen. Weight loss can help normalize blood sugar levels and medications can also be prescribed. Lifestyle changes, such as diet and exercise, are effective in preventing or delaying the onset of type 2 diabetes in individuals at risk.

Studies on migration have demonstrated that the rate of type 2 diabetes can rise significantly as a result of urbanization and the adoption of Western diet and lifestyle habits. For instance, the disorder is more common in the sedentary, high-fat Pima Indians of Arizona than it is in a closely related group of Pimas who lead a traditional lifestyle in a remote, mountainous region of Mexico. These Pimas are physically active, have lower body weights, and consume lower fat diets. For Native Americans and other ethnic minorities in the US, type 2 diabetes is a major health issue. Along with the rise in obesity, type 2 diabetes has become much more commonplace worldwide.

Individualized treatment plans for diabetics are created following a thorough medical evaluation and in collaboration with a registered dietitian or qualified nutrition expert. The updated therapeutic diet emphasizes complex carbohydrates, soluble fiber, and careful balancing of carbohydrate, protein, and fat intake. Due to the high risk of heart disease in diabetics, saturated and trans fats are limited, while physical activity and weight management are strongly recommended. Previous dietary guidelines restricted sugar consumption for diabetics, but new recommendations allow for moderate sugar intake as long as it is balanced with other carbohydrates in a meal. Diet and exercise are also essential in managing gestational diabetes, a condition that affects a small percentage of pregnant women and usually resolves after delivery, although these women are at increased risk of developing type 2 diabetes in the future.

The glycemic index is a new tool that was developed in the 1990s based on research that showed different foods high in carbohydrates had effects on blood glucose levels that could not be predicted based only on their chemical structure. For instance, compared to table sugar (sucrose), fruit, or milk, the simple sugars produced during the digestion of some starchy foods, like bread or potatoes, are absorbed more quickly and raise blood glucose levels more quickly. Practically speaking, though, a carbohydrate food's so-called glycemic effect is less significant if it is consumed as a component of a mixed meal. While the glycemic index can be a helpful tool when designing a diabetic diet, it does not replace the need for other well-established therapeutic approaches, such as controlling body weight and limiting total carbohydrate intake.

Chromium is a trace element that is necessary for glucose tolerance and an insulin cofactor. Additional chromium has been demonstrated to be beneficial for malnourished infants with impaired glucose tolerance; however, there is no proof that the majority of people with diabetes are deficient in chromium or require chromium supplementation. Hypoglycemia, or low blood sugar, is characterized by symptoms such as irritability, shakiness, headache, and confusion. It can also occur when a diabetic injects too much insulin. If left untreated, severe hypoglycemia can cause coma, seizures, and even death. Reactive hypoglycemia of non-diabetic origin is a unique disorder of carbohydrate metabolism characterized by excessive blood glucose drop (below 50 mg/dl) following an excess of insulin produced by the body in response to a meal high in simple sugars; hypoglycemia symptoms coexist. But this is not a common condition. Treatment with diet is effective for several hereditary metabolic disorders, also referred to as inborn errors of metabolism. Autosomal recessive traits are inherited by the majority of these relatively uncommon disorders (i.e., both parents must be carriers) and cause a particular enzyme or cofactor to either be completely absent or have decreased activity. The metabolism of fatty acids, carbohydrates, or amino acids through biochemical pathways may be impacted; each pathway may have several potential enzyme defects. Programs for screening new mothers and even prenatal diagnosis can sometimes enable early detection and effective intervention. The majority of these disorders have a poor prognosis and can lead to severe intellectual disability and other illnesses if treatment is not received quickly and aggressively. The most well-known of these illnesses is phenylketonuria (PKU), a condition in which phenylalanine is not correctly converted to tyrosine through metabolism. Tyrosine supplements and a lifetime restriction on phenylalanine intake are part of the treatment. Normal development and intellectual functioning are achievable with timely identification and careful management.

5.3.5. Obesity and Weight Control

The World Health Organization (WHO) has identified obesity as a global epidemic impacting over 500 million adults worldwide, while also coexisting with undernutrition in both developing and developed countries. There is a concerning rise in childhood obesity rates across the globe. Obesity, defined as excessive body fat for a person's height, is linked to various negative health outcomes including high blood pressure, abnormal blood lipids, heart disease, stroke, type 2 diabetes, gallbladder issues, arthritis, certain cancers, and a shorter lifespan. Genetics play a key role in determining body weight regulation, however, environmental factors such as high-calorie diets and lack of physical activity also play a significant role in how an individual's genetic predisposition unfolds.

Carbohydrates in the diet are not the cause of obesity. In certain Asian cultures, such as those where rice is a staple food, people tend to be thinner and have lower rates of heart disease and diabetes compared to Western cultures. The key to weight management is the balance between the number of calories consumed and the amount of energy expended over time. Body mass index (BMI) has replaced height-weight tables as a tool for determining healthy weight ranges. BMI is calculated by dividing weight in kilograms by height in meters squared: weight ÷ height^2 = BMI. In 1997, the World Health Organization recommended a healthy BMI range for adults as 18.5 to 24.9. A BMI below 18.5 is considered underweight, while 25 to 29.9 is overweight, and 30 or higher is considered obese. Determining overweight and obese status in children is more complex, as their BMI changes with age.

An effective weight loss plan for adults should aim for a gradual reduction of about one pound per week through a daily intake of 1,200 to 1,500 kilocalories (kcal) along with a balanced vitamin and mineral supplement. It is recommended to aim for an energy deficit of 500 kcal per day to achieve this goal. Consuming less than 1,000 kcal per day is not advised, instead, it is better to increase physical activity to help maintain lean tissue. For individuals who are severely obese and struggling to lose weight, medical consultation may be necessary to explore options such as weight-loss medications, appetite suppressants, nutrient absorption inhibitors, or even surgical procedures. It is important to note that fad diets or extreme restrictions, such as carbohydrate-restricted diets or very low-fat diets, may result in short-term weight loss but are often ineffective in maintaining long-term results. It is essential to establish healthy dietary and exercise habits to sustain weight loss success over time.

Establishing new habits is necessary for long-term weight management. These habits include eating less but healthfully, getting regular exercise, and modifying unhelpful behavior patterns like eating in front of the TV. Reducing consumption of foods high in fat, which are higher in energy, is also beneficial. Eating fewer portions and switching to water from beverages high in calories is another helpful strategy. Foods labeled as low-fat may actually contain more calories overall because the sugars used to replace the fat can also contain calories. People who use nonnutritive or artificial sweeteners do not always consume fewer calories overall. The *ob* gene was found in mice and humans as a result of studies conducted on genetically obese lab animals. The protein hormone leptin is secreted by adipose (fat) tissue cells, and it is controlled by this gene. The brain's regulatory center, the hypothalamus, receives a signal from leptin when fat reserves rise, encouraging a reduction in food intake and increased energy expenditure. Some genetic mutations cause the body to either produce insufficient amounts of functional leptin or to stop responding to the leptin signal. It is unknown whether leptin therapy will result in weight loss in individuals who are leptin-resistant or who do not have mutations in the *ob* gene, but treatment with leptin may be helpful for the small percentage of obese people who have an *ob* gene defect.

5.3.6. Eating Disorders

Eating disorders like anorexia nervosa and bulimia nervosa are critical health issues tied to an unhealthy obsession with body image. While girls and young women are often the most affected by societal pressures to be thin, boys and men can also struggle with these disorders. The consequences can be long-lasting and in some cases, deadly. The incidence of eating disorders has been on the rise over the past 50 years, particularly in the US and Western Europe.

Anorexia nervosa is characterized by low body weight, extreme undereating, a deep fear of gaining weight or being overweight (even when underweight), and a distorted body image. The effects can include weakened immune systems, anemia, and digestive problems. Without intervention, individuals may reach a state of semi-starvation similar to marasmus, requiring hospitalization and sometimes force-feeding to prevent death. Treatment typically involves a team approach, including a doctor, psychiatrist, nutritionist, and other health professionals. Bulimia nervosa is believed to be more common than anorexia nervosa, and it is possible for individuals to have both disorders. In bulimia nervosa, frequent episodes of binge eating are followed by purging behaviors such as self-induced vomiting, fasting, excessive exercise, or the use of laxatives, enemas, or diuretics. Treatment typically involves a structured eating plan.

Young athletes, particularly females, may restrict their food intake to adhere to weight guidelines and body image expectations of their sport. While females are most commonly affected, male athletes like gymnasts, wrestlers, boxers, and jockeys are also at risk. Intense training and low food intake in young female athletes can lead to amenorrhea (lack of menstruation for at least three months) and bone loss similar to that seen in menopausal women. Calcium supplements may be necessary.

5.3.7. Tooth Decay

Tooth decay, also known as dental caries, is caused by bacteria such as *Streptococcus mutans* metabolizing sugars and carbohydrates in dental plaque to produce acids that erode tooth enamel. Dental plaque is a combination of bacteria and sticky polymers that protect the teeth from saliva and the tongue, promoting decay.

All forms of sugar, including honey, molasses, brown sugar, and corn syrup, can contribute to tooth decay, as can fermentable carbohydrates found in grains, milk, fruits, and fruit juices. Consuming sugary or starchy foods between meals, especially those that stick to teeth, increases the risk of acid exposure. Artificial sweeteners are not decay-causing, and xylitol, found in some chewing gums, actually helps prevent new cavities by inhibiting plaque and bacteria. Putting a baby to bed with a bottle containing juice or sweetened beverages can result in "baby bottle tooth decay."

When enamel is forming in the jaws before the eruption of permanent teeth, fluoride is particularly effective at preventing decay. In areas where fluoride levels are not naturally high, fluoridating water is a safe and efficient public health measure. Fluoride-containing water that has one part per million or less prevents dental caries without causing teeth to become mottled as higher fluoride concentrations can. Children should take fluoride supplements if their area does not have fluoridated water.

Seaweed, brewed tea, and marine fish eaten with bones are important sources of fluoride in the diet. Regular oral hygiene practices, such as brushing and flossing the teeth and gums, as well as washing mouth after meals and snacks, are crucial for preventing dental caries and periodontal disease. A healthy immune system and a balanced diet are also essential for gum health. Important nutrients include calcium and vitamin D, which support strong teeth and a strong jawbone, and vitamin C, which helps prevent gingivitis (inflamed gums).

5.3.8. Heartburn and Peptic Ulcer

The burning sensation known as heartburn is caused by the lining of the esophagus becoming inflamed when hydrochloric acid-containing gastric contents flow backward from the stomach. Acid reflux, another name for occasional heartburn, is a common condition that is usually brought on by eating particular foods. GERD, or gastroesophageal reflux disease, is a condition that causes frequent heartburn in certain people. Limiting alcohol and caffeine intake, which relax the lower esophageal sphincter and actually cause reflux, as well as fat intake, which slows down the emptying of the stomach, is recommended for people with GERD. Although they don't seem to be the cause of the illness, foods like chocolate, citrus fruits and juices, tomatoes and tomato-based products, peppermint and spearmint oils, and some spices can make heartburn worse.

Losing weight can help overweight or obese individuals with GERD experience symptom improvement. Making lifestyle changes such as eating smaller meals, chewing food thoroughly, eating slowly, avoiding tight clothes, not smoking, and waiting at least three hours before lying down after eating can also help. It is advised to use antacids and acid controllers under medical supervision and not rely on them too heavily. Peptic ulcers are now known to be caused by the bacteria *Helicobacter pylori,* rather than stress or spicy foods. However, stress, coffee, caffeinated beverages, and alcohol can worsen an existing ulcer.

5.3.9. Bowel Conditions and Diseases

Constipation is when someone has trouble passing dry, hardened feces. It can be caused by not having enough fiber in the diet or other factors like taking calcium or iron supplements. Straining during bowel movements can also lead to diverticulosis, which are small pouches in the colon wall that can become inflamed and cause diverticulitis. Another result of straining is hemorrhoids, which are swollen veins in the rectum and anus that can cause pain, itching, and bleeding.

Increasing fiber intake through high-fiber foods like whole-grain breads and cereals, staying hydrated by drinking enough water, and regular exercise are effective ways to address constipation. Fiber, especially the insoluble type, helps soften and bulk up stool by drawing water into the colon consuming dried fruits like prunes, which have a natural laxative compound (dihydroxyphenyl isatin) and are high in fiber, can also promote bowel movements. While laxatives or enemas may offer temporary relief, they can disrupt fluid, mineral, and electrolyte levels and hinder vitamin absorption if used frequently. It's important to consult a doctor if there are persistent changes in bowel habits.

Unlike constipation, diarrhea is characterized by loose, watery stools and an increased frequency of bowel movements, which can be concerning. Acute bacterial diarrhea is a common condition that typically resolves on its own. Other causes of acute diarrhea include viral infections, parasites, food intolerances, medications, medical treatments, surgical procedures, and stress. It is important to stay hydrated when experiencing diarrhea, regardless of the cause. Severe and persistent diarrhea can lead to dangerous dehydration and electrolyte imbalances, requiring immediate medical attention, especially in infants and children. Prolonged vomiting carries similar risks.

Depending on which part of the gastrointestinal tract is affected, inflammatory bowel disease (IBD), such as Crohn's disease (regional ileitis) or ulcerative colitis, causes impaired absorption of many nutrients. Children who have IBD may not develop normally.

A diet that is low in fat and fiber, high in protein and easily digested carbohydrates, and free of lactose (milk sugar) is typically part of the treatment. It may also be advised to take supplements of fat-soluble vitamins, increase intake of specific nutrients like iron, calcium, and magnesium, and drink more fluids and electrolytes to replace fluid lost as a result of diarrhea.

A common gastrointestinal ailment called irritable bowel syndrome (IBS) is typified by an interference with intestinal peristalsis. The symptoms include constipation and diarrhea that alternate, as well as excessive gas, cramping in the abdomen, and discomfort. IBS does not harm the intestines, despite the fact that it can be quite uncomfortable. Dietary treatment entails recognizing and avoiding "problem" foods, such as legumes and other vegetables and dairy products that cause gas, as well as potentially cutting back on caffeine. A low-fat diet, smaller meals, and a gradual increase in fiber intake are beneficial for the majority of IBS sufferers.

5.4. FOOD-DRUG INTERACTIONS

Drugs can either improve or interfere with how nutrients are utilized, which can occasionally result in imbalances. One typical illustration is the increased potassium loss that happens when certain diuretics are used to treat high blood pressure. Drugs that modify the gastrointestinal tract's acidity, interfere with the function of the digestive system, or bind to nutrients themselves can also have an impact on the absorption of nutrients. For instance, consistent use of antacids, mineral oil, or laxatives can decrease the absorption of nutrients and eventually result in deficiency. Elderly people who take several medications are especially vulnerable to poor nutritional status. Conversely, food can change how a drug absorbs or interact with other drugs in an unfavorable way, which can make a drug toxic or ineffective. For instance, levodopa, a medication used to treat Parkinson's disease, is less effective when combined with protein and vitamin B6. When people taking monoamine oxidase (MAO) inhibitors for depression, tyramine, an amino acid derivative present in some aged cheeses and red wines, can raise blood pressure. Certain substances found in grapefruit juice have the ability to prevent some medications from breaking down, which can impact how well they are absorbed and work. These medications include calcium channel blockers, anticonvulsants, estrogen, antihistamines, protease inhibitors, immunosuppressants, antifungal medications, and certain statins that lower cholesterol. There may be unexpected consequences if a person eats grapefruit or drink grapefruit juice within a few hours or even days after taking these medications. Herbal products and vitamin and mineral supplements may also interact with medications. For instance, vitamin C, vitamin E, beta-carotene, and selenium are among the supplemental antioxidants that have been researched and may reduce the efficacy of some medications (e.g., when combined with statins, high-dose niacin can increase HDL cholesterol levels and enhance cardiovascular health. Additionally, the herbal remedy St. John's wort can affect how medications like antidepressants, protease inhibitors, and anticoagulants are metabolized. It can also lessen the efficiency of oral contraceptives.

5.5. FOOD ALLERGIES AND INTOLERANCES

A genuine food allergy is characterized by an aberrant immune reaction to a generally safe food ingredient, most commonly a protein. When a person has a food allergy that is antibody-mediated (also known as immediate hypersensitivity), their body releases histamine and certain immunoglobulin E antibodies within minutes or hours of being exposed to the allergen. This can cause symptoms that range from mild to fatal in the respiratory, gastrointestinal, or skin areas. Cell-mediated (delayed hypersensitivity) food allergies are far less common; symptoms, including a localized inflammatory process, may not manifest for up to a day. Food intolerances or sensitivities are adverse food reactions other than foodborne infection or poisoning that do not involve the immune system. The most prevalent of these is lactose intolerance, a genetically based insufficiency of the enzyme lactase, which breaks down lactose, the sugar found in milk.

While lactose intolerance and milk allergy are two different conditions, they are frequently confused. The true allergy rate to the protein in cow's milk is only around 1% of the population. Due to their developing immune and digestive systems, infants are most commonly affected by milk allergies. However, after early childhood, a large portion of the global population—aside from those descended from northern Europe—becomes partially intolerant to lactose. When undigested lactose enters the large intestine, it can result in diarrhea, flatulence, and discomfort in the abdomen. Small amounts of dairy products, particularly yogurt or other milk products containing the bacterium *Lactobacillus acidophilus*, can usually be consumed by lactose-intolerant people without much discomfort. Lactase tablets or drops, which break down lactose into easily absorbed sugars, are another option.

Wheat gluten and related proteins from rye and barley are intolerable to people with celiac disease, also known as nontropical sprue, celiac sprue, or gluten-sensitive enteropathy. According to recent research, oats might be safe if they aren't tainted with wheat. People with European ancestry are more likely to have celiac disease than people with African or Asian ancestry, who may also be affected by this type of cell-mediated food allergy. It is typified by inflammatory damage to the small intestine's mucosal cells, which results in nutritional malabsorption and symptoms like fatigue, diarrhea, weight loss, bone pain, and neurological abnormalities. There could be several nutritional deficits as a result, which would hinder children's growth. The disorder is frequently linked to autoimmune diseases, especially type 1 diabetes and autoimmune thyroid disease. Patients with celiac disease can recover if gluten is removed from their diet, even though the condition can be fatal if left untreated.

Drug side effects, such as those brought on by caffeine or alcohol, may also cause additional negative reactions to food or drink. Some people get headaches from certain foods, including chocolate, red wine, ripened cheese, and even ice cream. Among the food additives that can trigger reactions in vulnerable individuals are nitrate and nitrite preservatives found in processed meats, certain food colorants, especially tartrazine (also referred to as FD and C Yellow #5), and the flavor enhancer monosodium glutamate (MSG). Certain negative food-related reactions are solely psychological in nature and do not happen when the food is presented in a covert manner.

Although almost every food has the potential to cause allergies, cow's milk, eggs, wheat, fish, shellfish, soybeans, peanuts, and tree nuts (like almonds, walnuts, and cashews) are the foods that most frequently do so. Edible oils and other products made from these foods may still contain allergenic protein residues, depending on the processing techniques used. People with severe allergies may react even to minute amounts of a food allergen, such as inhaled vapors. The percentage of adults and children who actually have food allergies varies greatly among studies. All seem to agree, though, that slightly more children—roughly 3–8%—and fewer adults—roughly 2–5%—are affected. Food allergies usually pass in children, especially if the offending food is avoided for a year or two. Food allergies, however, can appear at any time and some allergies, like those to shellfish, peanuts, and tree nuts, can last a lifetime. Tightness in the throat, swelling of the lips or tongue, itchy lips, wheezing, breathing difficulties, headache, nasal congestion, skin rash (eczema), hives, nausea, vomiting, stomach cramps, diarrhea, and, in extreme situations, potentially fatal anaphylactic shock are typical signs of an antibody-mediated food allergy. It is recommended that those who are at risk of anaphylaxis always carry a syringe filled with adrenaline and seek emergency medical attention if an allergic reaction starts. Even medical professionals with expertise in allergy and immunology may find it challenging to diagnose food allergies. Skin tests, blood tests for antibodies to particular allergens, and even an elimination diet that involves removing suspect foods from the diet and then reintroducing them one at a time may not be conclusive. A double-blind food challenge is the most definitive diagnostic test because it keeps both the patient and the physician guessing as to whether a dangerous food or a harmless placebo is being given. Nevertheless, these controlled clinical tests are costly and time-consuming.

Although labels are frequently vague and should not be relied upon blindly, they are crucial for revealing hidden ingredients in packaged foods. For instance, the presence of milk protein is indicated by the inclusion of casein, caseinate, or whey in a product that is labeled as nondairy. Unusual foods like stew, chili, processed meats, oils, flours, cream-substitutes, and desserts can all contain peanuts.

Studies show that desensitization—exposing an allergic person to progressively higher concentrations of the antigen they are allergic to over time—may, at least occasionally, lessen the severity of an allergy. Different techniques are used to desensitize the body; sublingual immunotherapy uses tiny amounts of antigen administered as drops under the tongue, while allergy shots inject antigens. Building tolerance to food allergens is thought to be safe and effective, especially with sublingual immunotherapy.

5.6. TOXINS IN FOODS

The edible skins of fruits and vegetables are high in vitamins, minerals, and fiber, but the outer layers of these foods are usually more contaminated with pesticide residues and other environmental toxins. Additionally, pesticides have a tendency to build up in animal fat and skin. Eating a wide range of foods, washing fruits and vegetables with care, removing the skin from fish and poultry, and trimming fat from meat and poultry all help to lower the intake of harmful substances. Even organic produce needs to be thoroughly cleaned because, even though it doesn't contain artificial chemicals, it can still become contaminated during the growing process by rot, mold, feces, or other natural materials. Peeling aids in removing these undesirable substances and microorganisms, but it also results in the loss of important nutrients. Potatoes that have a slight green tint, even though it's just the innocuous compound chlorophyll, could have solanine, a naturally occurring toxin. When a potato is handled roughly, exposed to light or extreme temperatures, or gets old, solanine builds up. Solanine poisoning can cause cramps, headaches, and diarrhea, but consuming a large number of contaminated potatoes is necessary to produce a dangerous sickness. Despite the skin's high nutrient content, removing sprouts, peeling off green parts of it, or removing the skin altogether lowers solanine intake. Pregnant women should refrain from eating swordfish, shark, and tuna steaks due to the possibility of high levels of methylmercury (which is retained even after cooking). Nonbacterial toxins in seafood include ciguatera, which is present in some warm-water reef fish, dinoflagellates (microscopic algae) linked to the so-called red tide, which can cause paralytic shellfish poisoning when consumed, and scombrotoxin (histamine) in spoiled fish, which can cause a severe allergic reaction when eaten. Some species of mushrooms contain natural toxins that can cause symptoms that range from neurological effects to hallucinations and upset stomachs. The majority of mushroom deaths are caused by eating Amanita phalloides, also known as the death cap, which contains amatoxins that, while not fatal, can cause irreversible damage to the kidneys and liver. Eating wild mushrooms is not recommended because there are no treatments for mushroom poisoning and because amateur mushroom pickers frequently identify the wrong species.

5.7. FOODBORNE ILLNESSES

Any illness brought on by consuming food or drink contaminated with specific infectious or non-infectious agents is referred to as a foodborne illness. The majority of foodborne illness cases are brought on by pathogens like viruses, bacteria, or parasites. In addition, metals like lead, mercury, and cadmium can contaminate food through air, water, or soil pollution; organic pollutants like dioxin and polychlorinated biphenyls (PCBs), which are byproducts of certain industrial processes; and prions, which are abnormal forms of normally harmless proteins, are examples of other agents. Foodborne illness agents can cause a wide range of conditions, including neurodegenerative diseases like bovine spongiform encephalopathy (mad cow disease), disorders of the reproductive or developmental system, and gastroenteritis. Food poisoning is a term used occasionally to describe foodborne illnesses, especially those involving gastroenteritis.

On the other hand, food poisoning is a specific kind of foodborne illness that is brought on by toxins found in food. These toxins are usually produced by bacteria and manifest as symptoms shortly after the contaminated food is consumed. On the other hand, some other kinds of foodborne illnesses can take days to manifest because they are caused by ingesting the infectious organisms or chemicals. The burden of foodborne illness is believed to be high, despite the fact that it is difficult to determine the incidence because so many cases go unreported. For instance, diarrheal illnesses, a common stand-in for foodborne illness, affect tens of millions of people worldwide each year. Foodborne infections frequently start as outbreaks that could potentially infect a lot of people. For example, in 1988 there was a hepatitis A outbreak in China that affected over 300,000 people due to the consumption of contaminated clams, and in 1994 there was an outbreak of salmonellosis in the United States that affected 224,000 people due to the consumption of contaminated ice cream. Foodborne infections are often fatal. For instance, a listeriosis outbreak in 1985 that involved tainted cheese products in California resulted in 48 deaths out of 142 cases.

5.7.1. Common Foodborne Illnesses

There are numerous varieties of foodborne infections. This article discusses a number of common illnesses and their causes.

Numerous animals, including humans and cattle, have intestines that are home to a type of bacteria called Escherichia coli. Human health is not endangered by most strains of *E. coli*. When consumed by humans, a few strains, however, can have detrimental effects on health. For instance, *E. coli E. coli* 0157:H7, a bacteria found in cattle's intestines, is typically consumed through undercooked ground beef, but it can also spread through person-to-person contact, contaminated water, unpasteurized milk and fruit juice, and uncooked produce. *E. coli* 0157:H7 infections frequently cause severe cramping in the abdomen and bloody diarrhea. Hemolytic uremic syndrome is a rare condition that can lead to kidney failure and even death. Other strains of *E. coli* that can make people sick include O26:H11, O111:H8, and O104:H4.

The bacteria that cause salmonellosis are called Salmonella, and they are typically found in the intestines of birds, mammals, and reptiles. Humans typically contract salmonellosis from eating animal-based foods like milk, eggs, and meat. Salmonellosis can be fatal in those with compromised immune systems or poor health. Symptoms include fever, headache, nausea, vomiting, and diarrhea.

The bacterium *Listeria monocytogenes*, which causes listeriosis, is primarily spread through milk, soft cheeses, ice cream, raw vegetables, raw meat, and raw poultry. Extended refrigeration of certain foods increases the risk of transmission because *L. monocytogenes* can thrive at low temperatures. Listeriosis is characterized by flu-like symptoms, such as fever, lethargy, and muscle aches. Some people get disoriented, lose their balance, or even have convulsions. Listeriosis is particularly dangerous for expectant mothers because it can result in miscarriage and stillbirth. It can also cause meningitis, which is an inflammation of the membranes covering the brain and spinal cord, or septicemia, which is blood poisoning, in infants and people with compromised immune systems.

Bacteria called Campylobacter are the cause of *campylobacteriosis*. It is the most frequently found bacterial cause of diarrheal illness worldwide, and it is more common than salmonellosis in some countries. The main ways that *campylobacteriosis* is spread are through raw milk, undercooked poultry, and drinking water. It can be presumed that the majority of raw poultry is contaminated with Campylobacter bacteria due to its presence in the intestines of healthy birds. *Campylobacteriosis* manifests as fever, nausea, excruciating abdominal pain, and diarrhea. In a tiny percentage of cases, serious health effects like reactive arthritis and neurological disorders may arise.

5.7.2. Prevention and Control

Significant foodborne illnesses, such as salmonellosis and *E. coli* infections, are required to be reported to local, state, or federal health departments by doctors and medical laboratories. This is because these infections are considered reportable diseases in many jurisdictions. However, the reported number of cases is thought to be an underestimate of the actual number of cases because the majority of foodborne illness cases are mild and go undiagnosed. Several laws are in place to regulate the production, transportation, and handling of food with the goal of preventing foodborne illnesses and minimizing their impact. These laws address issues such as preventing the contamination of raw food, ensuring safe preparation and storage practices, and holding accountable any restaurants or food suppliers that are responsible for disease outbreaks or fail to adhere to proper food hygiene practices. Contamination of raw food can occur in various ways, including through the use of unclean water for irrigation or washing, the presence of fecal matter on meat and poultry during processing, the handling of food by workers carrying harmful bacteria or viruses, or the use of utensils and surfaces that are not properly sanitized. A high enough temperature during cooking eliminates a lot of bacteria and parasites. But some microbes might also be present in food that has been cooked, for example, if contamination happens during handling following pasteurization or cooking. If there are only a few microbes present, contamination that occurs after cooking or pasteurization may not always be harmful to human health. The majority of bacteria multiply quickly at room temperature, but *L. monocytogenes* is an exception that does not grow when exposed to refrigeration or freezing. Therefore, it is essential to promptly refrigerate even cooked foods to stop the growth of pathogenic microorganisms.

5.8. BOTANICALS AND FUNCTIONAL FOODS

Numerous herbal remedies are being investigated in thorough scientific investigations, including clinical trials, because they appear to have enough potential to prevent or treat disease. Consumers should use caution and knowledge when using the "botanicals" that are currently available on the market in many countries, as their safety and effectiveness have not been thoroughly investigated. Herbal products can have mild to severe side effects, just like pharmaceuticals do, and "natural" does not equate to "safe." Additionally, the amounts of active ingredients in supplements can differ greatly, and laboratory analyzes have shown that the potency listed on labels is frequently inaccurate. Certain preparations might even have unwelcome contaminants or contain none of the active ingredients specified on the label.

Herbal products with potential risks include ephedra (ma huang), which has been known to cause fatal reactions in certain individuals, particularly those with heart disease or high blood pressure, and comfrey and kava, which can harm the liver. Patients who are scheduled for surgery or other medical procedures may be advised to stop taking specific supplements days or even weeks before the procedure due to potential complications. Concerns about safety and effectiveness must also be addressed, since "designer foods" enhanced with herbs and bioactive ingredients are becoming more and more common.

The growing market for "functional foods," or foods with health benefits beyond simple nutrient value (like cholesterol-lowering margarine), is already blurring the lines between foods, dietary supplements, and medications. Furthermore, customized dietary recommendations to more successfully postpone or prevent disease may be made using genetic profiles to identify individual nutrient requirements, thanks to recent developments in molecular biology (Suskind et al., 2009; Kaur et al., 2019)

CASE STUDY

THE ROLE OF WATER QUALITY IN EARLY CHILD NUTRITION: A CASE STUDY IN UTTARAKHAND, INDIA

Undernutrition is a global public health problem especially in children below the age of 5 years. It can affect various aspects of children's development such as physical and mental development. It can be seen in the form of wasting, stunting or underweight. Water is an essential component of healthy growth and development, and ensuring access to clean water, adequate sanitation, and good hygiene is crucial for early childhood nutrition. Proper sanitation & hygiene and safe drinking water can reduce undernutrition and stunting in children by preventing diarrheal and parasitic diseases, and damage to intestinal development (environmental enteropathy).

Roughly 50% of all malnutrition is associated with repeated diarrhea or intestinal worm infections as a direct result of inadequate water, sanitation and hygiene (WHO). When children are undernourished, their resistance to infection is lowered and they are more susceptible and more likely to die from diarrheal disease and other infections. In fact, diarrhea is the second-leading cause of death globally in children under five years. The condition of malnutrition is further exacerbated in hills due to the limited availability and accessibility of micro-nutrient-rich staple crops around the year. Thus, as part of the Nutrition Security study, the team examined and collected scientific evidence on water quality in Mukteshwar, Nainital, Uttarakhand. This was investigated through on-site testing of drinking water samples and in-person consultations with medical experts.

The Importance of Water in Child Nutrition

Adequate hydration is crucial for a child's health and development. It helps regulate body temperature, transport nutrients and oxygen to cells, eliminates waste products, and supports the functioning of vital organs. Dehydration in children can lead to a range of health issues, including fatigue, headaches, constipation, urinary tract infections, and even impaired cognitive function.

In addition to providing hydration, water is also essential for nutrient absorption and digestion. It helps break down food and allows nutrients to be absorbed and transported to cells throughout the body. Without sufficient water intake, children may experience nutrient deficiencies, even if they are consuming a healthy and balanced diet.

It is, therefore, essential to ensure that children have access to clean and safe drinking water at all times. Parents and caregivers should encourage children to drink water throughout the day and offer fluids during meals and snacks. Water should be the primary source of hydration, with sugary drinks and fruit juices limited to occasional treats.

Water Quality and Child Health

Water quality is a critical public health concern. Poor water quality is an ongoing problem in India because the country lacks water treatment facilities that can handle the pollution caused by rapid industrialization and urbanization, and open defecation in many areas without sanitation infrastructure exacerbates the problem. Access to clean and safe drinking water is essential for preventing water-borne diseases and promoting good health. Inadequate or contaminated water can lead to a range of illnesses in children, including diarrhea (which is the second leading cause of death in children), cholera, typhoid fever, and other bacterial infections.

Water quality is determined by the presence or absence of harmful substances such as bacteria, viruses, and other contaminants, which can cause gastrointestinal diseases. Water can become contaminated due to various reasons such as natural sources, industrial activities, and poor sanitation practices. Children are particularly vulnerable to water-borne diseases as their immune systems are not fully developed, and they are more likely to get dehydrated from diarrhea and vomiting. Moreover, diarrhea increases malabsorption, in turn increasing the likelihood of malnutrition. Therefore, it can be hypothesized that water quality in the household has a significant impact on children's nutritional status in India. Therefore, diarrhea and malnutrition have a vicious cycle.

According to UNICEF, around 800 children die every day from diarrhea caused by poor water and sanitation. Lack of access to safe drinking water and sanitation facilities also affects the nutritional status of children due to the loss of nutrients through diarrhea and other illnesses. According to the latest National Family Health Survey (NFHS-5), 35.5% of children between the ages of 0 and 59 months are stunted, 19.3% are wasted, and 32.1% are underweight. The situation is slightly better in Uttarakhand, a state in northern India, with 27% stunting, 13.2% wasting, and 21% underweight. Around 7.3% of children under 5 years reported a prevalence of diarrhea 2 weeks before the survey at the national level compared to 4.4% in Uttarakhand. However, the region is highly vulnerable to the availability and accessibility of micro-nutrient-rich staple foods/crops. Under-five diarrheal diseases further exacerbate malnutrition by washing out the critical micronutrients.

Water Quality Assessment in Mukteshwar, Nainital, India

A study was conducted in Mukteshwar, Nainital district, Uttarakhand, to collect scientific evidence on water quality. The study included on-site testing of drinking water samples and in-person interactions with medical experts deployed in the region.

The water samples were collected and tested from different sources such as springs, naala (open well), water supplied in houses, and storage water. Physico-chemical and microbiological tests were conducted on the samples. The samples reported pH values within the acceptable range of Indian drinking water standards (IS 10500) of 6.5–8.5. Additionally, total dissolved solids (TDS) values were below 500 ppm, which is the acceptable limit for drinking water quality. However, 44% of the water samples had

TDS values below 50 ppm, which is too low for drinking water. Generally, water with TDS levels between 50 and 150 is considered most suitable for drinking. Only 13% of water samples tested had TDS values above 100 ppm. Such low TDS water does not need exhaustive water treatment such as reverse osmosis (RO) treatment.

Total coliforms were also tested as an indicator of the cleanliness of the water source. Total coliforms are a large group of different types of bacteria that share several characteristics. They are commonly found in the soil and intestines of animals, including humans, and their presence indicates that water is likely to contain other more harmful pathogens. About 50% of the water samples were contaminated with total coliforms. Although total coliform bacteria themselves do not necessarily cause harmful illnesses, their presence increases the probability of bacteria that can cause water-borne diseases (such as diarrhea, typhoid, and jaundice) in the exposed population.

The interaction with local medical experts emphasized that both viral and gastrointestinal diarrhea exist in the area with the increasing prevalence of the latter type of diarrhea. Water and personal hygiene-related disorders were widespread in 20–25 villages of Mukteshwar, strangely after the monsoon during July–October. The medical expert also suggested that human/animal open defecation is the prime cause of gastrointestinal diarrhea. The treatment methods such as boiling/disinfecting or filtering before drinking should be adopted by the population. Another way to prevent diarrhea and other water-borne diseases could be the maintenance of personal hygiene.

Water Availability in Hilly Areas

Besides water quality, water availability is another significant issue in most of the hilly areas. As per the Jal Jeevan Mission – Har Ghar Jal Scheme, 64% of households in Uttarakhand have tap water supply as of date. In rural areas, 46% of households have to travel more than 30 minutes to fetch drinking water. Many residents in Mukteshwar rely on storing rainwater from their rooftops during the rainy season for washing and cleaning purposes. However, this water may not be suitable for drinking due to potential contamination. The Uttarakhand Government has announced an ambitious scheme to provide tap water connections to households at a minimal cost of INR 1, which can improve water availability and quality in the future.

Way Forward

The study's findings highlight the need for proper management of water resources, particularly in the Himalayan region, where water scarcity is an issue. Investing in water quality and sanitation is an essential step towards ensuring child health and development. Governments and communities must prioritize safe drinking water and sanitation facilities, especially in areas where access is limited.

A few ways to reduce the prevalence of water-borne diseases include:

- The open wells are susceptible to microbial contamination, which may cause water-borne diseases (e.g., diarrhea, typhoid, etc.). So, point-of-use (POU) water treatment can significantly impact the health and nutrition of vulnerable populations (infants, children, and pregnant women). Hence, disinfection of open well water with bleaching powder or chlorine tablets is recommended for safe drinking water.
- The role of the village panchayats is critical for supplying disinfected water to the households.
- The Uttarakhand Har Ghar Nal Yojana should ensure adequate water supply and concrete measures to provide good quality water (free from microbes).
- Awareness campaigns or drives on water quality, quantity, sanitation and hygiene should be conducted to convey the information to villagers.
- Regular monitoring of water quality and testing for contaminants is also critical to ensuring that the water is safe for consumption.

CLASS ACTIVITY

PHYSICAL ACTIVITY CHARADES

- **Objective:-** Youngsters will understand how much daily physical activity they require as well as various approaches to achieving this goal.
- **Participants:-** Children of all ages.
- **Time:-** 10 to 20 minutes.
- **Supplies:-** Pieces of paper with lists of charade ideas on them, along with a drawing hat, bowl, or basket. Golfing, ice skating, soccer, badminton, tennis, dancing, bike riding, sledding, water skiing, horseback riding, canoeing, jogging, yoga, and so on are some activities to list on the slips of paper. Think outside the box!
- **Location:-** A classroom.
- **Pre-Activity Talking Points**
  1. Getting enough exercise improves both your appearance and sense of well-being. Do you know that children should aim for one hour of physical activity per day?
  2. How many of you play video games or watch TV? What is the recommended amount of "screen time" (video games and TV) for children per day, as determined by doctors? (two hours) We'll be playing "Activity Charades" today, so you can get up and guess some entertaining ways to get moving.

- **Directions**
    1. Give the group an explanation of the charades game. An individual "acts out" an activity without using words, and the audience attempts to guess what the "actor" is doing.
    2. Ask each volunteer to sketch a slip of paper with a task written on it, and then perform the task. Teams can play this as well. Having fun is the main goal.
- **Post-Activity Review**
    1. Can you think of a way to turn video games into "active games" where you have to stand and jump while playing? What other ways do you keep active on a daily basis?
    2. Let's all establish a goal for physical activity. – How many minutes do you think you spend exercising during the day? – Are you reaching the one-hour goal? – What are some strategies to increase your daily activity?

SUMMARY

- Undernutrition is a condition in which there is insufficient food to meet energy needs; its main characteristics include weight loss, failure to thrive, and wasting of body fat and muscle.
- Chronic undernutrition manifests primarily as protein-energy malnutrition (PEM), which is the most common form of malnutrition worldwide. Also known as protein-calorie malnutrition, PEM is a continuum in which people—all too often children—consume too little protein, energy, or both.
- Under most circumstances, there is no absolute dietary requirement for carbohydrates—simple sugars, complex carbohydrates such as starches, and the indigestible plant carbohydrates known as dietary fiber. Certain cells, such as brain cells, require the simple carbohydrate glucose as fuel. If dietary carbohydrates are insufficient, glucose synthesis depends on the breakdown of amino acids derived from body protein and dietary protein and the compound glycerol, which is derived from fat.
- Iron deficiency is the most common of all nutritional deficiencies, with much of the world's population being deficient in the mineral to some degree. Young children and premenopausal women are the most vulnerable.
- Dehydration may develop if water consumption fails to satisfy thirst; if the thirst mechanism is not functioning properly, as during intense physical exercise; or if there is excessive fluid loss, as with diarrhea or vomiting.
- The need for each nutrient falls within a safe or desirable range, above which there is a risk of adverse effects. Any nutrient, even water, can be toxic if taken in very large quantities. Overdoses of certain nutrients, such as iron, can cause poisoning (acute toxicity) and even death. For most nutrients, habitual excess intake poses a risk of adverse health effects (chronic toxicity).
- The relationship between diet and chronic disease (i.e., a disease that progresses over an extended period and does not resolve spontaneously) is complicated, not only because many diseases take years to develop but also because identifying a specific dietary cause is extremely difficult.
- Drugs may interfere with or enhance the utilization of nutrients, sometimes leading to imbalances. A common example is the increased loss of potassium that results from the use of certain diuretics to treat high blood pressure. Nutrient absorption can also be affected by drugs that change the acidity of the gastrointestinal tract, alter digestive function, or actually bind to nutrients.
- A true food allergy involves an abnormal immunologic response to an otherwise harmless food component, usually a protein. In the case of antibody-mediated (immediate hypersensitivity) food allergies, within minutes or hours of exposure to the allergen, the body produces specific immunoglobulin E antibodies and releases chemical mediators such as histamine, resulting in gastrointestinal, skin, or respiratory symptoms ranging from mild to life-threatening.
- Edible skins of fruits and vegetables are rich in vitamins, minerals, and fiber;

however, pesticide residues and other environmental contaminants are typically more plentiful in the outer layers of these foods. Pesticides also tend to accumulate in the fat and skin of animals.

- Foodborne illness, any sickness that is caused by the consumption of foods or beverages that are contaminated with certain infectious or non-infectious agents. Most cases of foodborne illness are caused by agents such as bacteria, viruses, or parasites.

REVIEW QUESTIONS

1. What are the most common nutrient deficiencies?
2. What is the connection between diet and chronic disease?
3. How can chronic diseases be treated with diet therapy?
4. What are food-drug interactions?
5. What is the difference between food intolerance and food allergy?
6. What is the cause of food allergies?
7. How to heal food allergies?
8. What are the symptoms of foodborne illness?
9. What is foodborne illness prevention?

REFERENCES

1. Coelho, V. A. T., Figueiredo, M. A., Rodas, C. L., Bastos, A. R. B., Coelho, L. C., & Carvalho, J. G. (2014). Mineral nutrition of ornamental pepper in nutritional deficiencies. *Nucleus, 11*(1), 169–177. https://doi.org/10.3738/1982.2278.875.
2. De, D., & Seshadri, D. (2012). Nails in nutritional deficiencies. *Indian Journal of Dermatology, Venereology, and Leprology, 78*(3), 237. https://doi.org/10.4103/0378-6323.95437.
3. Gimenez, M. S., Oliveros, L. B., & Gomez, N. N. (2011). Nutritional deficiencies and phospholipid metabolism. *International Journal of Molecular Sciences, 12*(4), 2408–2433. https://doi.org/10.3390/ijms12042408.
4. Hathcock, J. N. (1986). Drug-induced nutritional deficiencies. *Journal of Nutrition, 116*(11), 2311. https://doi.org/10.1093/jn/116.11.2311.
5. Kaur, D., Rasane, P., Singh, J., Kaur, S., Kumar, V., Mahato, D. K., Dey, A., Dhawan, K., & Kumar, S. (2019). Nutritional interventions for elderly and considerations for the development of geriatric foods. *Current Aging Science, 12*(1), 15–27. https://doi.org/10.2174/1874609812666190521110548.
6. Mehta, N. M., & Duggan, C. P. (2009). Nutritional deficiencies during critical illness. *Pediatric Clinics of North America, 56*(5), 1143–1160. https://doi.org/10.1016/j.pcl.2009.06.007.

7. Menon, A., Apthadevi, & Roshni, P. R. (2019). Nutritional deficiencies after bariatric surgery. *Research Journal of Pharmacology and Pharmacodynamics, 11*(3), 120. https://doi.org/10.5958/2321-5836.2019.00021.1.

8. Rahayu, S. F. (2022). Relationship of nutritional concerning families with the nutritional status of children in the working area of local public health center. *Healthy-Mu Journal, 4*(2), 93–98. https://doi.org/10.35747/hmj.v4i2.28.

9. Suskind, D. L. (2009). Nutritional deficiencies during normal growth. *Pediatric Clinics of North America, 56*(5), 1035–1053. https://doi.org/10.1016/j.pcl.2009.07.004

CHAPTER

6

Meal Planning

LEARNING OBJECTIVES

After studying this chapter, you will be able to:

- Understand food guides for selecting an adequate die
- Perform meal planning for the family
- Focus on meal planning for various age groups
- Explain the importance of special diets.

KEY TERMS FROM THIS CHAPTER

Adulthood	Anxiety	Arugula
Asparagus	Breastfeeding	Brussels sprouts
Colostrum	Family composition	Fiber
Flavor	Fluid intake	Infancy
Kale	Lactation	Menus
Osteoporosis	Pregnancy	

INTRODUCTION

Planning a meal for the week entails making a list of what a person will eat for each meal. A person can determine what supplies and ingredients they'll need to prepare their meals by organizing them. After that, they can make a shopping list to use when they make their Instacart order or goes to the grocery store. Planning a meal can make their life easier, which is why it's important. Everybody has a busy, hectic life, and there's already enough to worry about without having to decide on an impulse.

6.1. FOOD GUIDES FOR SELECTING AN ADEQUATE DIE

Strict restriction, maintaining an unattainable weight, or depriving oneself of favorite foods are not aspects of healthy eating. It is more important to eat balanced meals that make one feel fantastic, give them more energy, enhance their mood, and improve their health. Eating healthily does not have to be very difficult. One is not alone if one feels overloaded with contradicting diet and nutrition recommendations. It appears that for every expert who recommends a particular food, there are others who say the exact opposite. The truth is that although certain foods or nutrients have been demonstrated to improve mood, one's overall dietary pattern is crucial for overall health.

6.1.1. The Fundamentals of Healthy Eating

We all need a balance of protein, fat, carbs, fiber, vitamins, and minerals in our diets to maintain a healthy body and mind, despite what some extreme diets may suggest. It is not necessary to remove food groups from the diet; instead, choose a variety of foods from each group in moderation.

- **Both Mood and Cognitive Function are supported by Protein:** People with kidney disease may experience negative effects from eating too much protein, and it may replace other foods in the diet that are essential for certain nutrients. But studies indicate that a lot of us require higher-quality protein in our diets, particularly as we get older. That does not imply that a person must consume more animal products. Make sure the body gets all the necessary protein it needs by eating a range of plant-based protein sources every day.
- **Fat:** Aside from improving food taste and helping a person feel fuller after a meal, eating fats is beneficial for many biological processes. Still, not all fat is created equal. Saturated fats, for example, have been connected to a higher risk of developing certain diseases, but unsaturated fats are better for the heart and brain. The health of body and mind actually depends on omega-3 fats. Consuming a diet higher in unsaturated fat can both protect and elevate mood.
- **The Body and Brain Use Carbohydrates as Their Primary Energy Source:** The majority of carbohydrates should ideally come from complex, unrefined sources like fruit, whole grains, and vegetables as opposed to refined sources like sugars and donuts, white bread, and sugary drinks. A person can lessen the likelihood of abrupt blood sugar spikes, mood swings, and energy swings by converting from simple, refined carbohydrates to complex, unrefined carbohydrates and balancing their meals with protein and unsaturated fat.

- **Fiber:** Consuming high-dietary-fiber foods (vegetables, fruits, grains, nuts, and beans) can help to maintain a regular diet and reduce the risk of diabetes, heart disease, and stroke. It can also make skin look better and, by making feel fuller for longer, aid in weight loss.
- **Calcium:** Not getting enough calcium in the diet can cause osteoporosis, but it can also aggravate anxiety, depression, and trouble sleeping. Regardless of age or gender, it's critical to eat a diet high in calcium-rich foods, minimize those that deplete calcium, and obtain adequate magnesium and vitamins D and K to support calcium in its function.

6.1.2. Making the Switch to a Healthy Diet

Making the transition to a wholesome, well-balanced diet doesn't have to be all or nothing. It's not necessary to be flawless, to give up foods a person loves, or to make all the changes at once; doing so will almost always result in cheating or giving up on their new diet. A better strategy is to implement a few minor adjustments at a time. A person can accomplish more in the long run without experiencing deprivation or overwhelm from a significant diet change if they set modest goals. They should consider making small, doable dietary changes, such as including a salad once a day in their meals. As they get used to their minor adjustments, they can add more healthful options.

6.1.2.1. Setting Yourself Up for Success

Attempt to keep things simple in order to position for success. It is not difficult to eat a more balanced diet. Consider diet in terms of color, variety, and freshness, for instance, rather than obsessing over calorie counting. Try to cut back as much as a person can on packaged and highly processed foods and whenever they can, use more fresh ingredients.

- **Prepare More of Your Own Meals:** Taking greater control over what a person eats and keeping a closer eye on the ingredients in food can be achieved by cooking more meals at home. They might notice that the symptoms of anxiety, stress, and depression don't get worse and that they feel less drained, bloated, and agitated.
- **Add Balance to the Diet:** When making dietary changes, it's critical to concentrate on making adjustments that will enhance the general quality of the diet. A person's health may benefit from substituting healthy fats for saturated fats, such as grilled salmon for fried chicken. However, substituting refined carbohydrates for animal fats (like having a donut instead of bacon for breakfast) won't make them feel better or reduce their chance of heart disease.
- **Read the Labels:** It's critical to know what's in a person's food because manufacturers frequently put unnecessary ingredients in packaged foods that the body doesn't need.
- **Focus on How a Person Feels After Eating:** This will promote new, healthful tastes and habits. The better they feel after a meal, the more likely it is that their food is well-balanced and nutritious.
- **Drink Plenty of Water:** Although drinking water helps rid our bodies

of toxins and waste products, many people experience chronic dehydration, which can lead to headaches, fatigue, and low energy. Since it's common to confuse thirst with hunger, drinking plenty of water can also encourage you to eat less.

6.1.3. Moderation: Important to Healthy Eating

To put it simply, moderation is the practice of only consuming as much food as the body requires. After a meal, one should feel full but not overstuffed. Moderation is defined as cutting back on current eating habits. However, it doesn't entail giving up their favorite foods. For example, having bacon for breakfast once a week as opposed to every day could be regarded as moderation.

- **Try Not to Think of Certain Foods as "Off-Limits:"** When a person forbids a food item, it makes sense that they would crave it more and feel like a failure if they give in to temptation. To begin, practice being more mindful of those foods. What are your thoughts after eating different foods? Do you eat based only on habit or does your body demand a certain food?
- **Think Smaller Portions:** Recently, serving sizes have skyrocketed. When eating out, order a starter rather than an entree, share a dish with a friend, and avoid ordering anything supersized. Visual cues can assist with portion sizes at home. A standard light bulb is roughly the size of half a cup of mashed potato, rice, or pasta, and a portion of meat, fish, or chicken should be the size of a deck of cards. When a meal is almost over and a person is still hungry, assess what they need to feel full and try to eat more nutrient-dense foods like fruit or greens rather than highly processed ones.
- **Take Time**: It's critical to take a moment to appreciate food as fuel rather than as something to quickly consume between appointments or while en route to pick up the kids. Eat mindfully and stop eating before a person feels full, as it may take their brain a few minutes to signal their body that it is full.
- **Eat with Others Whenever Possible:** Eating by yourself frequently results in mindless overeating, especially when done in front of a TV or computer.
- **Be Mindful About Snacking.** Snacking can be a crutch when you're bored or stressed, but it can also help you stay full between meals throughout the day. When choosing snacks, aim to include a variety of carbohydrates, proteins, and fats. Some examples of this include cheese on an apple or crackers, or yogurt and granola.
- **Be Aware of Emotional Eating:** Hunger is not the only reason we eat. To decompress or deal with negative emotions like melancholy, loneliness, or boredom, a lot of us also turn to food. However, you can more effectively keep a balanced diet by learning more healthful ways to deal with stress and emotions.
- **It Matters When You Eat as Much as what You Eat:** Consume breakfast and smaller meals all day long. Healthy meals throughout the day will keep you energized, and

a hearty breakfast can speed up your metabolism.

- **Try to Avoid Eating Late at Night:** There is conflicting evidence, but some studies have connected eating late at night to gaining weight. This could be connected to eating without hunger. It's simple to eat mindlessly in the evening because so many of us see it as a time to decompress and unwind.

6.1.4. Add More Fruit and Vegetables to Your Diet

Fruits and vegetables are rich in vitamins, minerals, antioxidants, and fiber because they are nutrient-dense foods. You will naturally feel fuller if you concentrate on consuming the suggested daily intake of at least five servings of fruits and vegetables. Half a cup of raw fruit or vegetables, or a small apple or banana, for example, constitutes a serving. The majority of us should eat twice as much as we do now.

To increase your intake:

- Add antioxidant-rich berries to your favorite breakfast cereal
- Eat a medley of sweet fruit—oranges, mangos, pineapple, grapes—for dessert
- Use a colorful salad instead of your typical side of pasta or rice.
- Snack on veggies with peanut butter or a spicy hummus dip, such as carrots, snow peas, or cherry tomatoes.

6.1.5. How to Make Vegetables Tasty

Although simple salads and steamed vegetables can easily turn boring, there are many ways to infuse your vegetable dishes with flavor.

- **Add Color:** Brighter, deeper-colored vegetables not only have higher vitamin, mineral, and antioxidant concentrations, but they also add visual appeal and flavor variation to meals. Use roasted red cabbage wedges, yellow squash, sun-dried or fresh tomatoes, glazed carrots or beets, or colorful, sweet peppers to add some color.
- **Liven up Salad Greens:** Venture beyond the humble lettuce. Nutrient-dense foods include Chinese cabbage, broccoli, mustard greens, kale, arugula, and spinach. Try drizzling some olive oil over salad greens, adding a spicy dressing, or adding some almond slices, chickpeas, goat cheese, parmesan, or bacon on top.
- Experiment with cooking green beans, broccoli, Brussels sprouts, and asparagus in different ways. Rather than boiling or steaming these nutritious vegetables, consider grilling, roasting, or pan-frying them with chili flakes, garlic, shallots, mushrooms, or onion. You could also try marinating them in zesty lemon or lime before cooking (Skerrett & Willett, 2010; Marx et al., 2017).

6.2. MEAL PLANNING FOR THE FAMILY

Choosing what to eat for each meal of the day is part of meal planning. Every meal manager faces a challenge in providing the family with a healthy and nutritious meal because their well-being depends on it. When executed correctly, it turns out to be a fulfilling and gratifying experience. We eat three meals a day plus possibly one snack. Each family member participates in these meals. To make sure that every family member's needs are satisfied, we must arrange the meals. The plan can be adjusted to suit your family's needs and preferences as well as to take advantage of seasonal food sales at reduced prices. There must be a plan, though. The family can plan meals together as a cooperative project. Family members can collaborate to make the plans work by talking about the meal plans, the food budget, and the actual preparation. The meals will be more accepted and enjoyed as a result of this cooperation.

6.2.1. Objectives in Meal Planning

In planning meals your aim is:

1. To meet each family member's nutritional needs following their age and occupation.
2. Spend no more than your family's food budget allows.
3. To determine how much food should be bought from each food group.
4. To consider family size and composition.
5. To determine how often you need to buy different foods, take into account the space and storage conditions for food.

These extra actions must be done to convert the meal plan into meals that satisfy the needs of the family:

1. Make a grocery list by considering the dietary preferences of the people in your household.
2. Make use of preparation techniques that preserve nutrients without compromising flavor.
3. Serve scrumptious, esthetically pleasing meals that work with the members' schedules.
4. With the assistance of your family, efficiently manage your time, energy, and available materials.

You can save time, effort, and money by planning every meal for the week. It will also assist you in avoiding boring meals.

6.2.2. Nutritional Adequacy

A good meal plan must, first and foremost, provide for the nutritional needs of the entire family. You can accomplish this goal by making food choices based on the daily food guide. So, every meal needs to consist of:

- Cereals and their products: To provide energy and proteins.
- Dals, milk, egg, fish or meat; To supply proteins, B-vitamins and some minerals.
- Protective vegetables and fruits: To provide vitamins A & C, minerals and fiber.
- Sugars, oils and fats: To provide energy, satiety and improve palatability.

A person should be able to meet their needs for protein, vitamins, and minerals if they take the recommended amount of each group. The taste preferences of family members will dictate the variety selected within the group. Dietary recommendations include specific needs for different stages of the life cycle.

By consuming more units from the food groups, that supply these nutrients, these can be easily met. For instance, an adult may only require three of the group two units, but a teenager may require six to meet his needs for growth and maintenance in protein. A member who is overweight may receive a salad without oil seasoning, while other members may receive carrot sticks. An elderly member may receive a carrot salad that has been grate. These are a few doable strategies for tailoring family dinners to each member's specific dietary requirements.

One essential element of life is food. A sizeable portion of the family's financial resources must be set aside to cover their food needs. A low food allowance makes it more important to carefully plan meals and choose the right foods to guarantee that every family member receives an adequate diet.

6.2.3. Food Costs

A person has to keep in mind that the law of supply and demand governs food prices, not the nutritional value of the food. Although the economics of food consumption is a vast topic, individuals must budget their money and plan their meals to get the most out of their purchases. They must first determine how much money they can spend in total on food purchases. Next, they can choose which foods to purchase with the money they have available.

6.2.4. Factors Affecting Food Selection

Self-awareness plays a crucial role in life because it makes it simpler to set objectives and design a plan of action based on our unique requirements. Similarly, we must look at the factors influencing food choices if we hope to fully comprehend why we eat the foods that we do. We can start making plans to make better, healthier decisions once we are aware of these influences. Simultaneously, we are better equipped to steer clear of numerous traps that undermine healthy eating practices.

- **Food Acceptance:** Everybody has certain foods they enjoy and some they don't. But one can develop their taste in food. By sampling the cuisine of the places we visit, we can expand our knowledge of different foods. It makes our new surroundings easier to enjoy. Meal planning can also be made more varied by discovering new ways to use the foods that are readily available.

- **Tradition:** Tradition, whether it be family, national, or regional, has a big impact on what we eat. Many traditional selection methods rely heavily on experience, but occasionally bad practices spread like wildfire and have to be abandoned. It is a good idea to assess customary food choices in the context of current nutritional knowledge. As a result, we can maintain healthy eating habits and modify those that lead to bad choices.
- **Food Misinformation:** Food is a major topic of discussion and is covered in books, magazines, and newspapers. Additionally, a person hears and sees about it in commercials. While a good portion of this information might not be helpful, some of it might be. False information and misconceptions about food are widespread. But ignorance is the root of many of these. However, individuals will be able to use what they learn about food composition to inform their choice of foods. It is their responsibility to disseminate the knowledge they have acquired to help others get rid of myths about food that could be affecting their health.
- **Other Factors Affecting Meal Planning—Skill in Food:** Preparation: A tasty, palatable dinner requires skill in food preparation. Food preparation is a skill that is learned via experience. Creative pairings increase food's acceptability, provide diversity, and ultimately make eating more pleasurable. Certain dark green leafy vegetables, for instance, have a strong flavor that can be adjusted or diluted by adding other, milder foods or ingredients. Since fenugreek leaves have a bitter taste, adding ground dal or besan dilutes the bitterness and leaves behind an enjoyable mild flavor; adding grated coconut or jaggery also improves the flavor of the finished vegetable. This ability is what allows the finished vegetable to have a tolerable flavor. This ability is what elevates a hot soup made at home over a fine dining experience. Skill can be developed by practice and by seeing others who perform well. It's critical to keep honing the skills a person has learned until they become second nature.
- **Variety:** Learning how to cook foods from different parts of the world can improve one's enjoyment of food. Meal monotony is prevented if a new dish is made once a week.

6.2.5. Availability of Foods

Examining the seasonal variations in food availability is crucial. There are some foods from each group in each season. Cost is decreased by using seasonal foods. It is possible to grow some fruits and vegetables in the kitchen garden. Certain foods, such as wheat, rice, and beans, can be stored for the entire year and bought soon after harvest at a fair price.

- **Home Production:** In meal planning, fruits and vegetables that are grown on-site must be included. Certain fresh condiments, like curry leaves, mint, green chilies, coriander leaves, and various leafy vegetables, are simple to grow at home. Meal preparation helps determine how much food the family should realistically keep for consumption at home in rural areas where most staples are produced at home, ensuring that the family's needs are satisfied. Selecting fruits and vegetables that can be grown profitably to enhance the family's eating schedule without raising food costs is also helpful. If more fruits and vegetables are grown than the family consumes, the extra can be sold.

- **Schedules of Family Members:** When organizing a meal plan, it is important to consider the family members' schedules, including meal times and the proportion of meals consumed at home versus those consumed away from home. Menus should be adjusted so that items can be packed and the food tastes good even cold if packed lunches are made for the man of the house and kids at school.
- **Time:** The amount of time allotted for meal preparation and the assistance—whether from children or hired help—can have an impact on the menu and the selection of foods purchased.

6.2.6. Family Size and Composition

The foods that can be served depend on the size of the family. It is well known that, when family income stays the same, the amount of money spent on food per person drops as the size of the family grows. This is the situation in families with multiple children who live close to one another. Given that the family's financial resources remain constant regardless of the size of the family, a greater portion of it must be allocated towards meeting food expenses.

While staples like wheat, rice, and jowar are purchased in greater quantities, purchases of milk, vegetables, and fruits may decline. Dietary quality suffers as a result.

In severe circumstances, it might not be able to provide for the family's food needs, which would lead to partial starvation. Many families experience this when they move to cities in search of better living conditions, so it's critical to highlight the link between a family's size and food intake as well as their health and to support small family norms in the name of better survival and health.

- **Family Composition:** The types and quantities of food required, as well as the timing of meals, are influenced by the makeup of the family. For instance, since they are unable to consume large amounts at once, children under the age of five require more milk and more meals. As a child gets older, their eating schedule adjusts to fit the school schedule, so packing a lunch or snack may become necessary. Families with adolescents require larger amounts of food than adults do because it is necessary for their growth and daily activities. The adult members' dietary requirements will vary based on their level of activity. Due to dental issues or tooth loss, elderly family members may require a change in consistency in their diet, but they still require smaller portions. Festivals often require special menus, so keep that in mind when creating the menu.

6.2.7. Planning Process in Brief

Making monthly plans is imperative. Meal planning involves five key steps:

1. Make a list of all the foods that are available in the market that belong to each food group.
2. Based on the number of family members, determine which foods from each group fit within the food budget by looking at the prices.
3. Using the Daily Food Guide as a guide, estimate each family member's daily food requirements and use that information to compute the monthly food needs.
4. List the foods that need to be bought every day, every week, every fortnight, and every month.
5. Create menus based on the family's daily requirements.

To use up foods that might not be kept, it might be necessary to make some adjustments toward the end of the week. Occasionally, adding more guests may force a menu change. If the housewife is required to prepare meals, time and energy efficiency should be considered when creating the plan. Creating a grocery list in advance saves time when shopping. Certain dishes can be prepared in advance, such as preparing pastries or cake the day before the celebration. The facilities that are available for food preparation must be considered in the plan. Every family's meal plans have some degree of standardization, which facilitates effective meal management (Skerrett & Willett, 2010; Marx et al., 2017).

6.3. MEAL PLANNING FOR VARIOUS AGE GROUPS

Planning meals is a science and an art in and of itself. The homemaker chooses what to cook from the food that is on hand. However, there are a number of factors that affect meal planning, including season, budget, and nutritional needs which they have already studied. These differences exist between families. What was their typical lunch fare—rice, chapati, dal, cooked veggies, salad, curd, and occasionally fruits or desserts—do they recall? Overall, this is a well-balanced meal. Yes, because it combines protein, carbohydrates, and vitamins in appropriate proportions.

6.3.1. Adulthood

When a person reaches adulthood, their physical growth has reached a stable point in their life. The maintenance of bodily functions necessitates nutrition. Adults primarily require energy to maintain bodily processes and physical activity. The purpose of protein is to replenish the losses and wear and tear brought on by everyday bodily functions. As a result, requirements for other stages are discussed in relation to the adult stage, which is considered the norm. Additionally, adulthood is symbolic of the productive phase of life. Consequently, it's critical to meet an adult's nutritional needs to maintain vitality and a positive outlook on life, both of which are necessary for maximum productivity.

6.3.1.1. Changing Food Habits

Everybody has certain foods that they love to eat. These are served during celebratory events like birthdays and parties.

While it is important to consider family members' preferences when meal planning, our diets shouldn't be dictated by our personal tastes. Permitting someone to adopt a restricted eating pattern could be detrimental to their social standing and result in an inadequately balanced diet. Attempting to like new foods broadens one's diet, social circle, and enjoyment of food.

6.3.1.2. Food Selection

The daily food guide is used as a guide when choosing foods. The individual's body size and level of activity will determine how much of each food group is included. As a result, a laborer who works long hours might require more cereals, fats, and oils to meet their energy needs than a sedentary individual who works at a desk. The foods chosen must be used in the day's meals and fit into the person's daily routine.

6.3.1.3. Breakfast is an Important Meal

Since breakfast and dinner are served approximately 12 hours apart, the first meal of the day is crucial. Breakfast should consist of one-fourth to one-third of the food consumed during the day. Typically, rural farming families eat a substantial breakfast. However, the pattern differs from area to area in cities. It is traditional for many urban families to eat fluid and soft foods like rice, dal, salads, etc.) in the meal between 8:30 and 9:00 A.M. and prepare and transport the remaining meal components (chapati, bread, butter, sandwiches, veggies, pickles, etc.) to work in a lunchbox. For this reason, breakfast in the cities is defined as whatever is consumed before heading out of the house for work. Typically, it consists of preparing cereal, dal, fish, or meat, as well as curd or butter milk, vegetables, etc. Many people skip breakfast to eat less, especially housewives and girls. However, research indicates that eating a healthy breakfast improves performance at work and decreases the likelihood of overeating during the day. A person who skips breakfast is more likely to eat more at lunch and dinner or to take snacks, which increases their overall intake of calories. Snacks typically have low protein content and low satiety values. Additional comparative research has demonstrated that people who eat a healthy breakfast produce more work and are more alert. Since many people in cities and villages must pack their own lunches or choose from a small selection at snack counters, it's critical to schedule breakfast in order to provide enough food that is sufficient in terms of both quality and quantity.

6.3.1.4. Lunch Often a Poor Meal

Lunch or the noon meal should account for about one-third of the daily caloric intake. Lunch ought to be considered a meal in and of itself, rather than merely a midday snack to tide over until supper. Not just starchy foods, but also protective foods and protein should be included in a healthy lunch. A prevalent dietary habit among office workers involves consuming any accessible snack, along with bread, butter, and tea from the office canteen or a neighboring cafe. When milk or a milk product, such as a cold milk drink, is consumed in place of tea, this can occasionally get better. All too frequently, the lunch box from home includes chapati, bread, butter, or jam, and a cup of tea or coffee. These foods satisfy energy needs without offering much in the way of protein or protective foods, even though they might be more satisfying than the snacks bought at work. It's crucial to eat a portion of protein-rich foods at lunch. A variety of fruits and vegetables can help satisfy some of the requirements for protective foods.

Lunch should be carefully planned or selected, whether it is packed and brought to work, served at home, or purchased there. It is hard to make up for a meal missed. Neglecting this crucial meal on a regular basis can have a negative impact on a person's attitude toward life, behavior with coworkers, and performance at work. Lunch should be organized following the other requirements so that the foods chosen will provide a reasonable portion of the day's nutrients.

6.3.1.5. Dinner

This is the main meal of the day, and it can be served at night or at noon based on the schedule and traditions of the family. Planning a dinner menu that contributes to balancing the day's total intake of energy, protein, and other nutrients is a good idea. To increase the variety, one could cook an extra vegetable and salad and add a dessert or sweet. While the person who needs to watch his weight may eat more salads and vegetables, the active members may eat more heartily of the cereal dishes. Since the family is eating this meal together slowly, it greatly enhances the sense of community, enjoyment of the company, and relaxation.

6.3.1.6. Snacks

Having a tea break is now a common practice in many workplaces, where employees can enjoy a hot beverage along with snacks such as fried items. This break not only allows for a chance to drink and eat but also offers an opportunity to consume energy and nutrients. It is important to consider the nutritional value of snacks chosen during these breaks, as they can contribute to overall nutrient intake. While strict rules about snacking between meals may not be necessary, it is advisable to opt for items that are nutritious, such as milk, cereal-based dishes, salads, and fruits.

6.3.2. Nutrition During Pregnancy and Lactation

Two life stages in which an adult woman's needs are increased are pregnancy and lactation. During the nine months of pregnancy, she is responsible for providing internal support to the fetus and, later on, external support in the form of nursing the baby. For both the expectant mother and the nursing mother, adequate nutrition is essential because of the critical growth needs at the beginning of life. Several tissues develop to support and shield the developing fetus. The mother must take care of both her own needs and the developing fetus during pregnancy. Further development of related tissues and the accumulation of fat reserves are required to cushion the fetus before birth and to provide some of the energy required for the formation of milk during lactation. Pregnancy thus increases the requirement for all nutrients involved in tissue synthesis. Adolescent moms might require extra food to meet their own growth requirements because they haven't finished growing. Their health may be impacted if these are not met, which could have an indirect impact on the fetus's welfare. The mother might be better able to handle the demands of pregnancy if her diet was sufficient before becoming pregnant. Nobody wants to harm their unborn child's health by eating poorly. Nutritional studies have revealed, however, that a lot of women prioritize the needs of other family members over their own. During pregnancy, nothing changes in the situation. Pregnant women are therefore frequently the family members who eat the least. She does not take the time to sit down and eat because she is too busy taking care of other people. She can't eat when she's extremely exhausted. She is the most impacted if there is a shortage of food because she feeds the other members and consumes what is left over. The family must schedule the baby's arrival to prevent the expectant mother from going without food, both in terms of quantity and quality. To protect the health of the fetus, the expectant father should make an effort to make sure the expectant mother eats the correct kinds and amounts of food.

6.3.2.1. Dietary and Caloric Recommendations

It takes about 300 extra calories per day to keep a pregnancy healthy. A balanced diet rich in protein, fruits, vegetables, and whole grains should provide these calories. Limiting sweets and fats is a good idea. A nutritious, well-balanced diet can also aid in the reduction of certain pregnancy side effects, like nausea and constipation.

6.3.2.2. Fluid Intake During Pregnancy

Another crucial component of pregnancy nutrition is fluid intake. Pay attention to these guidelines when it comes to hydration during pregnancy:

- In addition to the fluids in juices and soups, a person can get adequate fluids by drinking multiple glasses of water each day. Consult a midwife or healthcare professional about limiting their intake of artificial sweeteners and caffeine.
- Avoid all forms of alcohol.

6.3.2.3. Ideal Foods to Eat During Pregnancy

The following foods are good for health and the development of the fetus while pregnant:

- Vegetables: veggies (for potassium and vitamin A): carrots, sweet potatoes, pumpkin, spinach, cooked greens, tomatoes, and red sweet peppers.
- Fruits: oranges, red or pink grapefruit, bananas, apricots, cantaloupe, honeydew, mangoes, prunes, and bananas (for potassium).
- Dairy: skim or 1% milk, fat-free or low-fat yogurt, and soymilk (for vitamins A and D, calcium, and potassium).
- Grains: cereals that are cooked or ready to eat (for folic acid and iron).
- Proteins: Lean beef, lamb, and pork; beans and peas; nuts and seeds; salmon, trout, herring, sardines, and pollock.

6.3.2.4. Foods to Avoid During Pregnancy

It is best to refrain from consuming the following foods while pregnant:

- Unpasteurized milk and foods prepared from it (soft cheeses such as feta, queso blanco and fresco, brie, Camembert, or blue-veined cheeses, unless they are identified as pasteurized milk-based products).
- Luncheon meats and hot dogs (unless they are served hot and steaming).
- Meat, eggs, and seafood that are raw or undercooked. Raw fish sushi should not be consumed; cooked sushi is safe to eat.
- Refrigerated pâté and meat spreads
- Refrigerated smoked seafood

6.3.2.5. Guidelines for Safe Food Handling

When handling and preparing food, abide by these general food safety guidelines:

- Wash. Before eating, chopping, or cooking, thoroughly rinse all raw produce under running water from the tap.
- Clean. Once a person has handled and prepared uncooked food, wash hands, knives, countertops, and cutting boards.
- Cook. Use a food thermometer to ensure that beef, pork, or poultry is cooked to a safe internal temperature.
- Chill. Keep everything perishable cold right away.

6.3.2.6. Prenatal Vitamin and Mineral Supplements

Before or soon after conception, the majority of medical professionals and midwives will recommend a prenatal supplement to ensure that all of the nutritional requirements are satisfied. Prenatal supplements, however, do not take the place of a balanced diet.

6.3.2.6.1. The Importance of Folic Acid

It is recommended by the US Public Health Service that women who are capable of bearing children take 400 micrograms (0.04 mg) of folic acid daily. The nutrient folic acid can be found in:

- Some green leafy vegetables
- The majority of citrus fruits, nuts, beans, berries, and fortified cereals for breakfast.
- Some vitamin supplements.

Birth defects of the brain and spinal cord known as neural tube defects can be prevented in part by taking folic acid. Different degrees of paralysis, incontinence, and occasionally intellectual disability can result from neural tube defects. Since most neural tube defects occur in the first 28 days after conception, folic acid is most beneficial during this time. Unfortunately, it's possible that a woman won't become aware of her pregnancy until 28 days. As a result, she should start taking folic acid before getting pregnant and keep doing so all the way through her pregnancy. Her midwife or doctor will advise her on the right dosage of folic acid based on her unique requirements. For instance, to prevent neural tube defects, women who take anti-epileptic medications may need to take higher doses of folic acid. When thinking about attempting to become pregnant, they ought to speak with their doctor.

6.3.3. Nutrition During Infancy

Making sure kids get enough nutrients to lay a solid foundation for the rest of their lives is one of the most significant parenting tasks. From infancy through adolescence, a child's development is greatly influenced by diet and nutrition. Emotional stability, mental capacity, growth, and immunity are all impacted by a nutritious diet. The significance of solid foods, bottle formulas, and breastfeeding will all be covered in this section.

6.3.3.1. Breastfeeding

All new mothers must make the decision to breastfeed. Family, friends, employers, and other supporters can be a big help when it comes to making decisions during pregnancy and continuing to breastfeed after the baby is born. Approximately 75% of babies in the US are breastfed from birth. However, according to the Centers for Disease Control and Prevention (CDC), only 15% of infants in the United States were still exclusively breastfed at six months of age, when solid foods should be introduced into a child's diet in addition to breast milk.

Healthcare professionals are usually the ones who start breastfeeding education programs. Doctors, nurses, and other healthcare professionals can clarify the advantages of breastfeeding and provide instructions on how to do it during prenatal care and shortly after a woman gives birth. In the US and Canada, hospitals are the site of almost all births. Breastfeeding mothers can benefit from knowledge about hospital labor, delivery, postpartum care, and discharge planning procedures. After a new mother leaves the hospital and returns home, she needs to have access to a trained person who can give her reliable information. Healthcare professionals who have received certification in breastfeeding management, such as registered nurses or dietitians, are known as International Board Certified Lactation Consultants (IBCLCs). They assist new mothers in resolving issues and educating their families about the advantages of breastfeeding. According to research, women who give birth in hospitals that provide IBCLCs to new mothers tend to breastfeed their babies at higher rates than women who give birth in facilities without these professionals on staff.

In addition, partners, spouses, and other family members can be very helpful in guiding a pregnant woman toward breastfeeding and in providing postpartum nursing support. A woman's decision to breastfeed or her capacity to continue the practice can also be influenced by her employment. Mothers who work have historically had lower breastfeeding initiation rates. Compared to new mothers who are unemployed or have extended maternity leaves, they typically breastfeed for a shorter period of time. The United States passed the Affordable Care Act (ACA) in 2010, which required employers to make accommodations for new mothers to pump breast milk in the workplace. This law mandates a clean, private area outside of the restroom for women to express milk during breaks, as well as enough time for breaks.

Communities can also encourage and support breastfeeding among its members. New mothers can seek support from other women in their community who have breastfed before, or they can join peer counseling groups. Additionally, community-based initiatives can offer assistance and education. For low-income families, the US Department of Agriculture's Women, Infants, and Children program offers breastfeeding information. The Loving Support program was introduced in 2004 with the goal of increasing breastfeeding duration rates throughout the United States by combining peer counseling with advocacy initiatives. La Leche League is a global initiative that offers breastfeeding education, support, and encouragement to mothers everywhere.

While it is important to recommend and encourage breastfeeding for new mothers, it is crucial to respect that it is ultimately a personal choice. No woman should be shamed for not being able to or choosing not to breastfeed. In some rare cases, breastfeeding may not be possible or in the best interest of the baby. The nutritional choices made by parents, whether it be breastfeeding or bottle-feeding, have a significant impact on early childhood development and the long-term health and well-being of a child. Therefore, it is essential to promote and support practices that prioritize the well-being of both infants and mothers.

6.3.3.1.1. Components of Breastmilk

In addition to offering babies enough nutrition, human breast milk shields them from harm. Additionally, breast milk has a lot of cholesterol, which is important for brain development. It is beneficial to understand the various forms and ingredients of breastmilk as well as the nutrients they offer for a baby's survival and development. Colostrum is produced right after birth, before the production of milk, and continues for a few days following the baby's arrival. Colostrum is yellowish or creamy in color and thicker than breast milk. An infant's nutritional needs during those early days are met by this protein-rich liquid. Colostrum is a small fluid, but it contains a lot of concentrated nutrition for babies. High in fat-soluble vitamins, minerals, and antibodies known as immunoglobulins, which are transferred from the mother to the child, is this particular milk. Neonatals protected against bacterial and viral infections by immunoglobulins, which offer passive immunity.

A few days after birth, colostrum transitions into transitional milk, which is a rich, creamy liquid that lasts for about two weeks. It contains high levels of fat, lactose, and water-soluble vitamins, as well as more calories than colostrum. When a new mother starts producing transitional milk, she may notice changes in the volume and composition of the liquid and experience an increase in breast size and weight. Mature milk is the final type of milk produced by a new mother, typically starting to be secreted around two weeks after giving birth. During feeding, two types of mature milk are present: foremilk, which is released first and contains water, vitamins, and protein, and hind-milk, which comes later and has higher fat content necessary for baby's weight gain. These different types of milk ensure that the baby receives the proper nutrients needed for healthy growth and development.

Breast milk is made up of approximately 90% water, which keeps babies hydrated. The remaining 10% contains carbohydrates, proteins, and fats, which provide energy and support growth. Lactose is the main carbohydrate in breast milk, similar to cow's milk. Essential fatty acids like DHA and ARA are found in breast milk. Whey is the main protein in breast milk, which is easier for infants to digest than casein found in cow's milk. Breast milk also contains all essential amino acids and proteins like lactoferrin, which helps absorb iron. Vitamins, minerals, and digestive enzymes are also present in breast milk, aiding in digestion. Additionally, breast milk provides hormones and growth factors necessary for a newborn's development.

6.3.3.1.2. The Mother's Diet

A mother's diet has a big influence on the quantity and quality of her milk. To enable breastfeeding, new moms must modify their calorie and fluid intake. Nursing their babies increases the recommended daily allowance (RDA) of almost all vitamins and minerals, just as it does during pregnancy. For the first six months of lactation, the RDA is 330 extra calories, and for the second six months, it is 400 extra calories. The extra energy consumed and fat reserves were used to support nursing. For instance, a lactating mother's daily calorie cost during the first six months after giving birth is 500 calories, of which 330 come from increased intake and 170 from maternal fat stores. This clarifies why breastfeeding could encourage new moms to lose weight. In order to sustain milk production, lactating women should also consume 3–1½ liters of liquids (roughly 13 cups) per day, according to the IOM.

Breastfeeding moms should abstain from cigarettes and illegal substances during pregnancy. It can be helpful to discuss some legal drugs and herbal products with a healthcare provider as they may be harmful. Certain foods that can cause gas in sensitive infants, like spicy foods, may need to be avoided by some mothers. Women who are nursing may consume alcohol, but they must wait to breastfeed until the alcohol has completely left their milk. Depending on a woman's body weight, this usually takes two to three hours for 12 ounces of beer, 5 ounces of wine, or 1.5 ounces of liquor. Because alcohol exposure can throw off an infant's sleep schedule, precautions must be taken.

6.3.3.1.3. Benefits of Breastfeeding

There are many advantages to breastfeeding for both the mother and the child. White blood cells, immunoglobulins, immune factors, and enzymes are all present in breast milk. Breastfeeding therefore boosts the baby's immune system and reduces the risk of respiratory illnesses, gastrointestinal issues, and ear infections in addition to diarrhea. Additionally, breastfeed babies have a lower chance of developing allergies and asthma, as well as a lower risk of SIDS. In addition, human milk promotes the development of beneficial bacteria in a baby's digestive system. After a baby is weaned off of breast milk, these advantages still apply. Other potential long-term effects are suggested by certain studies. Breast milk, for instance, may increase a baby's IQ and shield them from obesity and Type 1 diabetes, though these effects are still being studied. Numerous other important advantages of breastfeeding exist. Breast milk is easier for babies to digest than bottle formula because the latter contains cow's milk proteins that need time for the developing digestive systems of babies. Compared to infants bottle-fed, breastfed babies are less likely to get sick. Breastfeeding produces less garbage and plastic waste, making it a more sustainable practice. Families can also save money by breastfeeding because it eliminates the need to buy formula. The fact that breast milk is always available is another advantage. It doesn't need to be prepared, heated, or mixed. Furthermore, breast milk is always the proper temperature and is sterile. Additionally, nursing fosters a close bond between mother and child, which is a crucial emotional and psychological benefit of skin-to-skin contact. The mother's health benefits from the practice as well. Breastfeeding strengthens a woman's bones and guards against fractures in the future. Additionally, research has demonstrated that breastfeeding lowers the risk of ovarian and breast cancers.

6.3.3.1.4. Barriers to Breastfeeding

While almost all infants benefit greatly from breast milk, nursing moms may encounter certain difficulties in initiating and maintaining breastfeeding. These challenges include uncomfortable breast engorgement or fullness, painful and tender nipples, discomfort or insecurity in public, and accommodations for nursing or milk expression at work.

Acquiring the right technique is among the first challenges faced by nursing mothers. Development and growth can be slowed by improper latching, which can lead to insufficient intake. It could take a new mother some time to teach her infant to properly latch onto her nipples. OB nurses, registered dietitians, and International Board Certified Lactation Consultants (IBCLCs) are qualified to assist new mothers in learning the appropriate technique. Many nursing women and their newborns can benefit from breastfeeding facilitation measures, such as education, extended maternity leave, and laws protecting public breastfeeding.

6.3.3.1.5. Contraindications to Breastfeeding

While breastfeeding has many advantages, there are also risks that need to be taken into account in certain situations. In the developed world, a new mother who has HIV shouldn't breastfeed her child because the virus can spread through breast milk. These women can substitute breast milk with a safe bottled formula that is usually available. However, it can be difficult to find acceptable infant formula in developing countries with high rates of HIV infection. A lot of newborns wouldn't get the nutrients they require to grow and develop. Moreover, improper or tainted bottle formulas result in 1.5 million infant deaths annually. In order to reduce the risk of transmission, the WHO advises HIV-positive women in developing countries to breastfeed their children while taking antiretroviral drugs.

It is also not advised for women receiving chemotherapy or radiation therapy for cancer to breastfeed. In addition, a baby diagnosed with galactosemia—a condition in which the body is unable to process the simple sugar galactose—needs to be fed a diet free of galactose, which includes no breast milk. Still, only one in 30 to 60,000 newborns are affected by this rare genetic disorder. When it's not safe to breastfeed, parents and other caregivers can still provide for the newborn's nutritional needs by giving them formula.

6.3.3.2. Nutritional Requirements

During infancy, the body needs more macronutrients and micronutrients per kilogram than at any other point in the human life cycle. These requirements are impacted by the fast cell division that occurs during growth, which calls for nutrients for DNA synthesis as well as energy and protein. An adult's resting metabolic rate is half that of a baby. During this time, children are totally dependent on their parents or other caregivers to meet their needs. For almost all infants six months or younger, breast milk is the best source of nutrients. An infant may initially need to be fed eight or 12 times a day, if not more. Infants can progressively eat solid foods after six months to help meet their nutritional needs.

6.3.3.2.1. Energy and Macronutrients

The nutritional value of human breast milk serves as the foundation for infant dietary recommendations. About 45–65% of the calories in breast milk come from carbohydrates, meaning that an RDA of 130 grams is involved. Lactose, which newborns easily digest and tolerate, makes up the majority of the carbohydrates in human milk. Infants almost never experience lactose intolerance. 13 grams of breast milk a day, or 5–20% of total calories, are made up of protein. Protein is essential to an infant's growth and development. On the other hand, too much protein (which is only an issue when nursing) can result in dehydration, diarrhea, fever, and acidosis in premature babies. Breast milk contains between 30–40% fat in its calories. To promote neural pathways in the brain and other body parts, a high-fat diet is required. Saturated fats and trans fatty acids, however, prevent this growth. When a baby reaches six months old, it is no longer breastfed exclusively, so foods high in these fats should not be given to them.

6.3.3.2.2. Micronutrients

If babies get enough breast milk, they can get almost all the nutrients they need. However, there are a few outliers. Vitamin D, which is essential for building bone and absorbing calcium, is scarce in human milk. For this reason, vitamin D supplements in the form of drops are frequently required for breastfed children.

The babies who are most at risk of not getting enough vitamin D have darker skin and are not exposed to sunlight. Additionally, low levels of vitamin K, which is necessary for blood clotting and can cause bleeding or hemorrhagic disease, can be found in breast milk. Due to low vitamin K levels at birth, babies may need to be supplemented initially. In certain states, babies may also need to receive a vitamin K injection after delivery. Furthermore, babies absorb the iron in breast milk even though it is not a high concentration in it. Nevertheless, an infant requires a different source of iron than breast milk after four to six months.

6.3.3.2.3. Fluids

Compared to adults, infants require 10.5 milliliters of fluid for every calorie consumed, while adults only need 1.0 milliliters. This is because children's capacity for perspiration is lower and their body surface area is greater per unit of body weight. As a result, their chance of becoming dehydrated is higher. Breast milk or formula, however, can satisfy an infant's fluid requirements when provided by parents or other caregivers. When solids are introduced, parents need to make sure that young children stay hydrated all day long.

6.3.3.3. Bottle-Feeding

When given enough information and assistance, the majority of women can and should breastfeed. As was previously mentioned, some women decide not to breastfeed their children, while a tiny percentage are unable to do so.

A balanced nutritional intake is offered by infant formula to parents who opt for bottle feeding. But not all formulas are created equal, and parents and other caregivers need to take some crucial factors into account. The base for standard formulas is cow's milk.

Like breast milk, they have 20 calories per fluid ounce plus additional vitamins and minerals.

Infants with cow's milk protein allergies or those who experience diarrhea, constipation, vomiting, colic, or abdominal pain are typically given soy-based formulas. Infants allergic to soy and cow's milk are typically given hypoallergenic protein hydrolysate formulas. Hydrolyzed protein, which is divided into amino acids and tiny peptides to facilitate digestion, is used in this kind of formula. If breast milk is not available, low birth weight babies are given preterm infant formulas. Preterm infant formulas are fed until the infant reaches the target weight; each fluid ounce contains 24 calories.

Infant formula comes in three basic types:

1. Powder that needs to be combined with water. This kind of formula is the least costly.
2. Concentrates are liquids that need water to be diluted in them. This variety costs a little bit more.
3. Ready-to-use liquids that fit into bottles right out of the bottle. This kind of formula is the priciest. It does, however, require the least amount of planning. Traveling with ready-to-use formulas is also convenient.

The average baby requires 2.5 ounces of formula for every pound of body weight each day. Consequently, a baby should drink roughly 24 fluid ounces of breastmilk or formula each day on average. Due to their immature immune systems, parents and other caregivers should carefully follow the safety instructions when making formula for infants. Sterilize every piece of equipment used to prepare formula. To stop bacterial growth, a prepared, new formula needs to be refrigerated. To protect their children from foodborne illnesses, parents should never mix formula with tainted water. Pay close attention to the directions for the powdered and concentrated formula—a formula that is too thin won't contain enough calories or protein. A highly concentrated formula that gives too much protein and not enough water at the same time damages kidney function. It is noteworthy that the World Health Organization and the American Academy of Pediatrics both agree that breast milk is far superior to infant formula.

6.3.3.4. Introducing Solid Foods

The World Health Organization states that during the first six months of life, infants should only be breastfed or bottle-fed. Breast milk or bottle formula should be the only source of nutrition for at least the first four months, and ideally for six months, according to the American Academy of Pediatrics. Solid foods don't contain the proper nutrient mix for babies, so they shouldn't be given to them before they are six months old. Eating solids may also require consuming less bottled formula or breast milk. An infant might not ingest the appropriate amounts of different nutrients if that happens. An extrusion reflex occurs when parents attempt to feed a baby who is too young or not ready. This happens because the baby's tongue pushes the food out. After six months, eating solid foods becomes more feasible because the suck-swallow reflexes are weaker and the infant can move and hold up their heads.

Solid baby food can be made at home with a food processor, blender, food mill, or grinder, or it can be purchased commercially. Infants can chew soft foods and eat well-chopped or mashed solids by the time they are nine months to a year old. Early introduction of solid foods to infants increases the risk of food allergies. In order to detect allergic reactions or food intolerances, parents and other caregivers should only introduce one new food to their child at a time when introducing solids. At this point, taking an iron supplement is also advised.

6.3.3.5. Learning to Self-Feed

When solid foods are introduced, young children learn how to manage food and take care of themselves. Infants can pick up objects with their entire hand as early as six or seven months old (this is called the palmer grasp). While they can lift larger objects, it is challenging for them to pick up tiny food fragments. A child as young as eight months old may use a pincer grasp, which involves using fingers to pick up objects. After one, kids gradually start handling their food with utensils. Because very young children may play with or throw dishes and cups when they get bored with their food, it's important to have breakable items.

6.3.3.6. Food Allergies

In America 4–6% of young children suffer from food allergies. Cow's milk, eggs, shellfish, peanuts, and wheat are common food allergens. Breastfeeding women shouldn't alter their diets, though. According to research, nursing moms who try to prevent allergies in their babies by avoiding particular foods may end up doing more harm than good. The American Academy of Allergy, Asthma, and Immunology reports that mothers who abstained from specific dairy products had lower levels of an immunoglobulin specific to cow's milk in their breast milk. It is believed that this antibody shields kids from developing allergies. There is no proof that changing a mother's diet helps, not even in cases where her child is more susceptible to food allergies. Maintaining breastfeeding until solid foods are introduced to the infant's diet could be a preventative measure. As of right now, there is no scientific proof that introducing solid foods earlier—before four months of age—or later—after six months of age is a preventive measure. There is proof, though, that introducing solid foods after 17 weeks of age lowers the likelihood that a child will experience dietary allergies.

6.3.3.7. Early Childhood Dental Caries

Bottle-fed breast milk, formula, juice, and other drinks can cause early childhood caries, which is a disorder that affects primary teeth. The types of liquids given to an infant, as well as their frequency and duration of administration, are the primary causes of early childhood caries. A baby's mouth can become overflowing with liquids, and the natural or added sugars cause decay. Early childhood caries can also result from giving a child multiple bottles of juice or other sweet drinks throughout the day or from allowing a baby to suck on a bottle while they are awake or asleep for longer than mealtime. Moreover, the development and position of the jaw and teeth are impacted by this practice. The risk of early childhood caries persists into toddlerhood as kids eat more high-sugar foods. As a result, parents ought to refrain from providing their kids with sugar-filled snacks and drinks (Skerrett & Willett, 2010; Marx et al., 2017).

6.3.4. Meal Planning for School Children and Adolescents

When a child enters school, the parents' focus shifts to their child's coursework, making eating less important. As a result, school-age children's nutritional needs are frequently disregarded. When a child is well-nourished when they start school, the effects of neglect don't always become apparent right away. With the exception of a slow increase in food intake to meet the demands of the growing body, nutritional needs do not significantly change during the school year. However, attending school does require that activities and meals be scheduled to coincide with the academic calendar. One meal must be eaten away from home; this meal may be a box lunch or snacks purchased at school.

6.3.4.1. Food Needs of the School Child — 6–12. Years Old

At this point, it's critical to make sure that every meal contains enough protein, minerals, and vitamins. The amount of food required increases in proportion to the child's activity level and rate of growth. While the types of food required are the same as for younger children, the quantities required are higher.

With the exception of strongly spiced or fried foods, strong tea, and coffee, most foods served at family meals are acceptable to children of this age. At this age, the child is allowed to eat any fruit, even ones with seeds, like grapes and guavas, which were inappropriate for a younger child. He can eat many foods, which need biting and chewing such as popped cereals, roasted groundnuts, roasted corn on the cob, amla, bor, etc.

6.3.4.1.1. Spacing and Time of Meals

Typically, children require four or five meals. Encourage the youngster to wake up approximately 30 minutes before breakfast. The meal might take 20 minutes or so. Make sure the child has enough time to get ready for school by giving them approximately 30 minutes between breakfast and school time. The ideal window of time for lunch is between breakfast and dinner. A midmorning snack and a midafternoon snack are necessary for certain kids.

6.3.4.1.2. Breakfast the Neglected Meal

Either a small amount or none at all is eaten at first because of the need to get used to the school schedule. From the perspective of nutrition, this poses a risk. It's challenging to make up for a missed meal. A child's poor breakfast consumption can be attributed to a variety of factors. The child might not be able to eat breakfast and get to school in time if it is not ready in time. The breakfast may be rushed if the child doesn't wake up in time. It will be challenging for the child to wake up early enough to find time to prepare for school and have a healthy breakfast if he doesn't go to bed on time. If the child eats a poor breakfast, it may affect his ability to attend classes and learn his lessons during school hours.

6.3.4.1.3. Lunch

At this point, the child may have to eat this meal away from home, which would be a significant adjustment. Even when a child eats away from home, making sure they are well-fed takes a lot of preparation and supervision. The following items need to be included in the lunchbox: (a) cereal preparation; (b) pulse or milk product; and (c) vegetable or fruit preparation. A person can prepare a single dish that incorporates items from each of the three categories and may or may not have visible or hidden ingredients like sugar, ghee, oil, butter, or jaggery.

They can see that there are a plethora of additional variations that can be created using foods from the designated groups. Children's lunchtime enjoyment is enhanced when they eat with their friends and share it with them at school. Such sharing could increase the variety and exposure to food for them. When preparing a tasty packed lunch, there are a few things to keep in mind. They are as follows:

- To prevent spoiling, cool the preparation before packing.
- Food preparations that keep well in the refrigerator should be packed.
- To help moisten the mouth, include some fresh fruit or vegetables.
- To preserve their fresh flavor and texture, place the vegetable or fruit slices in a plastic bag.
- To help kids appreciate food and not waste it, only pack as much as they can eat contentedly.

6.3.4.1.4. Between Meal Snacks

A midmorning and a midafternoon snack might be necessary for certain kids. A variety of foods are typically available, including fruits like bananas and guavas roasted legumes, like cashews and groundnuts, etc. In addition to the snacks offered in the school canteen, these are accessible close to the school. These snacks can offer significant nutrients in addition to calories if they are carefully selected. When kids purchase candies, peppermints, soft drinks, or frozen fruit, they are primarily purchasing calories. These foods don't satisfy the body's needs; instead, they lessen or dull the appetite. These foods only provide energy. Consequently, it is claimed that these foods only offer "empty calories."

6.3.4.2. The Adolescent—12–16 Years

This is a phase of high activity and quick growth. In this age group, individual variation is also quite high. This stage of life is marked by a variety of physical, emotional, and mental transformations. Girls mature between the ages of 11 and 13, while boys go through their biggest changes between the ages of 13 and 15. Boys typically overeat at this age, especially if they enjoy outdoor sports. It is imperative that their nutritional needs be met with careful planning. As a result, it's critical to ensure that each meal includes enough proteins, minerals, and vitamins. The age of group activities is adolescence. Consequently, teaching nutrition as a group activity may aid in the improvement of eating habits.

6.3.4.3. Meeting the Food Needs of Adolescents

If more items from the daily guide are incorporated into the diet, the higher requirements of this time can be satisfied.

Boys might need to eat a lot of foods high in energy to meet their needs. Foods high in protein, iron, and other nutrients required for the synthesis and regeneration of red blood cells may require girls to pay particular attention to. All the foods in the food guide should be a part of the girl's diet, but she should pay particular attention to foods high in iron, like dals, leafy green vegetables, dried fruits, eggs, liver, and red meats, if they are allowed. Adolescents should gain weight that is appropriate for their height and body type. Any departure from the norm suggests a feeding issue that needs to be diagnosed and resolved with a dietician or nutritionist's assistance. Examining a three-day food journal can assist in pinpointing the precise deficiency or surplus, which can then serve as the foundation for a strategy.

6.3.5. Diet During 17–21 Years

At this stage of life, growth is typically slow. However, there is a certain level of tissue maturation that is supported by a healthy diet. When staying in hostels, boys and girls often try to cut back on their food consumption, particularly if the food costs are based on the number of meals consumed. This age group can find lodging and food at a reasonable cost from many establishments (colleges and universities); the government subsidizes a portion of the establishment's costs to lower the cost of food for college students. This group needs to be made aware of the connection between dietary intake and the body's need to store nutrients to be ready for adult responsibilities.

6.3.6. Nutrition for Middle Age Adults (40–60 yrs.)

The middle years are frequently a stable period of life. Typically, the individual is productive in their line of work, has a stable salary, and is no longer physically tasked with raising children. After age 30, the basal metabolic rate begins to decline. As a result, there is less need for energy, but there is still a need for other nutrients (proteins, minerals, and vitamins). Adults choose their meals following the Daily Food Guide. The amount of servings from each group will vary based on the individual's activity level and body size. Generally speaking, cutting back on sugar, sweets, oils, and fats will help to reduce energy intake while still maintaining a healthy body weight. To maintain optimal health, a reasonable exercise regimen that includes regular activities like yoga and walking is recommended.

6.3.7. Nutrition for Senior Citizens (60 + Years)

Individual differences exist in the rate of aging. Aging is influenced by several factors, such as genetics, psychosocial background, nutritional status, and lifestyle choices. No two people age the same way because of these differences. One encouraging finding is that those who maintain an active lifestyle both physically and mentally have been known to maintain their sense of taste and smell well into their eighties. Weightlifters are able to maintain the density of their bones as well as their lean muscle mass. When teeth are lost, eating a variety of textures of food is made possible by properly fitting dentures. Those with hearing impairments can continue to interact socially by wearing hearing aids. Following cataract surgery, lens implants allow for reading and unrestricted mobility, guaranteeing the capacity to pursue intellectual pursuits. Seniors are traveling and engaging in social and community activities at a higher rate than they were before the aforementioned technological advancements. Maintaining physical well-being throughout life is facilitated by healthy eating habits. Maintaining a regular fitness regimen of one's choosing helps to stop physical decline. Engaging in enjoyable social activities contributes to mental well-being. A positive outlook, consistent prayer and gratitude practices, and gratitude are assets that guarantee happiness in all ages, including old age. Make an effort to lead a life that requires less prescription drug usage. Attempt a quick nap or meditation if an individual is feeling drowsy rather than consuming stimulants like tea or coffee. Try getting enough exercise, drinking plenty of water, and consuming fiber on a regular basis if you have constipation.

6.3.7.1. Nutritional Requirements

All life long, normal nutritional needs persist. In general, a healthy diet can elevate mood, enhance quality of life, hasten healing from sickness, and even extend life. The degree to which aging has impacted a person's physiological capacities must be understood. All nutritional requirements remain unchanged, except for the need for energy resulting from a drop in metabolic rate. The average protein requirement per kilogram stays between 0.8 and 1.0. Except for a minor decline in the need for thiamin, riboflavin, and niacin, which is proportionate to the total energy intake, vitamin requirements do not change after the age of 50.

6.3.7.2. Food and Other Habits of Senior Citizens

Some dietary practices among the elderly population are malnutritious. These include eating an excessive amount of potatoes and sago on specific weekdays, like Monday, Tuesday, Thursday, Friday, or Saturday, in place of a meal. Some people choose to observe fasts by limiting their intake to tea, coffee, buttermilk, lemon sherbet, and so on. Meal consumption may be decreased by loneliness. Snacks like toast can be used in place of proper meals. Forgetting could lead them to believe they have eaten when they haven't. They occasionally ask for meals after they've already eaten, which annoys caregivers and family. To avoid having to use the restroom at night, they might cut back on their water consumption. The end effect is constipation. Reducing movement to prevent slips and falls has negative effects on digestion, circulation, and muscle tone. To meet the recommended allowances, a multivitamin tablet or capsule must be taken if the diet intake is lowered to less than 1,500 kcal per day. Consuming a range of foods high in nutrients ought to be promoted. Drinking enough water is essential to avoiding dehydration, maintaining urine volume, and preventing constipation. Food intake of vegetables and proteins may be reduced by missing or decaying teeth. Soups that are high in protein and vegetables can be used in these situations. Any changes to eating habits must be carefully considered and discussed with the individual in question.

6.3.7.3. Food Habits and the Elderly

The elderly are attention-seekers who are frequently tricked into purchasing special foods and needless dietary supplements. The salesperson convinces them to purchase pricey food supplements and specialty foods by acting as though they are concerned about their health. As a result, they spend money that could be spent on common foods that are suggested by the food guide and could offer more nutrients for less money. To put it briefly, a basic plan for promoting health and preventing disease involves eating a balanced diet that emphasizes eating a variety of foods in moderation, exercising frequently, not smoking, consuming less alcohol, and managing stress (Skerrett & Willett, 2010; Marx et al., 2017).

6.4. SPECIAL DIETS

Medical and nutrition experts use the term "special diet" to describe what kinds of foods or nutrients a person should include in their diet—or avoid—to promote healing, recovery, or improved health. Many people are first introduced to a special diet while they are hospitalized, where they may be given instructions to follow a clear liquids or soft foods diet, for example. However, special diets are not limited to the confines of hospitals; many people either try them or are curious about how they might benefit from them.

6.4.1. A List of Popular Special Diets

Although there are hundreds of unique diets, the following are some of the most well-liked ones at the moment. Every single one bans or restricts a particular food or food group.

- **Liquid Diets:** Liquid diets should only be used temporarily because they don't offer the body enough nutrients on a regular basis. In these kinds of situations, liquid diets are sometimes required.
  - After heart attacks
  - After certain kinds of surgery
  - With patients who have digestive problems
  - With patients who have acute infections
  - Before certain X-rays of the digestive tract
- **Soft Diets:** Foods that are easily chewed and digested are included in soft diets. For instance, foods that are hard to chew or digest, like nuts, shellfish, and spicy foods, are usually off limits in this kind of diet. Soft diets are commonly employed in the following circumstances:
  - After certain kinds of surgery
  - With patients who have infections
  - With patients who have digestive or chewing difficulty
- **Diabetic Diets:** A few adjustments are necessary for diabetic patients to guarantee that their health reaches its peak. The primary focus of this kind of diet is to stay away from foods that are high in sugar. Controlling the amount of refined carbohydrates in the diet is crucial, though, as they can have a hazardous effect on a diabetic's blood sugar level due to their rise and crash effect.
- **Low-Calorie Diets:** Diets low in calories are designed for patients who need to lose weight naturally or who are pursuing bariatric treatment. Combating obesity with the right diet is crucial for patient health because it is a major cause of

many health issues. Limiting high-calorie foods like butter, cream, and soft drinks is how these diets are carried out.

- **High-Calorie Diets:** Patients who are underweight for a variety of reasons, such as eating disorders, specific cancers, and hyperthyroidism, often require a high-calorie diet. High-calorie diets necessitate including more carbohydrates in addition to proteins. Contrary to popular belief, foods high in fat and/or cholesterol should not be included in a high-calorie diet because they digest slowly and suppress appetite.
- **Low-Cholesterol Diets:** Consuming foods high in saturated fats must be avoided when following a low-cholesterol diet. These foods include, among other things, cheese, eggs, and pork. Patients with atherosclerosis and heart disease need to follow this kind of diet to keep their hearts healthy.
- **Low-Sodium Diets:** Because sodium causes the body to retain water, it makes the heart work harder. Thus, a low-sodium diet is also necessary for patients with heart issues. In addition to avoiding salt addition, smoked meats, processed foods, and pickled foods should also be avoided when following a low-sodium diet. Individuals who suffer from the following ailments will probably have to eat low in sodium:
  - Hypertension
  - Heart disease
  - Edema
  - Kidney disease
- **High-Protein Diets:** All it takes to have a high-protein diet is to incorporate a lot of foods high in protein into a regular diet. Good sources of protein include foods like cheese, eggs, fish, and meat. A diet high in protein may be necessary for the following kinds of patients.
  - Children with difficulty growing
  - Pregnant/lactating women
  - Patients before/after certain types of surgery
  - Patients with acute burns
  - Patients with fever
  - Patients with infections
- **Low-Protein Diets:** Under certain circumstances, a diet that limits high-protein foods like meat, fish, or cheese may be necessary.
  - Patients who have certain allergies
  - Patients with certain types of kidney disease
- **Low-Residue Diets:** Avoiding high-fiber foods like raw fruits and vegetables, cereals, and nuts is known as a low-residue diet. Usually, this kind of diet is employed for:
  - Patients with colitis
  - Patients with diarrhea
  - Patients with other digestive or rectal disorders

6.4.2. Reasons People Follow a Special Diet

People may find themselves in situations where they are forced to adhere to a particular diet. For example, a person with lactose intolerance might require a dairy-free or minimally dairy-containing diet to be well. Another illustration would be someone who has a potentially fatal nut allergy and needs to adhere to a special diet that excludes all nuts and products made from nuts.

At times, people may discover that their dietary preferences have evolved over time or that they must adjust due to changes in their stage of life. For the benefit of her own health as well as the health of the developing fetus or nursing child, a pregnant or nursing woman may adhere to a special diet. A person may require or choose to adhere to a special diet while they heal from surgery or an illness. Additionally, a lot of people adhere to special diets to feel better than they do at the moment. Many people choose to experiment with special diets on their own, even though it is always best to work with a registered dietitian nutritionist who can guide them through special diets and help to decide which one is right for them (Skerrett & Willett, 2010; Marx et al., 2017).

CASE STUDY

DIABETES-NUTRITION-PHYSICAL ACTIVITY

Julie is a 55 year old woman with mild mental retardation. She lives in an independent support living arrangement and she works at a bakery in a supported employment arrangement. She weighs 205 pounds and was recently diagnosed with diabetes. Her doctor recommended that she reduce the amount of sugar in her diet, lose weight and monitor her blood sugar five times a day.

Julie attended a six-week course recommended by her doctor on diabetes in which she was instructed on how to monitor her blood sugar. Julie's goal is to monitor her own blood sugar twice a day and report it on a chart that is monitored by her residential staff. However, when the staff go to visit her, they usually find that there is nothing on the chart. Her staff usually responds by going over the steps for checking her blood sugar with her. They then have her check it and record it while they are there. Several times, Julie has become angry and thrown the glucometer across the room. On some of these occasions, her staff has had to leave without her blood sugar being checked. Frequently, Julie's blood sugar is too high when she checks it with her staff present.

In the six-week course, Julie also learned how to administer her own insulin shots. At first, Julie did not like the idea of giving herself shots, but she has learned how to it. Occasionally, Julie forgets to take her insulin when she needs it. She is supposed to take an insulin shot before every meal.

Julie's favorite food is chocolate. She likes to buy a chocolate donut or an éclair during her work break. After work, she likes to walk down the street to the convenience store and buy a Coke. On the weekends she likes McDonald's for a Chicken McNuggets meal deal. Julie has always picked out her own food at the grocery store. She really likes potato chips and ice cream. She buys these items every week.

Julie's doctor has recommended that she lose weight. If she did lose weight, she might have better control over her diabetes. Julie says that she is interested in losing weight, but has not yet made any efforts to change her diet. Her support staff thinks that she may not know what foods contribute to maintaining her high weight and what type of changes would help her reduce her weight.

Julie has never been involved in regular exercise. Her only regular exercise is her walk down the street to the convenience store and her two-block walk from the bus stop to the bakery where she works. One of the activities she does like is going to parks. Anything that involves being outside seems to be fun for her.

- What are the behaviors that you want changed?
- What health promotion actions would help?

CLASS ACTIVITY

MEAL FOR PREGNANT WOMEN

Visit a pregnant woman. Record the following–

Name -

Age -

No. of children -

Any specific information related to pregnancy

Food Eaten	Nutrients Present	Suggestions for Improvement

SUMMARY

- Healthy eating is not about strict limitations, staying unrealistically thin, or depriving yourself of the foods you love. Rather, it's about eating well-balanced meals that leave you feeling great, having more energy, improving your health, and boosting your mood.
- While some extreme diets may suggest otherwise, we all need a balance of protein, fat, carbohydrates, fiber, vitamins, and minerals in our diets to sustain a healthy body and mind.
- Moderation means eating only as much food as your body needs. You should feel satisfied at the end of a meal, but not stuffed. For many of us, moderation means eating less than we do now. But it doesn't mean eliminating the foods you love.
- Fruit and vegetables are nutrient dense, which means they are packed with vitamins, minerals, antioxidants, and fiber.
- Meal Planning involves deciding what to eat day at each meal. As the family's well-being and health are dependent on how well they are fed, it is a challenge to every meal manager to meet it. It proves to be a satisfying and rewarding experience, when done well. Each day we eat three meals and perhaps an additional snack.
- Self-awareness is important because when we understand ourselves better, it is easier to set goals and create an action plan according to our own individual needs. Similarly, in order to fully understand why we eat the foods we eat, we need to examine the factors that influence food choices.
- The family size affects the foods that can be served. It is known that the money spent for food per person decreases as the family size increases when the family income remains constant.
- Adulthood represents the steady state in life when a person has completed his/her growth in terms of body size. The nutritional needs are for the maintenance of body functions. The energy needs in adulthood are mainly to sustain body functions and activity. The protein requirement is to be made good the wear and tear and the losses, which occur as a result of normal life processes.
- Pregnancy and lactation are two stages of life when an adult woman's needs are increased. She has the responsibility of supporting the growth of the fetus internally during the nine months of pregnancy and later externally by nursing the infant.
- Diet and nutrition significantly impact a child's development from infancy into the adolescent years. A healthy diet affects growth and immunity, intellectual capabilities, and emotional well-being.
- Good nutrition can speed recovery from illness and generally can improve the spirit and quality of life and even prolong life. It is necessary to know the extent to which aging has affected a person's physiological capabilities. Except for energy needs, due to a decrease in the metabolic rate, the need for all the nutrients remains the same as before.

- The phrase "special diet" is a term used by medical and nutrition professionals to indicate the type of foods or nutrients that someone should add or eliminate to their diet to help them heal, recover, or feel better.

REVIEW QUESTIONS

1. What do you mean by meal planning?
2. What is the importance of meal planning?
3. How do you plan an adequate diet?
4. What are the essentials of an adequate diet?
5. Why a balanced diet?
6. What are the factors influencing food selection?
7. List the objectives in planning meals. How can you ensure nutritional adequacy in meals?
8. Discuss how to meet the nutritional needs at minimum food costs.
9. What are the non-nutritional factors, which affect meal planning? Discuss.
10. List the factors you will keep in mind while making a tiffin for school-going children.
11. Nutrition During Infancy

REFERENCES

1. Blatt, A. D., Roe, L. S., & Rolls, B. J. (2011). Increasing the protein content of meals and its effect on daily energy intake. *Journal of the American Dietetic Association, 111*(2), 290–294. https://doi.org/10.1016/j.jada.2010.10.047.
2. Conner, T. S., Brookie, K. L., Carr, A. C., Mainvil, L. A., & Vissers, M. C. M. (2017). Let them eat fruit! The effect of fruit and vegetable consumption on psychological well-being in young adults: A randomized controlled trial. *PLOS ONE, 12*(2), e0171206. https://doi.org/10.1371/journal.pone.0171206.
3. Hu, F. B., Manson, J. E., & Willett, W. C. (2001). Types of dietary fat and risk of coronary heart disease: A critical review. *Journal of the American College of Nutrition, 20*(1), 5–19. https://doi.org/10.1080/07315724.2001.10719008.
4. Hu, F. B., Stampfer, M. J., Manson, J. E., Rimm, E., Colditz, G. A., Rosner, B. A., Hennekens, C. H., & Willett, W. C. (1997). Dietary fat intake and the risk of coronary heart disease in women. *New England Journal of Medicine, 337*(21), 1491–1499. https://doi.org/10.1056/NEJM199711203372102.
5. Jakobsen, M. U., Dethlefsen, C., Joensen, A. M., Stegger, J., Tjønneland, A., Schmidt, E. B., & Overvad, K. (2010). Intake of carbohydrates compared with intake of saturated fatty acids and risk of myocardial infarction: Importance of the glycemic index. *The American Journal of Clinical Nutrition, 91*(6), 1764–1768. https://doi.org/10.3945/ajcn.2009.29099.

6. Krok-Schoen, J. L., Price, A. A., Luo, M., Kelly, O. J., & Taylor, C. A. (2019). Low dietary protein intakes and associated dietary patterns and functional limitations in an aging population: A NHANES analysis. *The Journal of Nutrition, Health & Aging, 23*(4), 338–347. https://doi.org/10.1007/s12603-019-1174-1.
7. Marx, W., Moseley, G., Berk, M., & Jacka, F. (2017). Nutritional psychiatry: The present state of the evidence. *Proceedings of the Nutrition Society, 76*(4), 427–436. https://doi.org/10.1017/S0029665117002026.
8. Masana, F., Tyrovolas, S., Kollia, N., Chrysohoou, C., Skoumas, J., Haro, J. M., Tousoulis, D., Papageorgiou, C., Pitsavos, C., & Panagiotakos, D. B. (2019). Dietary patterns and their association with anxiety symptoms among older adults: The ATTICA study. *Nutrients, 11*(6), 1250. https://doi.org/10.3390/nu11061250.
9. Morris, M. C., Tangney, C. C., Wang, Y., Sacks, F. M., Barnes, L. L., Bennett, D. A., & Aggarwal, N. T. (2015). MIND diet slows cognitive decline with aging. *Alzheimer's & Dementia: The Journal of the Alzheimer's Association, 11*(9), 1015–1022. https://doi.org/10.1016/j.jalz.2015.04.011.
10. Siri-Tarino, P. W., Sun, Q., Hu, F. B., & Krauss, R. M. (2010). Saturated fatty acids and risk of coronary heart disease: Modulation by replacement nutrients. *Current Atherosclerosis Reports, 12*(6), 384–390. https://doi.org/10.1007/s11883-010-0131-6.
11. Skerrett, P. J., & Willett, W. C. (2010). Essentials of healthy eating: A guide. *Journal of Midwifery & Women's Health, 55*(6), 492–501. https://doi.org/10.1016/j.jmwh.2010.06.019.
12. Veronese, N., Solmi, M., Caruso, M. G., Giannelli, G., Osella, A. R., Evangelou, E., Maggi, S., Fontana, L., Stubbs, B., & Tzoulaki, I. (2018). Dietary fiber and health outcomes: An umbrella review of systematic reviews and meta-analyses. *The American Journal of Clinical Nutrition, 107*(3), 436–444. https://doi.org/10.1093/ajcn/nqx082.
13. Vujovi⊠, N., Piron, M. J., Qian, J., Chellappa, S. L., Nedeltcheva, A., Barr, D., Heng, S. W., et al. (2022). Late isocaloric eating increases hunger, decreases energy expenditure, and modifies metabolic pathways in adults with overweight and obesity. *Cell Metabolism, 34*(10), 1486–1498.e7. https://doi.org/10.1016/j.cmet.2022.09.007.

CHAPTER 7

Effects of Food Additives on Health

LEARNING OBJECTIVES

After studying this chapter, you will be able to:

- Understand food additives and preservatives
- Know the health impact of food additives
- Identify common food additives to avoid
- Focus on regulation of use of food additives
- Perform safety evaluation of food additives
- Discuss recent developments and trends in the food sector.

KEY TERMS FROM THIS CHAPTER

Artificial food colors	Artificial sweeteners
Ascorbic acid	Bisphenols
Carrageenan	Consumers
Contamination	Flavor
Food additives	Guar gum
Indirect food additives	Monosodium glutamate
Nitrates	Perchlorate
Phthalate	Preservatives
Shelf life	Sodium benzoate

INTRODUCTION

Chemicals known as food additives and preservatives are added to food to raise its quality and lengthen its shelf life. Food additives are used to improve the food's flavor, texture, or color. For instance, many processed foods contain both sugar and salt as additives. Chemicals known as preservatives prolong the shelf life of food. For instance, ascorbic acid, or vitamin C, is added to some canned fruits and vegetables to prevent them from turning brown, and nitrates are used to prevent processed meats from spoiling. When ingested in excess over time, preservatives and additives can both be harmful to one's health.

7.1. UNDERSTANDING FOOD ADDITIVES AND PRESERVATIVES

Over time, a wide variety of food additives have been created to satisfy the demands of industrial food processing. To guarantee that processed food is safe and wholesome during its transit from factories or industrial kitchens to warehouses and retail locations, and ultimately to consumers, additives are added. Additionally, flavors, aromas, textures, and appearance of food can all be altered with the use of additives. Food additives can be chemically synthesized or derived from minerals, plants, or animals. Several 1,000 food additives are used, each with a specific purpose in mind. Based on their purpose, food additives can be divided into three major categories.

7.1.1. Indirect Food Additives

Indirect food additives don't have any technical purpose in food and aren't purposefully added. They can be considered contaminants if they accidentally contaminate food, if food is contaminated, or if food processing or storage leads to chemical and biochemical interactions that produce higher concentrations than are allowed. The significant health risks associated with high concentrations of indirect food contaminants in food and water make them a cause for concern. The outcome could be inflammation and intestinal infections brought on by eating contaminated food, along with the difficult task of curing or shielding people from the dangers of contaminated food. The term "chemical contamination" refers to the presence of a chemical in an unsuitable location. Among other things, a chemical hazard could come from the plate, the packing material, the air, water, soil, disinfectants, detergents, pesticides, or high concentrations of metals in or on contaminated fruits, vegetables, and meat. By adding unwanted pollutants and compounds created in the food during preparation—baking, roasting, canning, heating, fermentation, or hydrolysis—preservation, or transportation, food processing can create the conditions for food contamination. The migration of chemical compounds from plastic food containers or bags when heated, as well as printed closed packaging materials, glue, and adhesives, can result in indirect food contaminants. This is concerning because elevated concentrations in food and water can have detrimental effects on public health, particularly when migrant chemical levels are high. Globalization and the expansion of agribusiness have contributed to the rising risk and difficulty that has turned into a worldwide issue.

7.1.1.1. Bisphenols

In the 1960s, bisphenol A (BPA) was found to be an effective component in the production of polycarbonate plastics and polymeric metal can coatings, which led to an increase in the use of bisphenols as food additives. Recently, BPA was outlawed in baby bottles, and more plastic drinkware is marked as BPA-free. To stop metal from corroding in food and drink containers, polymeric resin coatings still contain BPA and similar substances.

BPA has received a lot of attention and research. It is categorized as an "endocrine disruptor" because it has the ability to bind to the estrogen receptor and induce bodily reactions as though estradiol is present. Human epidemiologic research and nonhuman laboratory research point to connections between BPA exposure and a variety of endocrine-related end points, such as altered puberty timing, decreased fertility, altered mammary gland development, and the emergence of neoplasias. BPA at environmentally relevant doses causes cells to turn into adipocytes, interferes with pancreatic β-cell function in vivo, and changes how adipocytes transport glucose. While longitudinal studies of prenatal exposure have produced less consistent relationships with postnatal body mass, cross-sectional studies have linked BPA to reductions in fetal growth, childhood obesity, and low-grade albuminuria. BPA exposure in utero has also been linked to adverse neurodevelopmental outcomes.

According to a thorough, cross-sectional study of preschool-aged children's exposure to dust, indoor and outdoor air, and solid and liquid food, dietary sources account for 99% of BPA exposure. Thermal copy paper and dental sealants are additional sources. African Americans have been shown to have higher urinary BPA concentrations, and there has been an inverse relationship found between BPA concentrations and family income. Sociodemographic differences in health may be partially explained by disproportionate exposure to endocrine-disrupting chemicals like BPA, as obesity is known to be more common among minority and low-income children. Due to consumer pressure, the FDA recently outlawed the use of BPA in baby bottles and sippy cups. As a result, many companies are voluntarily eliminating BPA from their products. However, it has frequently been replaced with closely related substitutes, like bisphenol S. These novel substitutes have been found in human urine and paper goods. The few studies that have examined bisphenol S have found that it is more resistant to environmental degradation than BPA and has comparable genotoxicity and estrogenicity. The only way that eliminating BPA from plastics and metal cans will benefit the economy and public health is if a safe substitute is found.

7.1.1.2. Phthalates

Phthalate esters are widely used in consumer products and fall into two categories: high-molecular-weight phthalates are used to make vinyl plastics for a variety of applications, such as flooring, clear food wrap, and flexible plastic tubing frequently used in food manufacturing, and low-molecular-weight phthalates are frequently added to shampoos, cosmetics, lotions, and other personal care products to preserve the scent. Di-2-ethylhexylphthalate (DEHP) is of particular interest within the high-molecular weight category because DEHP-containing plastic products are commonly used in industrial processes to produce food. There is ample evidence of racial and/or ethnic disparities in phthalate exposure. A large body of research, encompassing both human and animal studies, demonstrates that dibutyl phthalate, benzyl butyl phthalate, and DEHP are antiandrogenic substances that negatively impact the development of the male fetus' genitalia. These substances cause direct toxicity to the testicles, which lowers the body's levels of circulating testosterone and raises the possibility of hypospadias and cryptorchidism from birth. These phthalates are also linked to variations in sperm motility and quantity as well as variations in men's hormone concentrations. A DEHP metabolite called mono-(2-ethylhexyl)phthalate also interacts with three peroxisome proliferator-activated receptors, which are important for the metabolism of fats and carbohydrates. This interaction lends biological plausibility to the idea that DEHP metabolites contribute to childhood obesity and insulin resistance. Additionally, epidemiologic research has shown a connection between oxidative stress indicators and urinary phthalate metabolites. Research conducted in laboratories has revealed a connection between oxidative stress and phthalate metabolites. Oxidative stress seems to alter the endothelial relaxant nitric oxide, which promotes vasoconstriction, platelet adhesion, and the release of proinflammatory cytokines like interleukin-1. It also appears to lessen the insulin-dependent stimulation of insulin-signaling elements and glucose transport activity. Thus, phthalates may alter the results of metabolic health if they promote oxidative stress and cause inflammation. Additionally, emerging data from animals indicates that DEHP may alter metabolic profiles, cause cardiac myocyte dysfunction, and induce arrhythmias.

The National Health and Nutrition Examination Survey (NHANES) data show that between 2001 and 2010, there was a roughly 37% decrease in DEHP metabolites. These drops can be attributed to the substitution of diisodecyl (DIDP) and diisononylphthalate (DINP) for DEHP. These phthalates are becoming more commonly found in the population and have not been outlawed or restricted by regulatory bodies. In the 2009–2010 NHANES, urinary metabolites of DINP and DIDP were found in 94% and 98% of the population, respectively. The metabolites DINP and DIDP have been widely recognized as food contaminants. Cross-sectional data from the NHANES from 2009 to 2012 indicates that the metabolites' concentrations positively correlated with systolic blood pressure z scores and insulin resistance in children and adolescents.

7.1.1.3. PFCs

PFCs are artificially produced organic fluorinated compounds that have strong stability and thermal resistance due to their carbon–fluorine bonds. PFCs are widely used in nonstick cooking surfaces, fire-retardant foams, stain-resistant sprays for carpets and upholstery, and greaseproofing of paper and paperboard used in food packaging. >98% of US people have detectable levels of PFCs in their blood, including perfluorooctane sulfonic acid (PFOS), perfluorooctanoic acid (PFOA), perfluorohexane sulfonic acid (PFHxS), and perfluorononanoic acid (PFNA), according to the 2003–2004 NHANES. The majority of people are primarily exposed to PFOS and PFOA through the consumption of contaminated food, though exposure can also occur through inhalation and dermal contact. Research has linked exposure to PFOA and PFOS to many harmful health effects, including altered metabolism, a weakened immune response to vaccinations, and low birth weight. Concern over the potential of PFC to disrupt hormones is also growing; research has connected PFOA and PFOS to altered thyroid function and decreased fertility, among other consequences. These substances have half-lives in the human body ranging from two to nine years, making them incredibly persistent and bioaccumulative.

The United States phased out the production of PFOS and PFOA in 2002 and 2015, respectively, due to health and environmental concerns. These specific compounds, however, are just two of the parent family's more than a dozen members. For instance, closely related PFNA largely took the place of PFOA; rising PFNA concentrations were noted in the NHANES of 2003–2004 and have stayed steady since.

Three classes of long-chain PFCs were prohibited from being used as indirect food additives by the FDA in January 2016. However, short-chain PFCs with comparable structures, like PFHxS, might still be in use. Since NHANES 2003–2004, median levels of PFHxS have been measured, and they have stayed constant until NHANES 2009–2010. In addition to confirming increases in PFHxS concentrations (8.3% annually), a Swedish study of perfluoroalkyl acid trends between 1996 and 2010 also found increases of 11% annually in perfluoroalkylbutane sulfonate (PFBS), another short-chain PFC substitute for PFOS that is becoming more prevalent in food. According to NHANES 2011–2012, there were mild, sporadic (2%) detectable concentrations of PFBS in the US population. The structural similarity of short-chain PFCs to compounds that are prohibited implies that they may also be hazardous to human health, even though research has not adequately assessed the effects of exposure to these compounds on human health.

7.1.1.4. Perchlorate

The two most common ways that perchlorate gets into the food chain are as a component of nitrate fertilizers or as a contaminant in water. According to FDA-conducted exploratory studies, exposed crops may continue to contain high concentrations of the compound. Perchlorate is an indirect food additive as well. Food contamination can arise from the use of the substance as an antistatic agent for plastic packaging that comes into contact with dry foods (like sugar, flour, and starches) or from the breakdown of hypochlorite bleach, a cleaning solution used in the food manufacturing process. Because perchlorate interferes with the sodium iodide symporter (NIS), which permits essential iodide uptake in the thyroid gland, it is known to impair the production of thyroid hormones. Among other things, the thyroid hormone is essential for the development of the brain in the early years of life, and changes to normal hormone levels can have long-term effects on cognition. Because the developing fetus depends solely on the mother's thyroid hormone during the first trimester of pregnancy, exposure to perchlorate in pregnant women—especially those who are iodine deficient—raises special concerns. Pregnancy-related hypothyroidism in the mother has been linked to children's cognitive deficiencies. Another significant vulnerable group is infants, whose consumption of powdered formula may expose them to elevated levels of perchlorate from related packaging materials. The rise in neonatal hypothyroidism and other thyroid system disturbances that have been reported in the United States may be caused by perchlorate and other food contaminants that affect thyroid hormone homeostasis, such as polybrominated diphenyl ethers. Furthermore, the thyroid hormone is essential for healthy growth processes, and new research indicates that poor growth outcomes are linked to high exposure to various compounds that obstruct iodide uptake.

7.1.2. Direct Food Additives

Direct food additives are substances that are added to food on purpose to give it a desired quality or characteristic that consumers demand. These substances can be naturally occurring or artificially manufactured.

7.1.2.1. Artificial Food Colors

Artificial food coloring, also known as synthetic artificial food coloring (AFC), is added to food and drink for esthetic purposes. Young children find brightly colored products particularly appealing. AFCs can take the place of healthy ingredients in certain situations, like fruit juice drinks that have little to no real fruit in them. As of right now, nine AFCs—Blue 1, Blue 2, Green 3, Yellow 5, Yellow 6, Red 3, Red 40, Citrus Red 2, and Orange B—are authorized for use in the US. According to FDA data, between 1950 and 2012, the amount of AFCs used per capita per day increased by more than five times, from 12 to 68 mg.

Concerns have been raised by research over the past few decades about how AFCs affect kids' behavior and how they might exacerbate the symptoms of attention-deficit/hyperactivity disorder. Removing AFCs from the diet might help kids with ADHD (attention deficit/hyperactivity disorder). Blue 1 is one AFC that may be able to pass through the blood-brain barrier, although the exact mechanisms of action are still unknown. All things considered, more research is necessary to clarify the uncertainties in the scientific data and gain a deeper understanding of the effects of AFC exposure. Due to a variety of factors, including patent protection, including the lack of information regarding the ingredients, the literature that is currently available should be interpreted cautiously.

For each of the AFCs, the FDA has established acceptable daily intakes. These guidelines, along with the color additives' initial safety approval, are based on animal research that omitted neurologic or neurobehavioral endpoints. Considering that these effects have been noted in children, a comprehensive reevaluation of AFCs is necessary to ascertain whether they satisfy the agency's standard of safety, which entails providing strong proof that proves beyond a reasonable doubt that the color additive's intended use will not cause harm.

7.1.2.2. Nitrates and Nitrites

The use of nitrates and nitrites as preservatives in cheese, fish, and cured and processed meats has long been a source of concern. The American Medical Association stressed in a 2004 statement that due to the chemical makeup of their gastric tracts, infants are especially susceptible to methemoglobinemia from nitrates and nitrites. Nitrates and nitrites, while not inherently carcinogenic, can react with secondary amines or amides to form carcinogenic N-nitroso compounds (NOCs) in the body, which can lead to gastrointestinal or neural cancer, according to the American Medical Association statement. Ingestion of nitrates and nitrites in conditions that could cause endogenous nitrosation, or the production of NOCs, was categorized as "possible human carcinogens" (Group 2A) by the International Agency for Research on Cancer in 2006. The International Agency for Research on Cancer designated processed meat as "carcinogenic to humans" (Group 1) in 2015. This includes meat that has been salted, cured, or otherwise changed to improve flavor and preservation. There is strong evidence that eating processed meats increases the risk of colorectal cancer, and this processing can lead to an increase in the formation of NOCs. High maternal consumption of meats cured with nitrite has also been associated with a higher chance of brain tumors in the offspring during childhood, particularly astroglia tumors. According to current FDA regulations, final meat products may contain up to 500 ppm of sodium nitrate and 200 ppm of sodium nitrite. However, food made especially for newborns or young children cannot contain any nitrates or nitrites. Similar to perchlorate, nitrates can also cause thyroid dysfunction by obstructing the NIS and preventing the thyroid's necessary iodide absorption. Nitrate is still a major concern even though its relative potency is much lower than that of other common NIS inhibitors. This is because (1) NIS inhibitors may act additively, and (2) combined exposures from food and water may account for a larger proportion of NIS inhibition than from perchlorate exposure. Numerous physiological processes in the body, including normal growth, depend on thyroid hormones. According to new research, high exposure to NIS inhibitors, such as nitrate, is linked to a decrease in growth measures. Furthermore, because the fetus depends solely on the mother's thyroid hormone during the first trimester, as was previously mentioned about perchlorate, maternal thyroid disruption during pregnancy is particularly concerning. Thyroid hormone is essential for neurologic development, and deficits in this hormone during infancy can have long-term negative effects on cognitive function. Alternative sources of nitrate and nitrite preservatives, like celery powder, have been used more frequently in "natural" and "organic" product labels in recent years. These goods might have nitrate and nitrite concentrations that are on par with or greater than those in conventional goods made from sources based on sodium. Customers should therefore be informed that natural and organic products might not offer any advantages over conventional products when it comes to nitrates and nitrites alone.

7.2. THE HEALTH IMPACT OF FOOD ADDITIVES

Certain food additives may be harming human bodies even though they improve the taste and appearance of their food. Customers may be concerned about the long-term health effects of artificial food additives, even with the stringent safety regulations these ingredients go through.

7.2.1. Gut Health

There is currently little proof that artificial food additives that the FDA has approved for use are bad for digestion or gut health. Having said that, a lot of people are curious about whether artificial food additives and gut health are related because of the rise in gut disorders like IBS, Crohn's disease, and gut dysbiosis as well as the popularity of highly processed foods. Remember that the human microbiome is very different from that of animals, and this could result in variations in the metabolism of artificial sweeteners. Furthermore, the majority of these studies lack proper control and employ dosages that are much higher than what a typical person would take. Thus, additional research is required. Additionally, some studies on mice have demonstrated that the food coloring titanium dioxide and the emulsifier polysorbate 80 may change the diversity of the gut microbiome. For instance, the rise in bacteria that reduce sulfate could make IBS symptoms worse. Given that humans and mice have different digestive systems, it is unclear whether these effects also apply to humans. To properly comprehend the impact of artificial food additives on human health, long-term human studies are required. It is best to work with a healthcare provider who specializes in digestive disorders if a patient is sensitive to specific food additives.

7.2.2. Weight Gain

Popular food additives, and artificial sweeteners are used by some to counteract a high-sugar diet, but others think they might be more harmful than beneficial. Artificial sweeteners, also referred to as nonnutritive sweeteners, are sugar substitutes that don't add calories to food or drink. Aspartame, acesulfame K, saccharin, and sucralose are popular types. Artificial sweeteners have no calories, but it's believed that they can cause weight gain by changing the gut microbiome, which can affect hormone regulation and metabolism. But the majority of artificial sweeteners don't get to the lower intestine, which is home to the gut microbiome. As a result, it seems improbable that artificial sweeteners would alter the gut flora, and there is no proof in research that these sweeteners lead to weight gain. In fact, when human users consumed a normal amount of artificial sweeteners (within the recommended daily intake), one study found no difference in the colonies of gut bacteria. Furthermore, some speculate that, unlike sugar, artificial sweeteners may not trigger the release of hormones that control hunger, which would increase appetite and food intake. Nevertheless, correlations do not equate to cause and effect. Many times, additional confounding variables (such as total dietary intake, dieting history, and physical activity) can also affect an individual's weight gain or loss.

Furthermore, the study design and the external factors that are controlled for can affect the results. Many contradictory studies have been conducted as a result of these variances. For instance, several human cross-sectional studies indicate a dose-dependent relationship between the consumption of artificial sweeteners and body mass index (BMI); however, a large number of clinical trials and meta-analyzes have suggested artificial sweeteners may facilitate weight loss. Furthermore, a recent meta-analysis published in the British Medical Journal found no connection between BMI and the use of artificial sweeteners. The authors came to the conclusion that longer-term, higher-quality research is required. All things considered, more research is necessary.

7.2.3. Cancer

There may be a connection between some artificial food additives and an increased risk of developing specific cancers. A diet rich in nitrites and nitrates, which are frequently present in processed meats, has been specifically associated with a higher risk of colorectal cancer. Increased research demonstrating a dose-dependent relationship between processed meat consumption and colorectal cancer risk led the World Health Organization to classify processed meat as carcinogenic to humans in 2015. The additive known as bisphenol A (BPA), which is frequently present in food packaging, has also been connected to a higher risk of cancer and endocrine disruption. It is currently prohibited in baby products, but it is still present in some cans, water bottles, and other packaging. There is little evidence linking other food additives to a higher risk of cancer. It is unclear whether food additives have a role in this, but overall, a highly processed diet is linked to an increased risk of cancer. The best ways to lower the chance of developing cancer are to abstain from smoking, drink in moderation, maintain a healthy weight, engage in regular exercise, and eat a diet rich in fruits, vegetables, and fiber.

7.2.4. Children's Health

Customers sometimes question if artificial food additives, like food colorings, cause hyperactivity, attention deficit hyperactivity disorder (ADHD), or autism spectrum disorder (ASD) in kids. Many food products, particularly those marketed towards children, are enhanced in appearance and flavor by the use of artificial food colorings and flavorings. Artificial colors and flavors are typically used because they are less expensive and have greater vibrancy than natural options. Little research backs up the numerous organizations' and diets' claims—including the Feingold diet—that artificial food additives in children lead to behavioral problems or neurodevelopmental disorders. But BPA is an endocrine disruptor that can cause problems with development in developing babies. As a result, it can no longer be found in baby formula packages, sippy cups, or bottles. Seek for goods with the labeled phrase "BPA-free."

7.3. COMMON FOOD ADDITIVES TO AVOID

A person can probably find a food additive on the ingredients label of almost any food item in their kitchen pantry. They are used to increase a product's shelf life or to improve its flavor, texture, or appearance. While some of these substances are safe and should be consumed with little risk, others have been linked to negative health effects and should be avoided. The most popular food additives are listed here, along with suggestions for which ones to avoid consuming.

7.3.1. Monosodium Glutamate (MSG)

MSG, also known as monosodium glutamate, is a common food additive that is used to improve and intensify the flavor of savory foods. It can be found in many processed foods, including canned soups, frozen dinners, and salty snacks. It's frequently added to food at fast food and restaurant establishments as well. MSG has been the focus of intense debate ever since a 1969 study on mice revealed that high doses had detrimental effects on the nervous system and stunted growth. Because this additive cannot pass through the blood-brain barrier, it is unlikely to have any impact on the health of the human brain. In certain observational studies, MSG consumption has also been linked to weight gain and metabolic syndrome; however, other studies have not found this connection. Nevertheless, some people do react negatively to MSG and after consuming a lot of it, they may get symptoms like headaches, sweating, and numbness.

In one study, 61 participants who claimed to be sensitive to MSG received either a placebo or five grams of MSG. MSG sensitivity may be a real concern for some people, as it's interesting to note that 36% of respondents reported having a negative reaction to MSG while only 25% reported having a reaction to the placebo. It is advisable to exclude MSG from a person's diet if they encounter any unfavorable side effects after consuming it. Otherwise, MSG can be safely taken in moderation without running the risk of negative side effects if they can tolerate it. MSG is added to a lot of processed foods to improve their flavor. While some individuals may be sensitive to MSG, most people can safely use it in moderation.

7.3.2. Artificial Food Coloring

Everything from candies to condiments can be made to look more appealing by adding artificial food coloring. However, there have been a lot of worries in recent years regarding possible health consequences. Certain food dyes, such as Red 40, Yellow 5, Blue 1, and Yellow 6, have been linked to allergic reactions in certain individuals.

Furthermore, a review suggested that artificial food coloring might encourage children's hyperactivity, while another study indicated that certain kids might be more sensitive than others. Concerns have also been expressed regarding certain food dyes' possible link to cancer. Red 40 has replaced Red 3, also referred to as erythrosine, in most foods because it has been demonstrated in animal studies to increase the risk of thyroid tumors. Other food dyes, on the other hand, have not been linked to any cancer-causing effects, according to numerous animal studies.

To assess the safety and possible health risks of artificial food coloring for people, more investigation is still required. All the same, processed foods are the main source of food dyes and should be avoided in a balanced diet. Whenever possible, choose whole foods since they are naturally devoid of artificial food coloring and higher in vital nutrients. Artificial food coloring can trigger allergic reactions and encourage hyperactivity in kids who are already sensitive. Studies on animals have also demonstrated that red 3 raises the risk of thyroid tumors.

7.3.3. Sodium Nitrite

Sodium nitrite, a reddish-pink color and salty taste, is a preservative that stops bacteria from growing and is often found in processed meats. Nitrites can convert to nitrosamine, a substance with numerous detrimental health effects, when exposed to high heat and amino acids are present. According to one review, there is a correlation between a higher risk of stomach cancer and a higher intake of nitrites and nitrosamine. Numerous other studies have discovered a similar correlation, suggesting that consuming more processed meats may increase the risk of bladder, breast, and colorectal cancer.

Results from other studies are mixed and point to nitrosamine exposure possibly being associated with an increased risk of type 1 diabetes. However, it's advisable to limit consumption of processed meats and sodium nitrite. Try replacing processed meats with unprocessed meat and nutritious sources of protein in place of items like bacon, sausage, hot dogs, and ham. A person can replace processed meats in their diet with a variety of delectable high-protein foods, such as eggs, tempeh, beef, fish, and legumes. Processed meats frequently contain sodium nitrite, which has the potential to transform into the toxic substance nitrosamine. Increased consumption of processed meats and nitrites may increase the risk of developing various cancers.

7.3.4. Guar Gum

A long-chain carbohydrate called guar gum is used to bind and thicken food. It is a common ingredient in ice cream, sauces, soups, salad dressings, and other food products. Guar gum has been linked to numerous health advantages and has a high fiber content. For instance, a study found that it lessened constipation and bloating, two symptoms of irritable bowel syndrome. Additionally, taking guar gum with a meal increased feelings of fullness and reduced calorie intake from snacking throughout the day.

Guar gum may also help lower cholesterol and blood sugar levels, according to additional research. On the other hand, guar gum in excess can be harmful to one's health. This is due to its propensity to swell 10–20 times its original size, which could result in problems like small intestine or esophageal blockage. In certain cases, guar gum may also result in minor side effects like cramps, bloating, or gas. Guar gum is, however, usually regarded as safe when used in moderation. To further reduce the possibility of unfavorable side effects, the FDA has established stringent guidelines on the amount of guar gum that can be added to foods. A long-chain carbohydrate called guar gum is used to bind and thicken food. It has been linked to improved digestive health, decreased blood sugar and cholesterol, and elevated sensations of fullness.

7.3.5. High-Fructose Corn Syrup

Corn is used to make high-fructose corn syrup, a sweetener. It is commonly present in snack foods, juice, soda, candies, and cereals for breakfast. It contains a lot of fructose, a type of simple sugar that, in excess, can have major negative health effects. Specifically, there is a connection between weight gain and diabetes and high-fructose corn syrup. In one study, for 10 weeks, 32 participants drank a beverage sweetened with either fructose or glucose. By the end of the trial, compared to the glucose-sweetened beverage, the fructose-sweetened beverage significantly increased blood sugar and belly fat and decreased insulin sensitivity. Studies on animals and test tubes have also revealed that fructose can cause cellular inflammation. It is thought that a number of chronic illnesses, such as diabetes, cancer, and heart disease, are largely caused by inflammation. Furthermore, high-fructose corn syrup adds sugar and empty calories to food while leaving out vital vitamins and minerals that the body requires. It is best to avoid high-fructose corn syrup-containing foods and sugary snacks. Choose whole, unprocessed foods instead, and use stevia, yacon syrup, or fresh fruit to sweeten them. High-fructose corn syrup has been linked to inflammation, diabetes, and weight gain. It adds nothing but calories to the diet and has a high empty calorie content.

7.3.6. Artificial Sweeteners

Many diet foods and beverages use artificial sweeteners to increase sweetness while lowering calorie content. Artificial sweeteners that are frequently used are aspartame, sucralose, saccharin, and acesulfame potassium. Artificial sweeteners have been shown in studies to help control blood sugar levels and promote weight loss. According to one study, participants who took a supplement containing artificial sweeteners for 10 weeks consumed fewer calories and gained less weight and body fat than those who took regular sugar. Another study found that taking sucralose for three months had no impact on 128 diabetics' ability to control their blood sugar.

Be aware that some people may experience headaches from specific artificial sweeteners, such as aspartame, and that some people may be more susceptible to these effects than others. However, when used sparingly, artificial sweeteners are thought to be safe for the majority of people. However, carefully read ingredient labels and cut back on consumption if someone experiences any unfavorable side effects after using artificial sweeteners. Artificial sweeteners might support blood sugar regulation and weight loss. Though mild side effects like headaches are possible with some varieties, they are generally regarded as safe when used in moderation.

7.3.7. Carrageenan

Carrageenan, which is derived from red seaweed, is used in a wide range of food products as an emulsifier, thickener, and preservative. Almond milk, cottage cheese, ice cream, coffee creamers, and dairy-free goods like vegan cheese are common sources of carrageenan. Concerns regarding this popular food additive's safety and possible health effects have existed for decades. According to one study on animals, eating a high-fat diet in conjunction with carrageenan exposure elevated the subjects' levels of glucose intolerance and fasting blood sugar. Studies on animals and in test tubes have also revealed that carrageenan causes inflammation. Additionally thought to have a deleterious effect on digestive health, carrageenan has been linked to the development of growths and ulcers in the intestines.

According to a small study, patients with ulcerative colitis who were taking a carrageenan-containing supplement saw an earlier relapse than those who were taking a placebo. Regretfully, there is still a dearth of information on the effects of carrageenan, and more research is required to fully understand potential human health risks. Should an individual choose to restrict their consumption of carrageenan, an abundance of online resources is available to assist them in identifying brands and commodities devoid of carrageenan. Studies on animals and in test tubes have suggested that carrageenan may result in growths and ulcers in the intestines as well as elevated blood sugar. Additionally, a study discovered that carrageenan was linked to an earlier ulcerative colitis relapse.

7.3.8. Sodium Benzoate

Sodium benzoate is a preservative that is frequently added to acidic foods and beverages, such as fruit juices, salad dressings, pickles, and condiments. The FDA has declared it to be generally safe, but many studies have revealed possible adverse effects that need to be taken into account. For instance, a study indicated that giving 3-year-old children artificial food coloring and sodium benzoate together increased their hyperactivity.

In another study, 475 college students with ADHD symptoms were found to have higher intakes of beverages containing sodium benzoate. Sodium benzoate can also be transformed into benzene, a substance linked to the development of cancer, when combined with vitamin C. The highest concentration of benzene is found in carbonated drinks, and the formation of benzene is even more likely in diet or sugar-free beverages. In one study that looked at the amount of benzene in different foods, samples of cola and Cole slaw had benzene concentrations of over 100 parts per billion (ppb), which is more than 20 times the maximum amount of contaminants the EPA has set for drinking water.

Pay close attention to the food labels in order to reduce consumption of sodium benzoate. Foods containing benzoic acid, benzene, or benzoate should be avoided, particularly if they are combined with ascorbic acid or citric acid, which are sources of vitamin C. Sodium benzoate and elevated hyperactivity may be related. It may also result in the formation of benzene, a substance linked to the development of cancer, when combined with vitamin C.

7.3.9. Trans Fat

Trans fats are unsaturated fats that have been hydrogenated to lengthen their shelf life and enhance product consistency. Numerous processed food varieties, including biscuits, microwave popcorn, margarine, and baked goods, contain it. Trans fat consumption has been linked to a number of possible health hazards; in fact, the FDA recently decided to remove trans fat from its GRAS (generally recognized as safe) classification. Specifically, a number of studies have connected an increased consumption of trans fats to an increased risk of heart disease.

Eating foods high in trans fats was found to raise multiple markers of inflammation, one of the main risk factors for heart disease, in one study. Additionally, studies suggest that trans fats and diabetes may be related. Even a higher risk of type 2 diabetes by 40% was linked to a high intake of trans fat, according to a large study including 84,941 women.

The simplest and most efficient way to reduce trans fat intake is to eliminate processed foods from diet. A person can also make a few easy dietary changes, such as substituting vegetable oils with olive or coconut oil and using butter instead of margarine. Consuming trans fats has been linked to a number of detrimental health outcomes, such as diabetes, heart disease, and inflammation.

7.3.10. Xanthan Gum

Common food additives like xanthan gum are used to thicken and stabilize a variety of foods, including sauces, soups, salad dressings, and syrups. It's occasionally added to gluten-free recipes to enhance food texture. Xanthan gum has been linked to a number of health advantages. According to one study, eating rice with xanthan gum added lowered blood sugar levels compared to eating rice without it.

Additionally, a different study discovered that consuming xanthan gum for six weeks raised feelings of fullness while lowering cholesterol and blood sugar levels. There hasn't been much recent research on xanthan gum's possible advantages, though. Moreover, eating a lot of xanthan gum may also increase the production of gas, soft stools, and increased stool output.

However, xanthan gum is generally well tolerated and safe for the majority of people. It is advisable to cut back on xanthan gum consumption or think about removing it from the diet if a person experiences unpleasant symptoms after consuming it. Xanthan gum may aid in lowering cholesterol and blood sugar levels. When consumed in excess, it can lead to gastrointestinal problems like gas and soft stools.

7.3.11. Artificial Flavoring

Chemicals called artificial flavors are made to taste like other ingredients. They can be used to simulate a wide range of flavors, including fruit, caramel, popcorn, and more. According to research on animals, these artificial flavors may have some unsettling health effects. In one study, rats given artificial flavorings for seven days showed a significant decrease in red blood cell production.

Furthermore, it was discovered that some flavors, such as chocolate, biscuit, and strawberry, had a harmful effect on the bone marrow cells in those flavors. Similar findings from an animal study revealed that artificial flavorings flavored with grape, plum, and orange were toxic to mouse bone marrow cells and inhibited cell division. But bear in mind that the doses used in these studies were far more concentrated than what one would find in food, and more investigation is required to ascertain the potential health effects of artificial flavoring in food-based amounts. Check the ingredients label of foods if a person wants to limit their intake of artificial flavoring in the interim. Instead of "chocolate flavoring" or "artificial flavoring," look for the words "chocolate" or "cocoa" on the ingredients label.

7.3.12. Yeast Extract

Yeast extract, also known as hydrolyzed or autolyzed yeast extract, is used to enhance the flavor of some savory dishes, such as cheese, soy sauce, and salty snacks. It is prepared by mixing sugar and yeast in a warm environment, centrifuging the mixture, and then removing the yeast's cell walls. Glutamate is a naturally occurring amino acid that can be found in a variety of foods and is present in yeast extract. Similar to monosodium glutamate (MSG), consuming glutamate-containing foods can make those who are sensitive to its effects experience mild symptoms like headaches, numbness, and swelling. Furthermore, each teaspoon (8 grams) of yeast extract contains roughly 400 milligrams of sodium, which is a relatively high amount. It has been demonstrated that lowering sodium consumption can help lower blood pressure, particularly in high blood pressure patients.

Nonetheless, the majority of foods only have a small amount of added yeast extract, so most people are unlikely to experience significant issues from the glutamate and sodium in yeast extract. As of 2017, the Food and Drug Administration continues to deem yeast extract safe. If someone experiences adverse effects, they should think about consuming fewer processed foods containing yeast extract and increasing their intake of fresh, whole foods. Glutamate and high sodium content in yeast extract can cause symptoms in certain individuals. However, since yeast extract is added in small amounts to food, most people are unlikely to experience any issues.

7.4. REGULATION OF USE OF FOOD ADDITIVES

Strict rules controlling the licensing and management of food additives have been established by regulatory bodies and law enforcement organizations. Regulation is used to ensure that there are no dangers or malicious intent. Nevertheless, given the circumstances of use and the release of fresh scientific information, the use of food additives should continue to be closely monitored and evaluated when necessary. For example, very small amounts of humectants and anti-caking agents are used to protect food supply and preserve quality during shelf life.

According to studies, these additives are safe to use as long as you stick to the recommended dosages. Research and development are still being done in this field to find safer, natural substitutes and technologies that can improve food systems and public health even more. Food additives can occasionally be misused to cover up food adulteration or poor processing or preparation techniques. They can be used to cover up or mask deterioration, damage, subpar quality, or faulty ingredients, or they can take the place of a drawn-out, laborious procedure. To protect consumers, processing aids that show signs of allergies, intolerances, or religious concerns must follow food labeling laws regarding food additives.

7.5. SAFETY EVALUATION OF FOOD ADDITIVES

Food additives have become crucial in meeting consumer needs and ensuring a safe food supply since the introduction of food processing. Any material, natural or artificial, that is purposefully added to food for technological reasons during processing, packaging, transportation, or storage is referred to as a food additive. The following are just a few of the technological roles played by food additives.

1. enhancing the safety and quality by the inhibition of microbial growth;
2. extending the shelf-life by protection against any oxidative deterioration;
3. enhancing the flavor and odor;
4. stabilizing or retaining the color; and
5. improving the texture and consistency of a food, etc.

Food additives are not typically used as standard ingredients in meals or consumed as food on their own. Contaminants, additives added to food to preserve or enhance its nutritional value, and seasonings like salts, herbs, and spices are not included in this definition. Food additives come in a variety of forms, but some of the most widely used ones are thickeners, emulsifiers, sweeteners, coloring agents, antioxidants, and preservatives.

7.5.1. Safety and Public Health Significance

1. In general, food additives are not very toxic. In actuality, chronic exposure to food additives at levels above the safety reference is the main cause for concern regarding food safety.
2. The international food safety authority, known as the Joint Food Agriculture Organization/ World Health Organization Expert Committee on Food Additives (JECFA), is in charge of gathering and assessing scientific data on food additives and assigning a safety reference (i.e., the acceptable daily intake (ADI)) of the assessed food additives. JECFA also offers guidelines for acceptable usage levels.
3. The amount of a chemical, expressed on a body-weight basis, that can be consumed every day for the rest of one's life without posing a significant risk to health is known as the acceptable daily intake (ADI) of that chemical. A dietary intake above the ADI does not always indicate a danger to one's health. Since the emphasis of the ADI is a lifetime exposure, a brief deviation above the ADI would not have an adverse effect on one's health as long as the average intake over an extended period of time is not exceeded.
4. Certain food additives may cause acute effects in a small percentage of the population, e.g., for some individuals with allergy conditions, a small amount of sulfur dioxide may cause bronchoconstriction and an asthmatic reaction.

7.5.2. Principles for Using Food Additives

1. At the levels of usage, food additives shouldn't pose a health risk to the consumer.
2. Food additives should only be used when they are beneficial, do not pose a risk to consumers' health or deceive them, and fulfill one or more of the following technological needs. They should also only be used when no other practical, financially and technologically feasible means of achieving these goals can be found.

a. to preserve the nutritional quality of the food;

b. to supply components required for food made for customer groups with particular dietary requirements;

c. to improve the food's organoleptic qualities, stability, or keeping quality;

d. to help with food processing, packaging, transportation, or storage, as long as the additive isn't used to cover up the consequences of using subpar raw materials or unfavorable (including dirty) methods when performing any of these tasks.

3. The following are examples of good manufacturing practices (GMP) that apply to the use of all food additives–

a. The amount of food additive that is added must be kept to the lowest amount required to achieve the intended result.

b. the additive's quantity is minimized to the greatest extent practicable when used in food production, processing, or packaging to prevent any physical or technical effects from occurring in the food itself; and.

c. the additive is treated and prepared similarly to a food ingredient.

7.5.3. Advice to the Trade

1. It is advised that food manufacturers refrain from abusing food additives by taking into account the aforementioned Principles for Using Food Additives when creating new products.
2. Use caution when selecting food additives; only add the proper kind and quantity to the food to enable it to perform the intended technological function.
3. The Public Health and Municipal Services Ordinance, the regulatory requirements should be noted by the trade as well. 132, which deals with the application and marking of food additives.
4. Seek guidance from food scientists or the appropriate authorities if you have any questions about the use and labeling of food additives.

7.5.4. Advice to the Public

1. Buy foods from reputable sources.
2. Pay close attention to the ingredient list on prepackaged food labels to see if there are any added food additives that you can tolerate.
3. People who suffer from allergies, such as those who have asthma, may react hyper sensitively to certain food additives, such as sulfur dioxide, so they should be cautious when choosing what to eat. When necessary, medical professionals may be consulted for advice.
4. When selecting foods, steer clear of those with unusual color, texture, or smell as much as possible. If a food tastes strange, stop eating it.
5. Notify the appropriate authorities of any food anomalies so they can conduct an investigation and take further action.
6. It is recommended that the general public consume a well-balanced diet to prevent excessive exposure to food additives found in a limited variety of foods.

7.6. RECENT DEVELOPMENTS AND TRENDS IN THE FOOD SECTOR

Due to consumer preferences and the financial advantages they provide to food products—such as ease of processing, consistent quality, and longer shelf life—the need for food additives in the food processing industry has increased significantly. However, over the past few decades, there has been a significant advancement in food science and technology, leading to a growing range and quantity of food additives. The question of whether food additives are safe to use has come up more and more over time as a result of some controversies surrounding information about potential risks to the public's health.

Despite all of these achievements, food manipulation has had a significant impact on the distinct biochemical balance of human bodies. Notably, a number of food additives have been linked to unfavorable outcomes that raise questions about public health. Whether they are synthetic or natural, food additives have the same function when used in food processing. However, some consumers of foods containing additives—mostly synthetic food additives—have reported allergies and unfavorable side effects, which has left them unhappy. Numerous studies have connected certain food additives to a range of health risks, such as asthma, allergies, cancer, metabolic abnormalities, and behavioral disorders like childhood attention deficit hyperactivity disorder (ADHD) and multiple sclerosis (MS). Inattention, impulsivity, and hyperactivity are the hallmarks of ADHD, which is thought to include a range of neurobehavioral symptoms and intensities. Certain food additives cause hormonal imbalances that interfere with children's normal growth and development by upsetting their endocrine systems.

Food additives can also interfere with the body's normal metabolic processes, resulting in hormonal imbalances, chemical reactions, and/or physiological responses that can cause a host of health problems, including brain damage, nausea, heart disease, childhood obesity, and weakened immunity. Certain food additives and their metabolites may also inhibit specific growth hormones, leading to adverse effects that could potentially stunt a child's growth and development. While the majority of food additives that are sold commercially are thought to be safe, some are known to be toxic or carcinogenic. Knowing these substances' characteristics is crucial to ensuring their safe application or limits because they are purposefully added to food. Therefore, before being approved, food additives must undergo appropriate toxicity assessments or trials to rule out any potential negative effects.

Because it can bind to the estrogen receptor and cause the body to react as though there is estradiol present, BPA is categorized as an endocrine disruptor. Human epidemiologic studies have demonstrated a connection between BPA exposure and a range of endocrine-related consequences, such as altered puberty timing, decreased fertility, changes in mammary gland development, and neoplasia progression. The control system of the Acceptable Daily Intake (ADI), created by the World Health Organization's (WHO) Expert Committee on Food Additives and the Joint Food and Agriculture Organization (FAO) of the United Nations, supports the assessment of food additives globally (JECFA). The Codex Alimentarius Commission has established maximum usage levels for food additives, and only those that have undergone testing and certification as safe by JECFA—the International Joint FAO/WHO Expert Committee on Food Additives—may be used in internationally traded foods. As a result, all food additives have to be listed with other ingredients on the food label in the ingredients list.

Certain food additives, such as sugar, salt, spices, vinegar, and sulfites, have been used as preserving agents since ancient times. However, the use of additives and preservatives is necessary when food is to be kept in storage for an extended length of time in order to preserve its flavor and quality and stop bacteria and yeasts from spoiling it. When added, food additives—whether processed or fresh—become distinctive elements of the dish. For a variety of reasons, a large number of people consume prepared foods from the market, which may include additives and preservatives of some kind. Food labeling laws have been implemented to guarantee that consumers are aware of the ingredients in processed foods so they can make informed decisions. A number of food additives have been connected to various health risks, such as allergies, asthma, cancer, irritable bowel syndrome, mood swings, skin irritations or reactions, constipation, migraines, autism, sleep disturbance, and nasal congestion. This calls for more research and stricter regulation for their use in industry, even though their use is inevitable given that food additives are tailored to specific needs, low cost, functional foods, low calorie, and extended shelf life. For these reasons, to prevent negative effects, food safety monitoring authorities operating under the auspices of the World Health Organization (WHO) should constantly oversee and direct the regulation and control of national and international health authorities. Whenever possible, switching to organic foods has been a long-term solution to halt or reverse these effects (Qin, 2012; Kumar, 2020; Pepper et al., 2020).

CASE STUDY

PROLIFERATION OF ILLEGAL AND POTENTIALLY HAZARDOUS FOOD ADDITIVES IN PROCESSED AND PACKAGED FOODS IN AFRICA: A CASE STUDY AND HAZARD IDENTIFICATION IN GHANA

Food additives are normally added to food or feed to enhance their palatability, shelf life and attractiveness. There are several categories of food additives: acids, acidity regulators, antioxidants, anticaking agents, antifoaming agents, bulking agents, food coloring, preservative sweeteners, etc. The Joint FAO/WHO Expert Committee on Food Additives (JECFA) regulates food additives globally but individual countries have specific regulations, which are enforced by their food regulating bodies. There are several controversies surrounding the use of certain food additives and the acceptable daily intake. In most African countries, regulations and enforcement of the recommended food additives is weak due to inadequate experts, inadequate equipment for testing and lack of knowledge on the part of most consumers to read labels on especially processed and packaged foods. This situation is worsened by the proliferation of the African markets with products from especially countries that are noted to be food violators: India, China, Mexico, France, USA, Vietnam, Brazil etc. There is little information about food additives especially illegal ones in Africa as well as Ghana. Most people do not know about the designated numbers for the various food additives. This makes it easier for other countries to dump processed and packaged foods with illegal food additives in African countries. The objective of this study was to survey processed and packaged foods in Ghana and document all the illegal or potentially harmful food additives that are used to prepare them and their possible adverse effects on Ghanaians.

Materials and Methods

- **Sample Collection and Documentation:** A wide range of samples were randomly purchased on the Ghanaian Market from September 2013 to June 2014. In all, 63 items were purchased and including: Fruit juices, soft drinks, milk, ice creams, energy drinks, spices, noodles and margarine. Information documented on labels or packaging materials includes: trade name, type of food, company name, country of origin, additives/preservatives and whether they were certified by the GFDA.
- **Verification of the Products for Food and Drugs Authority Certification:** All the products sampled were checked for certification from the Ghana Food and Drugs Authority's registered products site (http://www.fdaghana.gov.gh/index.php?option=com_fdasearch&Itemid=36) to verify whether they were duly registered in the country.

- **Determination of Food Additives That Are Likely to Cause Adverse Effects:** The food additives in the various products purchased were first checked on the food intolerance network website (http://fedup.com.au/information/information/complete-lists-of-additives-6#code) to determine whether they were listed among the food additives that may cause adverse effects. Based on this list, calculate the percentages of the products that had the highest and lowest number of food additives likely to cause adverse effects. Do not, perform any chemical analysis on the products to check the levels of those potentially harmful additives.
- **Data Analysis:** The results were calculated using Microsoft Excel and presented with graphs and tables with descriptive statistics.

Results

- **Countries of Origin of Products Sampled:** Figure 7.1 shows the country of origin of all the processed or packaged food sampled in this research. The majority of the products were manufactured here in Africa (46%) followed by Europe (30%), Asia (21%) and the USA (3%). Out of the percentage manufactured in Africa (46%), 62% was manufactured in Ghana by either foreign or local companies followed by Nigeria (15%), Algeria (10%), Togo (7%), Ivory Coast (3%) and South Africa (3%). Among the European countries, the UK recorded the highest number of products (26%) followed by Bulgaria (21%), Germany (17%), Netherlands, Austria (11%) and Romania, Spain (7%). Among the Asian countries, the highest number of products was from China (24%), followed by Dubai (19%), and Lebanon (17%) while Bangladesh, Turkey, Thailand, Taiwan and Vietnam had 8%.
- **Categories of Foods Sampled:** The highest category of food sampled was soft drinks (38%), followed by energy drinks (18%), fruit juice (13%), ice cream (11%), milk (5%), noodles (6%), spices (3%) and biscuit and margarine (1.6%) each. Among the soft drinks, 46% originated from Europe (Bulgaria, Germany, Netherlands, Northern Ireland and the UK), 38% were produced in Africa (Algeria, Ghana Nigeria and Togo) and 17% originated from Asia (China, Turkey and Dubai).

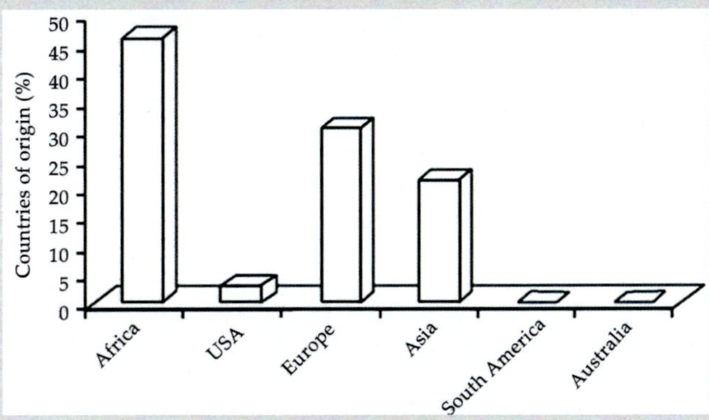

Figure 7.1. *Percentage of countries of origin represented by the various continents from which the sampled products originated.*

Only 8% of the soft drinks sampled originated from Ghana. Most of the energy drinks originated from Europe (46%) followed by Asia (27%), the USA (18%) with only 9% coming from Africa. The only energy drink produced from Africa in this study was from Ghana. Most of the fruit juices were produced in Africa (50%) followed by Europe (25%) and Asia (25%). All the ice creams sampled originated from Ghana (100%). Forty percent (40%) of the milk originated from Africa (all from Ghana), 40% from Europe and 20% from Asia. Most of the noodles were from Asia (75%) and Africa 25%. All the spices sampled originated from Africa (Ghana and Cote de Ivoire) while the biscuit and margarine originated from Ghana.

- **Verification of the Products for Food and Drugs Authority Certification:** Upon verification from the Ghana Food and Drugs Authority's product register, 37% of all the products sampled on the market were not registered with the GFDA at the time of collecting samples for this research. The Ghana Food and Drugs Authority certified only 63% of the products. Of all the uncertified products, the majority of the products originated from Europe (52%), of which Bulgaria represented 34%, followed by the UK (17%), Germany (17%), Austria, Netherlands, Northern Ireland and Romania represented 8% each. 26% of the non-certified products originated from Asia, of which China represented 50% followed by Thailand, Turkey and Bangladesh which represented 17% each. The uncertified products from Africa represented 17% and originated from Algeria, Ghana, Nigeria and Togo representing 25% each. Only 5% of the illegal products came from the USA. The results show that most of the illegal products sampled came from European countries followed by Asia, Africa and the USA. Only 1 Ghanaian product out of the total of 17 from all the 63 products sampled was not registered. The majority of the uncertified products were soft drinks (50%), noodles (18%), energy drinks (14%), fruit juice and milk represented 9% each. Of all the uncertified products, 82% of them contained 1 or more food additives that are potentially hazardous.
- **Food Additives in Products Sampled:** Out of the total number of 63 products sampled, 6 (10%) indicated there were no additives added on their labels. The majority (50%) of products with "No additives" labels were fruit juices followed by milk (33%) and a soft drink (17%). Ninety percent (90%) of all products sampled contained 1 or more food additives. All the food additives found in this survey are presented in the Table 7.1. 71% of all the products sampled contained one or more additives that are likely to cause adverse reactions when consumed.

Table 7.1. Categories of Potentially Harmful Food Additives in Sampled Products

Category	Food Additive
Artificial colors	*Tartrazine (E102), sunset yellow (E110), Amaranth (E123), brilliant scarlet or Ponceau (E124)* and *Allura red (E129)*
Natural colors	*Annatto extracts (E160b)*
Preservatives	*Sodium benzoate (E211), potassiumsorbate (E202), tert-Butylhydroquinone, tBHQ (E319), butylated hydroxytoluene, BHT (E321), butylated hydroxyanisole, BHA (E320)* and *sodium metabisulphite (E223)*

Flavour enhancers	*Monosodium L-glutamate (E621), disodium guanylate (E627)* and *disodium inosinate (E631)*
Artificial sweeteners	*Aspartame (E951), sucralose (E955), cyclamates (E952), saccharin (E954)* and *acesulphame-K (E950)*
Miscellaneous (1)	*Carrageenan (E407), anthocyanins (E163), paprika oleoresins, (E160c), cochineal or carmines red (E120)* and *disodium 5'-ribonucleotides (E635)*
Miscellaneous (2)	Sodium phosphates (E339), potassium phosphates (E501), lecithin (E322), guar gum (E412), sodium alginate (E401), locust bean gum (E410), Mono and di-glycerides of fatty acids (E471), sodium carboxymethylcellulose (E466), glycerin (E422), starch sodium octenylsuccinate (E1450), sucrose acetate isobutyrate (E444), phosphoric acid (E338), citric acid (E330), malic acid (E296), caramel i (E150a), caramel iii (E150c), caramel iv (E150d), magnesium carbonates (E504), acetylated distarch adipate (E1422), calcium disodium EDTA (E385) and gum Arabic (E414)

Italicized words are food additives that are likely to cause adverse reactions

Source: Courage Kosi Setsoafia Saba, 2015. Proliferation of Illegal and Potentially Hazardous Food Additives in Processed and Packaged Foods in Africa: A Case Study and Hazard Identification in Ghana. *Journal of Food Resource Science, 4: 73–81.*

- **Potentially Harmful Artificial and Natural Colors Found in the Products Sampled:** With the artificial colors: Tartrazine (E102) was found in 43% of the ice creams, 13% of soft drinks, 25% of fruit juice, 25% of noodles sampled, sunset yellow (E110) was found in 43% of ice creams, 18% of energy drinks and 13% of soft drinks sampled, amaranth (E123) was found in only 29% of the ice creams sampled, brilliant scarlet (E124) was found in 14% of the ice creams, 9% of the energy drinks and 50% of the spices sampled and Allura red (E129) was found in only 9% of the energy drinks.
- **Potentially Harmful Preservatives in Products Sampled:** Sodium benzoate (E211) was found in 46% of the soft drinks, 27% of the energy drinks and 38% of the fruit juices. Potassium sorbate (E202) was found in 29% of the soft drinks, 36% of the energy drinks, 25% of the fruit juices and 50% of the spices. Sodium metabisulphite (E223) was found in only 1 of the fruit juices (13%). tert-butylhydroquinone, tBHQ (E319) was found in both spices (50%) and noodles (25%) only. Butylated hydroxytoluene, BHT (E321) was found in only one noodle (25%) while Butylated hydroxyanisole, BHA (E320) was also found in only one noodle (25%).
- **Potentially Harmful Artificial Sweeteners in Products Sampled:** Aspartame (E951) was found in 21% of the soft drinks, sucralose (E955) was found in 8% of the soft drinks, cyclamates (E952) was found in 21% of the soft drinks, saccharin (E954) was found in 13% of the soft drinks, acesulphame potassium (E950) was found in 17% of the soft drinks while saccharin was found in 4% of the energy drink.
- **Potentially Harmful Flavor Enhancers in Products Sampled:** Monosodium L-glutamate (E621) was found in 50% of the spices and 100% in the noodles, disodium guanylate (E627) was found in 75% of the noodles while disodium inosinate (E631) was also found in 75% of the noodles.

Discussion

- **Countries of Origin of Products Sampled:** Most of the products sampled originated from Africa (46%) and most of the products from Africa originated from Ghana (62%). However, 54% of the products came from outside Africa. This results show that what is happening in other countries in terms of regulation of food additives must be of concern to us in Ghana and Africa since most of the products were imported from outside. What this study designed but was constrained by funding, was to verify whether these additives with potentially adverse effects are consumed in the countries of origin. This is to ascertain whether some of the products that contain potentially hazardous food additives were intentionally dumped into Africa. There is a tendency to dump products that contain illegal or banned additives in Africa because most of the people in countries like Africa are either ignorant of the E numbers or do not know the adverse effects of some food additives.
- **Categories of Foods Sampled:** Out of the top 4 most common categories of products sampled, 3 categories of products (soft drinks, fruit juice and ice cream) are mostly consumed by children and women. This means any illegal or adverse food additives found in those products are likely to have deleterious effects on children and women.
- **Verification of the Products for Food and Drugs Authority Certification:** It was clear that some products entered the Ghanaian market without the approval of the Food and Drugs Authority. The number of potentially hazardous food additives (82%) that are present in the uncertified products is really a cause for concern. The influx of these products in the country may pose potentially hazardous adverse effects to the people if not properly regulated. The situation is more alarming because 50% of the uncertified products were soft drinks, which are mostly consumed by the vulnerable (children and women) in the population. This clearly shows the loophole in the food regulatory body and the risk of contracting any disease if all trust is put in the regulatory bodies alone. The government must, therefore, encourage other researchers, especially in the universities to carry out research to complement the efforts of the regulatory bodies.
- **Food Additives in Products Sampled:** The result of this study is really a course of concern. Some of these products especially soft drinks, fruit juice and ice creams are taken daily. With the little knowledge that exists in the population about the Acceptable Daily Intake (ADI) of these additives, most people are likely to take them above the acceptable limit. This may lead to toxic effects in consumers and may lead to severe complications. The worrying aspect of this study is that artificial colors or a sodium benzoate preservative (Table above), either mixed or individuals are known to increase hyperactivity in 3-year-old children in the general population. The additive, sodium benzoate (E211) was found in 46% of the soft drinks, 27% of the fruit juice and 27% of the energy drinks sampled. Some of these products are given to children in Ghana and

this may lead to hyperactivity in Ghanaian children. It is worth mentioning that 2 energy drinks from Europe (Germany and the UK) provided cautions on the labels of their products that Sunset Yellow (E110) and Ponceau 4 R (E124) may have an adverse effect on the activity and attention in children. It was, therefore, surprising and scary that most of the products sampled on the Ghanaian market are not have such labels warning consumers about the health effects of consuming such additives.

- **Potentially Harmful Artificial and Natural Colors Found in the Products Sampled:** Patients with allergic disorders have been found to have high leukocyte hypersensitivity to tartrazine in 10.8%, sunset yellow in 4.8% and ponceau or brilliant scarlet in 13.2%. In a similar study, some of these food colorants (amaranth and tartrazine) had a toxic potential to human lymphocytes and have been found to bind directly to DNA. This means that either children or adults who take these products regularly or in excess are at risk of allergic disorders and possible DNA damage, which may result in cancer. The alarming revelation in this study is that tartrazine and other artificial colors are prohibited in several countries including Iran and strictly regulated in the UK by the Food and Standard Agency because of their possible effects on children (http://www.food. govuk/science/additives/foodcolors/). Tartrazine (E102) has been reported to produce several harmful effects in humans and mice of which hyperactivity, asthma and urticaria are most common. However, some studies have also debunked the claim that there are no adverse effects of these additives. Sunset yellow (E110) has also been reported to produce adverse effects in people especially children and has been regulated in most countries. The other artificial colors Allura red (129), amaranth (E123) and ponceau (E124) have also been reported to have adverse effects in several studies. The only natural color found to be potentially harmful among the sampled products was annatto extracts (E160 b), which was found in one of the spices. Annatto extracts (E160 b) have been implicated in causing allergy reactions in certain products. Even though annatto extracts (E160 b) cause allergy reactions, the risk of its allergenicity in Ghana is very low since it was found in one out of the 63 products sampled. But it effect cannot be overemphasized since there is the proliferation of spices in the Ghanaian market. In order to be on the safer side in light of these findings and counter findings, consumers need to be careful when consuming these additives.
- **Potentially Harmful Preservatives in Products Sampled:** Sodium benzoate has been found to cause allergic reactions and other medical conditions when included in food and drugs. The cytotoxicity and genotoxicity of tert-butylhydroquinone (E319) have been demonstrated. Butylated hydroxyanisole, BHA (E320) has also been reported to have carcinogenic effects on epithelial cells. Liver toxicity has been linked to the overconsumption of butylated hydroxytoluene, BHT (E321). Even though there seems to be a very low amount of these food additives in the products sampled, they are present in one of the spices that is widely used in Ghanaian homes and restaurants, so it is a cause for concern in the general public. It is worth noting that one of the spices (popularly

known in Ghana as a cube) in the research contained the highest number of additives that are likely to cause adverse effects. It contains 8 potentially harmful additives [Potassium sorbate, (E202), ponceau 4R (124), monosodium L-glutamate (E621), disodium 5'-ribonucleotides (E635), potassium metabisulphite (E224) butylated hydroxytoluene (E321) butylated hydroxyanisole (E320) and tert-butylhydroquinone, (E319) out of the 11 additives used in its preparation.

- **Potentially Harmful Artificial Sweeteners in Products Sampled:** Artificial sweeteners have generated a lot of controversies. Long-term aspartame exposure could alter the brain's antioxidant status and can induce apoptotic changes in the brain. Aspartame has also been reported to have a possible effect on cognitive functions. Artificial sweeteners are also implicated in certain types of cancers. Studies have also reported that there is no adverse effect of the consumption of artificial sweeteners with regard to their carcinogenic potentials. Long-term use of artificial sweeteners have also been linked with a significant increase in body weight. Acesulphame potassium has also been reported to trigger allergy reactions as well as affect cognitive functions. There are several controversies on the consumption of artificial sweeteners. Considering the long-term exposure to artificial sweeteners, consumers must be educated in Ghana to know the possible side effects of products that contain them. There is more cause for concern especially in children who take soft drinks in Ghana indiscriminately.
- **Potentially Harmful Flavor Enhancers in Products Sampled:** A study has reported that high doses of monosodium L-glutamate (E621) may lead to partial infertility in males. Monosodium L-glutamate-treated rats have also been reported to be more susceptible to developing anxiogenic and depressive like behavior. Oral consumption of Monosodium L-glutamate appears to cause alkaline urine and may increase the risks of kidney stones with hydronephrosis in rats. With the percentage of monosodium L-glutamate that is found in noodles products sampled (100%), there is a serious cause for concern due to the high craze for noodles especially by children in Ghana. In recent times these noodles have flooded the Ghanaian markets. This means that consumers in Ghana are at risk of all the harmful effects of these products.

Conclusion

This study is to survey processed and packaged foods in Ghana and document all the illegal or potentially harmful food additives that are used to prepare them. The study shows that most of the products sampled (71%) contain one or more food additives that could potentially cause adverse effects in the Ghanaian population. The potentially adverse effects that these food additives could trigger in the Ghanaian populace range from allergy, hyperactivity, hypersensitivity, asthma, urticaria and several types of cancer effects on cognitive functions to partial infertility in males. Most of the products that contain potentially harmful additives (soft drinks, fruit juice and ice creams) are mostly consumed by children, who have been reported in several studies to be the most vulnerable to the consumption of potentially harmful additives. The serious concern is the long-term effects these additives have on consumers. The long-term effect of these additives, if not properly monitored will have a serious repercussion on the future generations in other countries. In the course of this study, most people seem to have little knowledge about food additives and could care less about the labels to look for any potentially hazardous additives. This means that there is generally low public knowledge about food additives in Ghana. The Food and Drugs Authority must be resourced very well to carry out extensive tests on products for some of these harmful additives whether they are local or foreign. There is a need for research into the levels of inclusion of these additives to establish if they meet the acceptable daily intake recommended by JECFA.

CLASS ACTIVITY

UNDERSTANDING LACTOSE INTOLERANCE

Table 7.2. Diagnostic Approaches for LI with their Benefits and Limitations

Diagnostic Approach	**Type of Sample**	**Analyte or Focus of Detection**	**Expected Variation in Lactose Intolerant Individuals**	**Benefits**	**Limitations**
Lactose Tolerance Test	Blood	Blood glucose levels at 30, 60, and 120 min post-lactose consumption	Glycemia remains steady	Minimally intrusive; economical	Results affected by subject-specific factors (e.g., gastric clearance)
Genomic analysis	DNA from blood	LCT gene mutations	Existence of known LNP-associated mutations	Can verify or rule out primary intolerance; minimally intrusive	Does not detect secondary intolerance
Lactase quantification in duodenal biopsies	Biopsy of post-bulbar duodenal mucosa	Enzyme lactase levels	Diminished or absent enzyme levels	Accurate diagnosis of lactase deficiency	Invasive; requires technical expertise during endoscopy; costly
Hydrogen Breath Test	Breath	Hydrogen levels before and after ingesting 25 g lactose challenge	Minimum 20 ppm rise in breath hydrogen from the baseline	Affordable; non-invasive; highly sensitive and precise; straightforward execution and interpretation	Time-consuming (3–6 h); requires avoiding a large array of foods and behaviors prior to testing to avoid false positives
Gaxilose Test	Blood and urine	Xylose concentration after gaxilose ingestion	Xylose levels not elevated indicating reduced lactase activity	Non-invasive; direct assessment of lactase activity; patient comfort	Relatively new test; lower specificity than lactose tolerance test

Table 7.3. Clinical studies on lactose supplementation to individuals with lactose intolerance.

Study	Intervention	Subject Count	Condensed Findings
[15]	Study 1: lactose (incrementally increasing from 0.3 to 1.0 g/kg of body weight over 17 days vs. dextrose control periods (crossover) Study 2: Lactose dose increased from (0.6 to 1.0 g/kg of body wt/day) vs. dextrose control in a 10-day crossover study	Study 1: 9; Study 2: 20	Study 1: Increased fecal β-galactosidase. Non-significant decrease in breath hydrogen post-lactose challenge. Study 2: No severe symptoms; significant reduction in flatulence and breath hydrogen during lactose phase in response to a lactose challenge.
[14]	17 g of lactose twice a day vs. sucrose control for 14 days	46 (lactose n = 24; sucrose n = 22)	Increased fecal β-galactosidase, decreased breath hydrogen after lactose challenge; non-significant decrease in symptoms.
[93]	Incremental lactose doses from 3 g to 12 g twice daily for 12 weeks	25	Bifidobacterium and fecal β-galactosidase activity increased; reduced post-lactose challenge breath hydrogen. Lactose well-tolerated.
[97]	33 g of lactose from 4 servings dairy food per day for 21 days	14	Significant decrease in breath hydrogen after lactose challenge; minimal LI symptoms without significant change over time.
[16]	15 g of lactose for six days	24	Decrease in Bacteroides and Clostridium; increase in Lactobacillus; no data on LI symptoms or hydrogen breath concentration.
[98]	25 g of lactose twice daily for two weeks	41 (23 LI, 18 LP)	Significant increase in Bifidobacteria for LI group; non-significant increase in Lactobacilli; non-significant reductions in symptoms and hydrogen breath after a lactose challenge.

Source: Angima G, Qu Y, Park SH, Dallas DC. Prebiotic Strategies to Manage Lactose Intolerance Symptoms. Nutrients. *2024; 16(7):1002. https://doi.org/10.3390/nu16071002.*

SUMMARY

- Additives are added to ensure processed food remains safe and in good condition throughout its journey from factories or industrial kitchens, to warehouses and shops, and finally to consumers. Additives are also used to modify the sensory properties of foods including taste, smell, texture and appearance.
- Indirect food additives are not intentionally added and do not perform any technical function in food. They may be regarded as contaminants as they get into food unintentionally, through food contamination, or arise from chemical and biochemical interactions in food during processing or storage resulting in higher concentrations than is permitted.
- Direct food additives are naturally occurring, or artificially manufactured substances that are intentionally added to food for a specific desirable quality or characteristic whether physical, chemical, biological, or sensorial that consumers demand.
- Whilst enhancing taste and appearance, some additives in our food may be harming our bodies. Despite the strict safety protocols that artificial food additives undergo, consumers may worry about these ingredients' long-term health effects.
- Artificial sweeteners are popular food additives, but while some use them as a solution to a high-sugar diet, others believe they may do more harm than good. Also known as nonnutritive sweeteners, artificial sweeteners contain zero calories but add sweetness to food and beverages. Popular varieties include aspartame, acesulfame K, saccharin, and sucralose.
- Artificial food colorings and flavorings are used to enhance the look and taste of many food products, especially those catered to children. Though natural options exist, artificial colors and flavors are usually used due to their vibrancy and low cost.
- Some of the food additives have been associated with adverse health effects and should be avoided, while others are safe and can be consumed with minimal risk.
- Regulatory Authorities and law enforcement agencies have established restrictive regulations governing the licensing and control of food additives. Regulation is used so that no risks or ill intentions happens.
- With the advent of food processing, food additives play an important role in providing a safe food supply as well as meeting the consumers' needs. Food additive means any substance, either natural or synthetic, intentionally added to food for a technological purpose in the processing, packaging, transport or storage of such food.
- The need for food additives in the food processing industry has risen dramatically due to consumer preferences, and the economic benefits they bring to food products such as ease of processing, consistent quality and longer shelf life.

REVIEW QUESTIONS

1. What are food additives as food preservatives?
2. Give some examples of food additives as food preservatives.
3. How do food additives affect health?
4. What are the positive impacts of food additives?
5. What are the most harmful food additives?

REFERENCES

1. Babbel, J., Ramos, C., Wangberg, H., Luskin, K., & Simon, R. (2021). Adverse reactions to food additives. *Journal of Food Allergy, 3*(1), 8–23. https://doi.org/10.2500/jfa.2021.3.210004.
2. Etwaroo, D. H., Goburdhun, D., & Ruggoo, A. (2019). Classes of food additives in food products sold in Mauritius. *British Food Journal, 121*(5), 1125–1139. https://doi.org/10.1108/bfj-10-2018-0705.
3. Kumar, A. (2020). Consumer awareness about food additives. *Journal of Advanced Research in Quality Control & Management, 5*(2), 21–28. https://doi.org/10.24321/2582.3280.202001.
4. Lück, E., & Jager, M. (1997). *Antimicrobial Food Additives*. Springer Berlin Heidelberg. https://doi.org/10.1007/978-3-642-59202-7.
5. Martyn, D. M., McNulty, B. A., Nugent, A. P., & Gibney, M. J. (2013). Food additives and preschool children. *Proceedings of the Nutrition Society, 72*(1), 109–116. https://doi.org/10.1017/s0029665112002935.
6. Nabavi, S. M., Nabavi, S. F., Loizzo, M. R., Tundis, R., Devi, K. P., & Silva, A. S. (2020). Food additives and human health. *Current Medicinal Chemistry*. https://doi.org/10.2174/9789811446139120010.
7. Pepper, A. N., Sriaroon, P., & Glaum, M. C. (2020). Additives and preservatives: Role in food allergy. *Journal of Food Allergy, 2*(1), 119–123. https://doi.org/10.2500/jfa.2020.2.200014.
8. Qin, X. (2012). Food additives. *Journal of Pediatric Gastroenterology and Nutrition, 54*(4), 564. https://doi.org/10.1097/mpg.0b013e3182464b0e.
9. Xu, B. (2013). Safety and management of food additives in the United States. *Advanced Materials Research, 781–784*, 1328–1331. https://doi.org/10.4028/www.scientific.net/amr.781-784.1328.
10. Zhong, Y., Wu, L., Chen, X., Huang, Z., & Hu, W. (2018). Effects of food-additive-information on consumers' willingness to accept food with additives. *International Journal of Environmental Research and Public Health, 15*(11), 2394. https://doi.org/10.3390/ijerph15112394.

CHAPTER

8 Diet Therapy

LEARNING OBJECTIVES

After studying this chapter, you will be able to:

- Understand basic concepts of diet therapy
- Identify types of diet therapy
- Discuss diet therapy for weight management
- Focus on diet therapy for diabetes management
- Perform diet therapy for heart disease prevention and management
- Explain diet therapy for food allergies and intolerances.

KEY TERMS FROM THIS CHAPTER

Calorie	Diabetes
Diet plan	Dietary assessment
Dietitian	Electrolytes
Food allergies	Gastrointestinal disorders
Genetic traits	Kilojoule
Legumes	Nutritionist

INTRODUCTION

Dietary interventions or specific diets are used in diet therapy to treat or manage a variety of medical conditions. It is frequently combined with other medical interventions to enhance general health outcomes. The aim of diet therapy is to limit or stay away from foods that might make certain conditions worse while simultaneously giving the body the nutrients it needs to support optimum health. Typically, a diet therapist will collaborate with a patient to pinpoint their unique health issues and create a customized diet plan. Numerous medical conditions, such as obesity, diabetes, heart disease, gastrointestinal disorders, and food allergies or sensitivities, can be treated with diet therapy. Diet therapy may occasionally be utilized to assist in managing the symptoms of other chronic illnesses or specific cancers.

Depending on the needs and health status of the patient, there can be a wide range of dietary interventions used in diet therapy. A patient with diabetes, for instance, might benefit from a diet low in sugar and carbs, while a patient with heart disease might benefit from a diet high in fiber and low in saturated fats. All things considered, diet therapy can be a useful strategy for managing a variety of medical disorders and enhancing general health outcomes. To create a diet plan that is both safe and effective, it is crucial to collaborate with a trained healthcare professional.

8.1. BASIC CONCEPTS OF DIET THERAPY

We need food to survive on a basic level. An individual's health and nutritional status are determined by the type and quantity of food they consume. Getting enough nutrients from food is the foundation of a person's nutritional health. In addition to nutrients, food also contains some non-nutrients that can occasionally be crucial in the fight against illness.

Only when food is consumed in the proper amounts, at the appropriate times, and in the proper ways will it nourish and improve health; otherwise, it will devour us. Our eating habits are the most significant factor among the many others. Food and health are directly related because it is the only factor that can be changed for the better and still yield positive results. Nutrients found in food provide the body with nutrition and serve a variety of physiological, regulatory, and body-building purposes. The human body is like a machine; the kind of fuel or nourishment it receives and the materials it is made of determine how well it functions.

The fundamental part of the human body and the state of physical fitness are dependent upon:

1. Genetic traits
2. The quality of material used in its construction
3. Regular maintenance

The biggest overall influence on longevity and health is lifestyle. Lifestyle behaviors such as drug use, smoking, diet, lack of physical activity, psychological stress, and so on; of these, diet is the most significant and controllable factor affecting human health.

8.1.1. Purpose

The main health issues of today are largely caused by acute and chronic conditions stemming from inadequate nutrition and excessive consumption. \

Many cases of heart disease and cancer can be linked to unhealthy dietary choices and obesity. The prevalence of chronic illnesses is on the rise, driven by the increasing rates of obesity among the American population. A balanced food intake is encouraged by diet therapy as it is essential for overall health. One can help maintain the best possible health by eating foods that are appropriate for each individual and making sure they are getting enough water. For the health of individuals, populations suffering from malnutrition as a whole, and groups with specific medical and nutritional needs, eating a healthy diet is essential. A customized diet plan can educate the patient about food-related diseases and the effects of different nutrients (protein, carbohydrates, fats, and alcohol) on ailments, diseases, and obesity. Dietary therapy can be customized to treat patients for specific illnesses, help with side effect and/or complications reduction, and enhance overall health.

8.1.2. Precautions

For every individual, a customized diet plan is recommended. People who are sensitive to certain foods or have medical conditions must be extremely watchful and conscientious about what they eat. People shouldn't embark on a "fad" diet without first seeing a doctor or registered dietitian. Popular low-carb diets, for instance, may deprive the body of the glucose required for brain and central nervous system functions, which can be dangerous at times.

8.1.3. Methods of Feeding

Nutrition intervention aims to provide the patient with enough nutrients in the most physiologic, safe, and economical way possible. Sometimes a disease prevents a person from eating enough or any food at all. There could be a problem with the stomach or bowel, or the organs may have been removed surgically, partially or completely. Nutrition must be administered under those circumstances in a different way, such as enterally and parenterally.

- **Enteral:** Enteral nutrition is defined as nutrition that is given through the gastrointestinal tract. It can be fed to you orally through a tube.
- **Oral:** Nutrition support products known as oral nutritional supplements (ONS) give consumers a safe, efficient means of increasing their dietary intake or meeting their needs for specific nutrients. Even though ONS users may be able to eat a regular diet, they still need supplemental nutrition because their regular diet isn't enough to meet all of their needs. In other cases, when a patient needs a liquid-based diet, ONS may be helpful. A doctor or registered dietitian will frequently prescribe or suggest ONS products. Some individuals get all of their nourishment from ONS.
- **Tube Feeding:** Enteral nutrition therapy (EN) can be given as a tube feeding directly into the gastrointestinal tract if a person has a disease or condition that restricts or impairs oral intake. Life-sustaining nutrients are provided by enteral nutrition through tube feeding, which is frequently necessary as a first-choice feeding technique for individuals who are unable to swallow food or who have digestive system impairments. Specialized liquid meals with protein, carbs, fats, vitamins, minerals, and other life-sustaining nutrients are part of EN therapy. These dietary supplements are designed to address specific requirements for a range of illnesses and ailments.
- **Parenteral:** Parenteral nutrition (PN) is the intravenous (IV) feeding of nutrients into a vein, thereby avoiding the gastrointestinal tract and going straight into the systemic circulation. It is a unique liquid mixture that includes fats, proteins, carbohydrates, vitamins, minerals, and other essential nutrients for life. When oral or enteral routes are insufficient or contraindicated for meeting nutritional needs, parenteral nutrition (PN) offers an additional or alternative approach to nutrition intervention.

8.1.4. Benefits of Diet Therapy

Diet therapy offers the following advantages:

- It helps in weight reduction.
- It guarantees that the body's systems run smoothly.

- It increases vitality and focus.
- It helps with the detoxification plan.
- It controls high blood pressure.
- It improves the immunity system.
- It helps in balancing hormones.

8.1.5. Side Effects of Diet Therapy

The following are the side effects of diet therapy.

- It lowers resting energy expenditure.
- It demonstrates that unfavorable changes in body composition include hormone fluctuations, a loss of bone density, irregular menstruation, and a reduction in resting energy expenditure.
- It leads to irritability.
- It can lead to fatigue (Traverso-Yepez & Hunter, 2016; Downs et al., 2020).

8.2. TYPES OF DIET THERAPY

A menu plan that restricts the consumption of specific foods or nutrients is known as a therapeutic diet. It is usually advised by a physician and coordinated by a dietician as part of the treatment of a medical condition. A therapeutic diet is usually just a modified version of a regular diet. It is changed or tailored to fit the person's dietary needs. The following are the various forms of diet therapy.

8.2.1. Calorie-Modified Diet

Diets with altered calories are recommended to address weight issues while maintaining a healthy diet. For those who are overweight or obese, low-calorie diets are recommended as a means of losing weight. People with significantly higher energy needs, such as athletes in training or those battling illnesses like cancer, AIDS, or cystic fibrosis, are advised to follow high-calorie diets.

Anorexia nervosa is also treated with high-calorie diets. Dietitians create calorie-modified diet plans, which should be recommended after a thorough physical examination, dietary assessment, or dietary history. Enough energy is provided by a low-calorie diet to meet an individual's energy requirements and level of activity. It has a wide range of well-balanced foods, but it also restricts alcohol and carbs. A low-calorie diet shouldn't be designed to help someone lose more than one to two pounds (500 grams to one kilogram) every week. It is generally not advisable to lose more than 1 lb (or 500 grams) per week for someone who is slightly overweight. A high-calorie diet typically adds 500–1,000 extra calories, which causes most people to gain 1 pound (500 grams) of weight per week. It emphasizes foods that are high in calories per unit volume and has a moderate fat content along with a high protein content. It is advised to eat snacks in between meals to increase calorie intake. Limit your daily intake of wine, beer, or liquor to no more than two drinks when calorie constraints permit.

8.2.2. Fiber-Modified Diet

Vegetables and fruits are great providers of dietary fiber. Important nutritional advantages of fiber include its ability to help move food through the digestive tract more easily and reduce the risk of constipation. According to research, a low dietary fiber intake may be linked to colon cancer as well as an increased risk of diverticulosis.

Diets high in fiber are advised for the following reasons: whole grains (especially bran), raw fruits and vegetables, nuts, and seeds; and unpeeled fresh fruit.

- increase fecal bulk
- increase intestinal movement
- avoid or manage diverticulosis, Crohn's disease, irritable bowel syndrome, and constipation.
- help lower cholesterol
- help those who are overweight lose weight and help diabetics with their sugar tolerance.

Avoiding raw fruits and vegetables, whole grains, nuts, and seeds is a key component of low-fiber diets, which instead focus on consuming soft and easily digestible foods. These diets are typically recommended for:

- decrease fecal bulk
- slow intestinal movement
- decrease stomach acid secretion
- treat a range of conditions, such as heart attacks, diarrhea, indigestion, and bowel inflammation.

8.2.3. Protein-Modified Diet

A person following a high-protein diet should consume approximately 0.05 oz (1.5 g) of protein for every kilogram of body weight. Half to two thirds of the daily requirement for protein should come from complex proteins like those found in meats and milk. Diets high in protein are advised for those who.

- possess a greater need for protein as a result of malnutrition low in protein calories, extreme stress, or diseases like AIDS, cancer, or burns with high metabolic rates that cause substantial protein loss.
- possess celiac disease, malabsorption syndromes, or other conditions involving inadequate food absorption.

On a low-protein diet, which forgoes meats and dairy, approximately three-fourths of the daily allotment of protein must come from high-value protein sources. It is possible to prescribe supplements to avoid amino acid deficiencies. Kidney disease and cirrhosis are treated with low-protein diets.

8.2.4. Low-Cholesterol Diet

The first line of defense against high cholesterol, which is a contributing factor to atherosclerosis and heart disease, is dietary modification.

Low-fat diets are recommended to lower blood pressure, lower serum cholesterol (which can be inherited and may also need cholesterol-lowering medication), treat atherosclerosis, diabetes, and lower the risk of heart disease. Most people must adhere to a low-cholesterol diet for the remainder of their lives because it is not a cure for the conditions it is intended to treat.

According to the American Heart Association eating plan, total fat consumption should not exceed 30% of total calories and total cholesterol should not exceed 0.01 oz (300 mg) per day.

Less than 10% of calories should come from saturated fats; individuals with high blood cholesterol or heart disease should limit their consumption of saturated fats and cholesterol to 7% of total calories.

The AHA eating plan also suggests:

- 8–10% of calories should come from polyunsaturated fatty acids.
- The remaining fat, up to 15% of total calories, should come from monounsaturated fatty acids.

These recommendations are applicable to everyone in good health who is older than two.

8.2.5. Low-Fat Diet

Too much fat is a part of most American diets. Typically, fat accounts for 40% or more of the total calories ingested daily. Dietary guidelines advise consuming no more than 20–35% fat in daily calories because consuming too much fat has been related to heart disease, obesity, and various cancers. An extremely low-fat diet restricts daily fat consumption to 0.88–1.05 oz (25–30 g), while a low-fat diet typically caps daily fat intake at 1.76 oz (50 g). The grams of fat in a person's diet are determined by their caloric requirements. Diets low in fat are advised for:

- help prevent heart disease
- help prevent colon, prostate, and breast cancers
- aid in the treatment of numerous illnesses, such as heartburn, gout, AIDS, gallbladder disease, liver disease, celiac disease, and inflammatory bowel disease.
- lose or control weight

To prevent essential fatty acid deficiencies, some fat must be consumed in the diet; however, most people eat more than enough fat to meet their needs. People who reduce their fat intake are likely to consume fewer calories, which will help them lose weight and lower their risk of coronary heart disease.

8.2.6. Gluten-Free Diet

Certain grains and products containing grains contain the proteins gluten and gliadin. These proteins impair an intolerant person's ability to absorb food because they are harmful to the intestinal tract's cells. An intolerance to these proteins is the root cause of celiac disease. Patients with celiac disease experience bloating, diarrhea, weight loss, and malnutrition as a result of this intolerance. Reducing gluten-containing foods from the diet can relieve symptoms, stop additional intestinal damage, and improve nutritional malabsorption. All foods containing wheat, rye, barley, and malt are off limits on a gluten-free diet, which needs to be adhered to for life.

8.2.7. Low-Purine Diet

This diet limits foods that make the body produce uric acid, like eggs, liver, and sardines. It is typically prescribed as a part of a comprehensive treatment plan that also includes exercise and medication for kidney stones and gout, a condition that is typically brought on by an excess of uric acid in the body. This diet stresses drinking approximately 2 quarts (1.89 liters) of water and fruit juice daily to promote the excretion of uric acid, as well as eating fruits and vegetables that increase urine alkalinity and the solubility of uric acid. Other purine-containing foods that must be avoided include sweetbreads, liver, kidney, and some types of fish, such as anchovies, sardines, and mackerel, as well as limiting the amount of other purine-containing foods like shrimp, meats, and dairy products.

8.2.8. Low-Salt Diet

Americans ingest 0.17 oz (5 g) of salt or sodium on average per day.

Even in the absence of any heart disease symptoms, 0.08 oz (2.4 g) of sodium is the recommended upper limit according to dietary guidelines. Less than 0.07 oz (2 g) of sodium should be consumed daily by most heart disease sufferers, and some low-salt diets only allow 0.008 oz (250 mg) of sodium per day. For those with congestive heart failure or high blood pressure, the amount of salt in their diet is crucial.

Some experts, particularly in Western countries, think that consuming excessive amounts of salt is a major cause of high blood pressure. Over sodium promotes fluid retention in the body, which raises the amount of fluid the heart pumps and circulates in the blood. People with congestive heart failure may also be adversely affected by diets high in salt because the excess fluid can cause congestion in the lungs.

One common ingredient in salt substitutes is potassium chloride. However, those who have renal issues may be adversely affected by excessive potassium. Using flavored vinegar, herbs, spices, or lemon juice can improve food flavor without adding salt.

8.2.9. Low-Phenylalanine Diet

Phenylketonuria (PKU) is typically treated with a low-phenylalanine diet. If untreated, PKU, a rare genetic disorder that affects the breakdown of dietary phenylalanine, can cause severe, progressive mental retardation. To lessen or avoid mental retardation, strict dietary compliance is required due to the extremely restrictive nature of the diet. It is essential to have a doctor or registered dietitian provide close supervision.

People with PKU are unable to tolerate a typical diet. Dietary treatment requires staying away from foods that are high in protein. Merely the necessary quantity of phenylalanine for the body is present in a prescribed diet. The following are the fundamentals of the PKU diet:

- Foods high in protein and consequently phenylalanine, such as meat, fish, cheese, eggs, milk, and nuts, are prohibited.
- Other foods which contain moderate amounts of protein (e.g., potatoes, and cereals) are provided in tiny, precisely measured portions. To maintain stable phenylalanine levels, these foods are divided up throughout the day's meals.
- In moderation, the majority of fruits, certain vegetables, and salads are safe to consume, but overindulgence should be avoided.
- Use of sugar, jam, syrups, and fats like butter, lard, and cooking oil is quite permissive.
- Many manufactured low-protein foods are available with prescriptions. You can take any of these at any time to add some diversity to your diet. Foods include spaghetti, biscuits, low-protein bread, pasta, and flour.
- Infants can be fed phenylalanine-free formulas.

Because they consume more carbohydrates than normal, people with PKU have a higher risk of tooth decay.

8.2.10. Diabetic Diet

Generally speaking, controlling one's diet is essential to managing diabetes. No single diet can satisfy the requirements of every diabetic. Diabetics should follow the general guidelines for healthy eating that were previously covered. There are numerous dietary strategies for managing blood sugar levels. All adults over the age of two, including those with diabetes, should adhere to the Dietary Guidelines for Americans. Although there are more intricate dietary techniques for blood sugar control, some experts think that these dietary recommendations may be adequate for diabetics. Nonetheless, the complexity of these techniques discourages a lot of diabetics from utilizing them. The most widely used method of blood sugar control is the Diabetic Exchange Lists, which were created by the American Diabetic Association. Other nutrition experts point out that the food pyramid has certain shortcomings, such as not placing enough emphasis on meal planning, and instead advise switching to a Mediterranean diet. While carb counting programs can be helpful, they can also be complex and demanding on a dedicated learner. Additionally, some people may find it challenging to comprehend the ideas behind the Diabetic Exchange Lists. Controlling blood glucose levels is crucial for type 1 and type 2 diabetics using insulin or oral medication. This can be achieved by adjusting blood glucose-lowering factors like exercise and food intake in conjunction with insulin administration or medication.

Dietary practices that help with glucose regulation include:

- Stick to a meal plan.
- Appropriately treat hypoglycemia (low blood sugar).
- Quickly respond to hyperglycemia (high blood sugar).
- Maintain consistent snacking habits.

Eating more complex carbohydrates is the American Diabetes Association's recommendation. This goes against the advice given in previous years. More grains, beans, and starchy vegetables are the healthiest foods for everyone to eat in order to reduce fat and cholesterol, according to recent studies. The total amount of carbohydrates consumed has a bigger effect on blood glucose regulation than the type of carbohydrates. That being said, consuming complex carbohydrates (e.g., beans, whole grain bread, etc.) are preferable to consuming meals high in sucrose, or simple sugar.

For overweight individuals with type 2 diabetes who are not taking medication, blood sugar control and weight loss are especially crucial. After initial weight loss, health effects are at their peak. The progression of type 2 diabetes can be slowed down by 10% in body weight reduction. Other crucial matters include blood pressure regulation and lipid (triglyceride and cholesterol) level control.

Diabetes increases the risk of cardiovascular disease by approximately twice as much as non-diabetics, so reducing fat intake is crucial. Based on available data, individuals with diabetes mellitus have equivalent protein needs as healthy individuals. However, when nephropathy sets in, adults should consume no more than 0.8 grams of protein per kilogram daily, of which 80% should come from high-value biological protein.

8.2.11. Traditional Diets May Offer Some Health Advantages

Compared to the typical Western diet, which is frequently excessively high in fat and cholesterol, certain populations follow far healthier eating practices. Studies have demonstrated the health benefits of the hunter-gatherer diet, the traditional Mediterranean diet, and the Japanese diet. With a low incidence of coronary heart disease, the dietary practices common to Mediterranean countries that emphasize the consumption of olive oil (monounsaturated fatty acid) as the primary source of fat appear to offer optimal health benefits. The Mediterranean diet is characterized by a high consumption of fruits, vegetables, pulses, nuts, cereal products, and fish, and a low consumption of meat and dairy products (Traverso-Yepez & Hunter, 2016; Downs et al., 2020).

8.3. DIET THERAPY FOR WEIGHT MANAGEMENT

The best way to reach and stay at a healthy weight is through diet therapy for weight management. Diet therapy presents a valuable solution to the issue of weight-related problems that are plaguing an increasing number of people worldwide. By using this strategy, people can establish a calorie deficit, which will aid in weight loss and enhance general health outcomes. A variety of dietary interventions that are customized to meet the unique needs and objectives of each individual are part of diet therapy for weight management. Calorie control is one of the most crucial parts of weight management. People must consume fewer calories per day than their body expels in order to lose weight. This can be accomplished by cutting back on high-calorie foods, choosing low-calorie foods and beverages, and controlling portion sizes.

Another essential element of diet therapy for weight management is macronutrient balance. While preserving muscle mass, a balanced diet with a moderate intake of protein, carbs, and healthy fats can aid in weight loss. Additionally, a balanced diet can lower the chance of developing chronic illnesses like diabetes, heart disease, and some forms of cancer. Consuming fiber is a crucial component of diet therapy for managing weight as well. Consuming a diet rich in fiber can encourage feelings of fullness, which can help you consume fewer calories overall. Nuts, seeds, whole grains, fruits, and vegetables are examples of foods high in fiber.

Another essential component of diet therapy for weight management is meal timing. Regular mealtimes throughout the day can aid in weight loss by reducing the risk of overindulging. A nutritious meal plan could consist of three meals a day plus one or two snacks as needed. It's crucial to eat breakfast because it can speed up your metabolism and keep you from overindulging later in the day.

8.3.1. The Role of Macronutrients in Diet Therapy

Carbohydrates, proteins, and fats are examples of macronutrients, and they are important for controlling weight. The distinct qualities of each macronutrient influence weight gain and loss. The body uses carbs as its main energy source. Excessive consumption of them may lead to weight gain. But not every carbohydrate is made equal. Blood sugar spikes and crashes brought on by simple carbohydrates, like those in refined sugars, can induce hunger and overindulgence in food. Whole grains, fruits, and vegetables contain complex carbohydrates, which digest more slowly and give longer-lasting energy. Therefore, by decreasing hunger and preventing overeating, a diet high in complex carbohydrates can aid in weight management.

In addition to being involved in numerous metabolic processes, proteins are necessary for the development and maintenance of tissues. Additionally, protein keeps you feeling full, which can help you control your weight. A high-protein diet has been linked to decreased appetite, increased satiety, and preservation of lean muscle mass during weight loss, according to studies. Although fats are necessary for energy and metabolic functions, they are frequently vilified as being unhealthy and a cause of weight gain. But not every fat is made equally. In addition to being vital for general health, healthy fats—like those in nuts, seeds, and fatty fish—can also help with weight loss by encouraging fullness and lowering inflammation. Conversely, unhealthy fats like trans and saturated fats should be avoided since they can cause weight gain. To control weight, a diet rich in all three macronutrients must be balanced. Appropriate portions of lean proteins, complex carbohydrates, and healthy fats will support overall health and weight loss goals while providing sufficient energy and satiety.

8.3.2. The Importance of Portion Control

One of the most important aspects of weight management is portion control. Eating a lot of food can lead to overindulgence in food and weight gain. Conversely, eating in moderation can support both achieving and maintaining a healthy weight. Here are some pointers on controlling portion sizes:

1. **Use Smaller Plates:** Opting for a smaller plate can aid in consuming smaller food portions without the sensation of being restricted. A smaller plate simplifies the process of determining appropriate portion sizes.
2. **Measure Food:** To measure food, use a food scale, measuring cups, or other portion control devices. This guarantees that they are eating the right portions and prevents overeating.
3. **Eat Slowly:** Eating slowly can help you avoid overindulging and feel fuller sooner. Enjoy your food slowly, chew it thoroughly, and take a break in between bites.
4. **Use the Hand Rule:** Using your hands is an easy method of estimating portion sizes. A serving of protein, for instance, should be the size of a palm; a serving of vegetables, a fist; and a serving of carbohydrates, a cupped hand.
5. **Pack Leftovers:** If someone goes out to dine, pack the leftovers to prevent eating too much at once.
6. **Be Mindful of Snacks:** Remember that snacks can quickly accumulate calories, so watch how much you eat. To prevent overindulging, use portion control containers or divide snacks ahead of time into baggies.
7. **Read Labels:** To learn about calorie counts and serving sizes, read the labels on packaged foods. Consider serving sizes and make necessary adjustments.

One of the most important aspects of weight management is portion control. To control portions and reach weight loss goals, try using smaller plates, measuring food, eating slowly, applying the hand rule, packing leftovers, keeping an eye on snacks, and reading labels.

8.3.3. Fiber in Weight Management

One important element of a balanced diet that can help with weight management is fiber. Foods high in fiber promote fullness, which can help people eat less and lose weight. Furthermore, fiber can lessen the risk of heart disease and some types of cancer, help control blood sugar levels, and avoid constipation. Soluble and insoluble fibers are the two different varieties. In the digestive tract, soluble fiber gels and dissolves in water to slow down the absorption of carbohydrates and help control blood sugar levels. Contrarily, insoluble fiber does not dissolve in water and gives stool more volume, which helps to ward off constipation.

Sources of high-fiber foods include:

1. **Fruits:** High-fiber foods include kiwi, oranges, bananas, pears, apples, and berries.
2. **Vegetables:** Vegetables high in fiber include spinach, broccoli, carrots, sweet potatoes, and Brussels sprouts.
3. **Legumes:** High in fiber are split peas, kidney beans, black beans, chickpeas, and lentils.
4. **Whole Grains:** Good sources of fiber include oats, barley, quinoa, brown rice, and whole wheat products like pasta and bread.
5. **Nuts and Seeds:** High-fiber foods include pumpkin seeds, flaxseeds, almonds, and chia seeds.

To prevent stomach discomfort, it's crucial to gradually increase fiber intake and drink lots of water. For women, aim for at least 25 grams of fiber daily, and for men, aim for 38 grams.

8.3.4. The Impact of Hydration

Being hydrated is crucial for managing weight because water is necessary for numerous body processes, such as controlling metabolism, digestion, and body temperature. Water consumption can also aid in weight loss by lowering hunger, raising satiety, and speeding up metabolism. Our bodies may confuse thirst for hunger when we are dehydrated, leading us to overeat.

Staying hydrated can assist in lowering the risk of overindulging in food and pointless snacking. Water consumption before meals can also help us eat less, which eventually contributes to weight loss.

Throughout the day, carrying a reusable water bottle is a good way to stay hydrated. This makes it possible to drink water consistently and makes tracking fluid intake easier.

Moreover, eating a diet high in water-rich foods, such as fruits and vegetables, can help maintain proper hydration. Drinking soups and broths is another healthy way to stay hydrated, especially during the winter months. People should be aware of the symptoms of dehydration, which include fatigue, dry mouth, and dark urine. These signs may suggest that you should be drinking more water. A minimum of eight glasses of water should be consumed each day, though the exact amount will depend on your age, activity level, and general health.

8.3.5. The Effect of Processed Foods

Because processed foods are high in calories, contain added sugars, unhealthy fats, and have a low nutrient density, they can significantly contribute to weight gain. They are frequently made to be incredibly convenient and palatable, which makes it simple to overeat and take in more calories than one needs.

The following advice can help you consume fewer processed foods:

1. **Choose Whole, Unprocessed Foods:** Choose whole foods like vegetables, fruits, whole grains, lean meats, and healthy fats. These foods can aid in feeling full and content because they are high in nutrients, fiber, and water.
2. **Read Food Labels:** When perusing the nutritional label of packaged foods, try to find those with the least amount of added sugar, sodium, and unhealthy fats. Select goods with easily identifiable, basic ingredients.
3. **Cook at Home:** You can manage the ingredients and portion sizes when you cook at home. To make eating healthily more convenient, try to prepare meals ahead of time.
4. **Snack on Healthy Options:** Pick for healthier snacks like fresh fruit, veggies with hummus, or nuts instead of processed foods like chips and candy.
5. **Limit Intake of Sugary Drinks:** Sugar-filled beverages, such as juice and soda, can cause weight gain. Instead, go for low-fat milk, unsweetened tea, or water.

An individual can help minimize the intake of processed foods and lower the risk of weight gain by adhering to these suggestions and concentrating on whole, unprocessed foods.

8.3.6. The Benefits of Mindful Eating Practices

A practice known as mindful eating entails being totally present and involved in the eating experience. Numerous advantages of this eating strategy for managing weight have been demonstrated, including:

1. **Increased Awareness of Hunger and Fullness Cues:** Gaining greater awareness of your body's hunger and fullness cues can be achieved through mindful eating. This can aid in improving appetite control and preventing overeating.
2. **Reduced Emotional Eating:** To recognize and control the emotions that lead to overindulgence in food or unhealthy eating habits, mindful eating is helpful. One can learn to deal with emotions in a healthier way by increasing their awareness of them.
3. **Improved Digestion:** Eating slowly and with more chewing is possible when one eats mindfully. This can lessen gastrointestinal problems like indigestion and bloating.
4. **Increased Enjoyment of Food:** Eating mindfully promotes experiencing all of the senses—taste, texture, and aroma of food—when consuming food. This can encourage greater meal enjoyment and greater satisfaction from smaller portions.
5. **Reduced Stress:** Through relaxation and a decrease in overwhelming or anxious feelings related to food and eating, mindful eating can help people feel less stressed.

The concept of mindful eating is derived from mindfulness, which is a widely accepted and centuries-old practice found in many different religions.

An intentional focus on one's feelings, ideas, and bodily experiences in the present is known as mindfulness. The goal of mindfulness is to increase awareness of one's surroundings and decisions rather than to react to them. Eating mindfully entails experiencing and appreciating the foods one chooses by using all of their sensory faculties, both physiological and emotional. This fosters a greater sense of appreciation for food, which enhances the dining experience in general. Making decisions while eating mindfully encourages one to choose foods that are both gratifying and healthy for the body. But since there are various kinds of eating experiences, it discourages "judging" one's eating habits. With increased awareness of our eating patterns, we can work to modify our behavior for the good of both the environment and ourselves.

Promoting a more pleasurable mealtime and awareness of the dining environment is the aim. Four guiding principles for mindful eating were outlined by Fung and colleagues: what to eat, why we eat what we eat, how much to eat, and how to eat.

Mindful eating:

- takes into account the meal's larger components, such as the food's origin, preparation method, and chef.
- recognizes both external and internal cues that influence our food intake
- observes the appearance, flavor, aroma, and physical sensation of the food as we consume it.
- recognizes the physical sensations that a meal causes in the body.
- expresses gratitude for the meal
- may practice meditation or deep breathing either before or after the meal.
- considers the impact that the food we eat has on the environment both locally and globally.

8.3.7. The Role of Physical Activity

It takes physical activity to control weight. People can burn calories when they exercise, which produces a calorie deficit that is essential for weight loss. Physical activity is essential for maintaining a healthy weight even for people who are not actively trying to lose weight. Increasing muscle mass through regular exercise can also help a person burn more calories even when they're not exercising. Furthermore, being physically active can aid in the prevention or management of diseases like heart disease, high blood pressure, and type 2 diabetes, all of which are linked to being overweight or obese.

It's imperative to understand that weight control is not solely reliant on physical exercise. A nutritious diet plays a crucial role in managing weight as well. By combining a healthy diet with regular physical exercise, individuals are more likely to achieve and sustain a healthy weight. The amount of physical activity required for weight control varies based on factors like age, gender, weight, and fitness level. Generally, adults are advised to engage in at least 150 minutes of moderate-intensity aerobic activity or 75 minutes of vigorous-intensity aerobic activity per week, in addition to muscle-strengthening exercises at least twice a week.

In general, physical activity is crucial for managing weight, and it's critical to identify enjoyable physical activities for an individual to incorporate into their routine on a regular basis.

8.3.8. The Impact of Stress

The control of weight can be significantly impacted by stress. Stress causes the body to release cortisol, a hormone that can boost appetite and lead to increased fat storage, especially in the abdominal area. Furthermore, stress can result in emotional eating, a behavior in which people turn to food as a comfort or a coping mechanism.

There are a number of strategies to help with stress management and weight loss.

1. **Exercise:** Engaging in physical activity helps you burn calories and manage stress. Endorphins, which are released during exercise, have been shown to elevate mood and lower stress.
2. **Mindfulness and Meditation:** Stress and emotional eating can be lessened with the use of mindfulness and meditation practices. In order to prevent overeating, mindful eating practices can also help one become more conscious of their body's hunger and fullness cues.
3. **Yoga:** Yoga incorporates breathing exercises, meditation, and relaxation methods. Yoga can help you become more flexible and strong while lowering your stress levels.
4. **Sleep:** Sleep is crucial for stress management and for promoting weight loss. Sleep deprivation can exacerbate stress, appetite, and weight gain.
5. **Social Support:** Having friends and family as a support network can lower stress levels and hold people accountable for their weight loss objectives.
6. **Time Management:** Effective time management can lower stress levels and make it simpler to give physical activity and a healthy diet top priority.

Although stress can significantly affect how one manages their weight, there are strategies one can employ to both reduce stress and promote weight loss. To manage stress and reach weight loss goals, incorporate physical activity, mindfulness practices, yoga, adequate sleep, social support, and efficient time management (Traverso-Yepez & Hunter, 2016; Downs et al., 2020).

8.4. DIET THERAPY FOR DIABETES MANAGEMENT

Diabetes is a long-term illness that impairs the body's capacity to utilize and store glucose, leading to elevated levels of blood sugar. Diabetes is managed with a combination of diet therapy, exercise, and medication. Diet therapy is a crucial component of diabetes management as it can enhance blood sugar regulation, lower the risk of complications, and enhance general health. In addition to providing for nutritional needs, diet therapy for diabetes management aims to reach and maintain healthy blood sugar levels. This is usually accomplished by eating a balanced diet that contains a range of foods in sensible portions from each food group. A certified dietitian can assist in creating a customized meal plan that takes into account each person's requirements and preferences.

A popular method for meal planning for diabetics is carb counting. Since carbohydrates have the biggest effect on blood sugar levels, it's critical to pay attention to how much of them you eat. A person's ability to consume carbohydrates is influenced by several variables, including age, gender, weight, degree of physical activity, and use of medications. A person with diabetes should aim to consume 45–60 grams of carbohydrates per meal, with a typical carbohydrate serving size of about 15 grams. Carbohydrate-rich foods include fruits, dairy products, grains (like bread, rice, and pasta), and sweets. To slow down the digestion and absorption of carbohydrates, these foods can be included in a diabetic meal plan, but only in moderation and in combination with other foods that offer protein, fat, and fiber.

Maintaining muscle mass and encouraging fullness require protein. Lean meat, poultry, fish, eggs, tofu, and legumes are good sources of protein. Fat is a necessary component of the diet as well, but it's crucial to select healthy fats like those in nuts, seeds, avocados, and fatty fish. Given its ability to enhance blood sugar regulation and increase feelings of fullness, fiber is a crucial nutrient for the management of diabetes. Whole grains, legumes, fruits, and vegetables are all excellent providers of fiber. It's crucial to pay attention to the frequency and timing of meals in addition to monitoring the amount of carbohydrates consumed. Consuming meals on a regular basis can help control blood sugar levels and avoid crashes and spikes. Low blood sugar and overeating at the next meal can result from skipping meals or waiting too long between meals.

8.4.1. Develop a Healthy-Eating Plan

A dietitian can help people with diabetes or prediabetes create a healthy meal plan. This is usually advised by their healthcare provider. This program is intended to help control blood sugar levels, help you maintain a healthy weight, and help manage heart disease risk factors like high blood pressure and elevated blood fats. An adverse rise in blood glucose levels is produced by the bodies of people who consume excess fat and calories. Not controlling blood sugar levels can lead to serious problems, such as hyperglycemia, which can lead to long-term consequences like heart, kidney, and nerve damage if left untreated. By making healthy food choices and keeping an eye on eating patterns, one can maintain a safe blood glucose range. Losing weight has several other health advantages and can help most people with type 2 diabetes better control their blood sugar. A diabetes diet offers a structured, nourishing approach for people who need to lose weight to safely achieve their objective.

8.4.2. Recommended Foods

Eating three meals a day at consistent times is the foundation of a diabetes diet. This aids people in making better use of the insulin that is either produced by their bodies or obtained via medication. A registered dietitian can help create a diet plan based on a person's preferences, lifestyle, and health objectives. Additionally, they can offer advice on how to eat better, such as choosing portion sizes that suit one's size and degree of activity. It is advised to eat a diet rich in fiber-rich foods, fish, healthy fats, and carbohydrates to ensure that you are getting enough calories.

- **Healthy Carbohydrates**: Blood glucose is produced during digestion from the breakdown of simple carbohydrates (sugars) and complex carbohydrates (starches). Pay attention to good carbs like:
  - Fruits
  - Vegetables
  - Whole grains
  - Legumes, such as beans and peas
  - Low-fat dairy products, such as milk and cheese

Steer clear of less nutritious carbs, like those found in meals or beverages that have extra sodium, sugar, or fat.

- **Fiber-Rich Foods**: All plant food components that the body is unable to digest or absorb are considered dietary fiber. Fiber helps regulate blood sugar levels and moderates how the body breaks down food. High-fiber foods consist of:
  - Vegetables
  - Fruits
  - Nuts
  - Legumes, such as beans and peas
  - Whole grains
- **Heart-Healthy Fish**: Consume heart-healthy fish at least twice a week. Choices like salmon, mackerel, tuna, and sardines are packed with omega-3 fatty acids, known to help reduce the risk of heart disease. Stay away from fried fish and varieties like king mackerel that may contain high levels of mercury.
- **'Good' Fats**: Foods high in polyunsaturated and monounsaturated fats can assist in lowering cholesterol. These consist of.
  - Avocados
  - Nuts
  - Canola, olive and peanut oils

Don't go overboard though, as fats all contain a lot of calories.

8.4.3. Foods to Avoid

Maintaining blood glucose levels within a target range and giving the body enough nutrition to meet its needs are the main objectives of diet therapy for diabetes. Foods that are high in fiber, vitamins, and minerals and low in fat, salt, and sugar make up a healthy diet for people with diabetes. Eating the healthiest foods in moderation and adhering to regular mealtimes are the fundamental components of a diabetes diet. A diabetes diet is a nutritious eating plan that is naturally low in calories and fat. Whole grains, fruits, and vegetables are essential components. For most people, a diabetes diet is actually the best eating plan.

Diabetes accelerates the clogging and hardening of arteries, increasing the risk of heart disease and stroke. Foods that include the following may hinder your efforts to maintain a heart-healthy diet.

- **Saturated Fats:** Steer clear of animal proteins like butter, beef, hot dogs, sausage, and bacon as well as high-fat dairy products. Don't use too much coconut or palm kernel oil either.
- **Trans Fats:** Steer clear of the trans fats present in stick margarines, shortening, baked goods, and processed snacks.
- **Cholesterol:** High-fat dairy products, high-fat animal proteins, egg yolks, liver, and other organ meats are among the foods high in cholesterol. Limit the amount of cholesterol you consume to no more than 200 mg per day.
- **Sodium:** Don't consume more than 2,300 mg of sodium per day. In the event that a patient has high blood pressure, the doctor might advise even lower goals.

8.4.4. Putting it All Together: Creating a Plan

Create a diabetes diet using a few different strategies to assist a person in maintaining blood glucose levels within normal limits. A person may discover that one or more of the following strategies, with the assistance of a dietitian, is effective for them.

- **The Plate Method**: There is an easy meal-planning method provided by the American Diabetes Association. It basically emphasizes increasing vegetable consumption. To prepare the plate, follow these steps:
  - Use non-starchy vegetables like tomatoes, spinach, and carrots to cover half of the plate.
  - A protein, such as tuna, lean pork, or chicken, should take up one-fourth of the plate.
  - Use a whole grain product, like brown rice, or a starchy vegetable, like green peas, to fill the final quarter.
  - Incorporate healthy fats like nuts and avocados into your diet in moderation.
  - Include a serving of dairy or fruit, along with a glass of water, unsweetened tea, or coffee.
- **Counting Carbohydrates**: Carbohydrates have the largest effect on blood glucose levels because they are broken down into glucose. People may need to learn how to calculate their carbohydrate intake to properly control their blood sugar levels, which will allow for any necessary adjustments to insulin dosage. It

is critical to keep an eye on how many carbohydrates are included in each meal and snack. A dietitian can teach people how to accurately portion food and develop their food label reading skills. Additionally, they can impart knowledge on how to recognize and comprehend the sugar content.

- **Choose Your Foods**: Dietitians can provide recommendations for specific food choices to assist with meal and snack planning. selecting a variety of foods from lists that contain groups like fats, proteins, and carbs.
- **One Serving in a Category is Called a "Choice**:" A meal choice has roughly the same quantity of calories, fat, protein, carbs, and other nutrients as a serving of every other food in the same category, as well as the same impact on blood sugar. On the list of options for starch, fruits, and milk, for instance, there are options with 12 to 15 grams of carbohydrates.
- **Glycemic Index**: Some diabetics choose their foods based on the glycemic index, particularly when it comes to carbohydrates. This system assigns a ranking to foods that contain carbohydrates according to how they affect blood glucose levels.

8.4.5. Carbohydrate Counting

To help diabetics maintain stable blood sugar levels, a diet therapy called "carbohydrate counting" involves keeping track of the number of carbohydrates consumed during meals and snacks. To avoid dangerously high or low blood sugar, the aim of carb counting is to balance the amount of carbohydrates consumed with insulin and prescription drugs. Numerous foods, such as fruits, vegetables, grains, and dairy products, contain carbohydrates. After consumption, carbohydrates are converted to glucose, or sugar, which enters the bloodstream and elevates blood sugar levels. If a person has diabetes and their body cannot make enough insulin or use it correctly to control blood sugar, this can lead to complications.

Understanding how to estimate the amount of carbohydrates in various foods is essential for putting carbohydrate counting into practice. You can accomplish this by reading food labels or by using resources that list carbohydrates, such as reference books, websites, or apps. Personalized carb counting plans can also be developed with the assistance of a registered dietitian. When estimating the amount of carbohydrates in a meal or snack, a diabetic can use this information to calculate how much insulin or medication is needed to control their blood sugar levels. To keep blood sugar levels steady, the objective is to eat a steady amount of carbohydrates throughout the day. All things considered, carb counting may be a useful diet therapy for managing diabetes; however, to be successful, it needs constant supervision, education, and support. In order to meet specific needs and goals, it is crucial to develop and modify a carbohydrate counting plan in collaboration with a healthcare team that includes a registered dietitian.

8.4.6. Glycemic Index

Consuming foods with a lower glycemic index (GI) is one diet therapy used to help diabetics control their blood sugar levels. The GI is a system of numbers that ranks foods according to how quickly they raise blood sugar in relation to a reference, like white bread or glucose. Foods that have a high GI (above 70) digest quickly and raise blood sugar levels quickly, whereas foods that have a low GI (below 55) digest more slowly and raise blood sugar levels more gradually. Foods with a moderate GI (56–69) are in the middle of the spectrum.

Selecting foods with a lower GI is the aim of the glycemic index diet therapy in order to help maintain stable blood sugar levels. Foods with a low GI include the following:

- leafy greens, broccoli, and cauliflower are examples of non-starchy vegetables.
- Legumes such as beans, lentils, and chickpeas
- Whole grains such as brown rice, quinoa, and oatmeal
- Fruits such as apples, berries, and pears
- Nuts and seeds

Contrarily, foods with a high GI that should be restricted or eliminated include:

- Sugary beverages such as soda and fruit juice
- processed foods like crackers, pasta, and white bread.
- Candy and sweets

It's crucial to remember that other factors also influence blood sugar levels besides the glycemic index.

Blood sugar levels can also be influenced by the quantity of carbohydrates, fat, protein, and fiber content of a meal, as well as by portion sizes.

Thus, it's critical to collaborate with a healthcare team, which should include a registered dietitian, to create a customized glycemic index plan that addresses each person's needs and goals.

8.4.7. Mediterranean Diet

The Mediterranean diet, which focuses on whole grains, fruits, vegetables, nuts, seeds, and healthy fats like olive oil and fatty fish, has been shown to be beneficial for diabetes management.

The Mediterranean diet can help control diabetes in the following ways:

1. **Helps Regulate Blood Sugar Levels:** The Mediterranean diet is low in simple carbohydrates and high in fiber. This lowers the risk of blood sugar spikes and helps to regulate blood sugar levels.
2. **Improves Heart Health:** Heart disease is more common in those who have diabetes. The Mediterranean diet is linked to a lower risk of heart disease and is high in heart-healthy fats.
3. **Promotes Weight Loss:** It is crucial to maintain a healthy weight in order to manage diabetes. The Mediterranean diet is focused on consuming less processed foods and more fiber, which can aid in weight loss efforts.
4. **Reduces Inflammation:** Type 2 diabetes and insulin resistance are linked to chronic inflammation. Anti-inflammatory foods like fruits, vegetables, and nuts are abundant in the Mediterranean diet and can help lower inflammation.
5. **Supports Overall Health:** Numerous chronic diseases, such as diabetes, heart disease, and specific cancers, are less likely to occur with the adoption of the Mediterranean diet. This diet offers a diverse array of nutrients that promote overall health.

If someone has diabetes and is thinking about switching to a Mediterranean diet, it's crucial to collaborate with a medical professional or a registered dietitian to create a personalized plan that suits their individual requirements and preferences. Diets rich in olive oil and fish have been linked to lower rates of high blood pressure and cholesterol, two conditions that increase the risk of heart disease.

For those who have diabetes, a Mediterranean-style diet can be very beneficial because it may lower their chance of developing certain complications from their condition. Additionally, there is evidence that the Mediterranean diet can help type 2 diabetics lose weight and better control their blood sugar levels.

Since this diet is mostly plant-based, it contains a lot of whole grains, beans, pulses, nuts, and seeds, as well as fruits and vegetables. Olive oil is also used. In moderation, it also contains some dairy products (milk and yogurts) and lean proteins (fish, eggs, and chicken). Wine is usually only consumed in moderation, along with red meat and processed foods, which are typically consumed in much smaller amounts.

1. The nutritional information in this meal plan is sourced from our recipes and the sixth edition of Carbs and Cals, unless specified otherwise.
2. Although whole milk and semi-skimmed milk have been combined, feel free to use your preferred type. Any substitute for dairy products needs to be calcium-fortified and unsweetened.
3. These weekly meal plans provide the recommended daily intake of fiber for an individual.
4. Although the daily food intake for one person is specified in this meal plan, it's still crucial to remember to stay hydrated. This includes unsweetened tea or coffee, plain water, and unsweetened milk.

8.4.8. DASH Diet

To help control diabetes, the DASH (Dietary Approaches to Stop Hypertension) diet therapy emphasizes eating foods low in processed sugars, saturated fat, and sodium. Rich in fruits, vegetables, whole grains, lean protein, and low-fat dairy products, the DASH diet can help lower blood pressure, increase insulin sensitivity, and lower the risk of heart disease. Additionally high in fiber, the DASH diet can help lower blood sugar and promote better digestive health. Furthermore, the DASH diet places a strong emphasis on eating foods high in minerals like calcium, magnesium, and potassium, which are vital for good health overall and can lower the risk of complications from diabetes.

Foods that are part of the DASH diet include the following:

- **Fruits and Vegetables:** They can aid in blood sugar regulation and are high in fiber, vitamins, and minerals. Try to consume five or more servings of fruits and vegetables each day.
- **Whole Grains:** These have a high fiber content and can aid in blood sugar regulation. Quinoa, brown rice, and whole wheat bread are a few examples.
- **Lean Protein:** Opt for low-fat protein options like lean chicken, fish, and legumes as they are packed with essential nutrients and can aid in controlling blood sugar levels.

- **Low-Fat Dairy Products:** These can aid in blood sugar regulation and are high in calcium. Opt for low-fat cheese, yogurt, and milk.
- **Nuts and Seeds:** These can aid in blood sugar regulation and are high in protein, fiber, and good fats. Limit your intake of unsalted nuts and seeds.
- Limit or stay away from processed foods, sugar-filled drinks, and foods heavy in trans and saturated fats.

8.4.9. High-Fiber Diet

Because it lowers cholesterol, supports a healthy digestive system, and helps regulate blood sugar and cholesterol levels, a high-fiber diet may help manage diabetes.

The following advice can help you include more high-fiber foods in your diet:

- **Choose Whole Grains:** Select whole-grain foods like brown rice, quinoa, whole wheat pasta, and whole-grain bread in place of refined grains.
- **Eat Plenty of Fruits and Vegetables:** Including a variety of fruits and vegetables in your diet is important for getting enough fiber, vitamins, and minerals. Try to have a minimum of five servings of fruits and vegetables each day.
- **Include Legumes:** Chickpeas, black beans, and lentils are examples of legumes that are high in protein and fiber. Toss them into stews, soups, and salads.
- **Snack on Nuts and Seeds:** Nuts and seeds are an excellent source of wholesome fats and fiber. Select unsalted options like flaxseeds, chia seeds, walnuts, or almonds.
- **Drink Plenty of Water:** It is crucial to stay hydrated by drinking an adequate amount of water to support digestion and maintain regular bowel movements.

8.4.10. Low-Carb Diet

Diabetes can be effectively managed with low-carb diets as they lower blood sugar levels and minimize the need for medication. Before beginning a low-carb diet, it's crucial to speak with a healthcare professional, particularly if you take medication for diabetes or have other medical issues.

For diabetics, the following general recommendations apply to a low-carb diet:

1. **Limit Carbohydrate Intake:** Typically, a low-carb diet limits daily carbohydrate intake to 20–50 grams. This can change based on personal preferences and needs.
2. **Focus on Protein and Healthy Fats:** Make sure your diet is rich in sources of high-quality protein, such as fish, poultry, meat, eggs, nuts, and seeds. Include foods high in healthy fats, such as nuts, avocado, and olive oil, in your diet.
3. **Choose Low-Carb Vegetables:** Broccoli, cauliflower, spinach, and kale are examples of low-carb and high-fiber vegetables that can be included in a diet.
4. **Avoid or Limit Sugary Foods:** Sugar-filled foods ought to be restricted or avoided, including soda, sweets, and desserts.
5. **Monitor Blood Sugar Levels:** When beginning a low-carb diet, it's critical to regularly check blood sugar levels and, with the assistance of a healthcare professional, modify medication as necessary.

6. **Stay Hydrated:** Staying hydrated by drinking an adequate amount of water is essential for maintaining good health and managing blood sugar levels.
7. **Consider Working with a Registered Dietitian:** A certified dietitian can offer advice on how to choose healthful foods and assist in developing a customized low-carb meal plan (Traverso-Yepez & Hunter, 2016; Downs et al., 2020).

8.5. DIET THERAPY FOR HEART DISEASE PREVENTION AND MANAGEMENT

A key component of heart disease management and prevention is diet therapy. Heart disease is a general term for a number of conditions affecting the heart and blood vessels. It is also referred to as cardiovascular disease. Heart failure, arrhythmias, coronary artery disease, and heart attacks are some of these ailments. Diet therapy is a crucial part of managing and preventing heart disease, which is the world's leading cause of death. Over the past few decades, a lot of research has been done on the connection between diet and heart disease. Based on available data, some dietary patterns may raise your risk of heart disease, while other dietary patterns may lower it. A heart-healthy diet rich in fruits, vegetables, whole grains, lean protein sources, and healthy fats is advised by the American Heart Association. This diet is low in sodium, added sugars, and fats—both trans and saturated.

8.5.1. Dietary Fats and Cholesterol Levels

Cholesterol is an indispensable lipid involved in numerous metabolic processes and a fundamental component of every cell membrane in the body. It is created in the liver by the body using the food we eat. Low-density lipoprotein (LDL) and high-density lipoprotein (HDL) are blood lipids, or fats, that contain cholesterol. HDL (or "good") cholesterol helps to eliminate cholesterol from the body and prevents plaque from accumulating in the arteries, whereas LDL (or "bad") cholesterol can cause plaque to form in the arteries.

- **Saturated Fats:** Saturated fats that are saturated. Known colloquially as "bad fats," saturated fats have a tendency to raise blood levels of LDL (or "bad") cholesterol. Animal products (butter, coconut oil, meat fat, including lard and dripping, beef, lamb, chicken skin, and palm oil) and processed foods, such as pastries and biscuits, are common sources of saturated fats. Full-fat or reduced-fat dairy products: While full-fat dairy products, like milk, cheese, and yogurt, do contain saturated fat, the evidence suggests that this kind of fat has no discernible effect on heart health.

The Heart Foundation suggests that while reduced fat varieties of milk, yogurt, and cheese are better for individuals who need to lower their LDL cholesterol, unflavored varieties are okay for the general public to consume.

- **Eggs**: Eggs naturally contain cholesterol, which was formerly believed to be detrimental to heart health. Nonetheless, studies indicate that eggs do not affect heart health; that is, they don't raise or lower the risk of heart disease in the general population. The Heart Foundation suggests consuming no more than seven eggs a week for those with type 2 diabetes or those who need to lower their LDL cholesterol.
- **Trans Fats**: Similar to saturated fats, trans fats also tend to lower levels of HDL (good) cholesterol in the blood while raising levels of LDL (bad) cholesterol. Consequently, they are more harmful to our health and may raise our chance of developing cardiovascular conditions (like heart disease and stroke). When monounsaturated or polyunsaturated vegetable oils are "hydrogenated" and solidified to create margarines, deep-frying oils, or shortening for baked goods, trans fatty acids are created. The food industry uses these tougher vegetable fats and shortenings in processed foods (like cakes, biscuits, and deep-fried takeout meals). Additionally, some meats, butter, and dairy products naturally contain some trans fatty acids. Saturated fat-containing butter is preferred over table margarines, which are mostly monounsaturated and polyunsaturated and have very low levels of trans fatty acids.
- **Monounsaturated and Polyunsaturated Fats:** Lower your chance of developing heart disease by substituting unsaturated fats (also known as "good fats") for trans and saturated fats in your diet. Use oils derived from seeds or plants (such as olive, avocado, sunflower, canola, safflower, peanut, soybean, and sesame) in place of butter, coconut and palm oil, lard, dripping, and copha. Avocados, unsalted nuts, and seeds (such as chia, tahini, and linseed) are additional sources of unsaturated fats.

8.5.2. Blood Pressure and Salt (Sodium)

Consuming a diet high in salt is associated with hypertension, which can elevate the risk of heart disease and stroke. The majority of individuals consume over 10 times the necessary amount of salt to meet their sodium needs. It's important to note that salt contains sodium and chloride. The main source of sodium in our diet is not from table salt, but from packaged and processed foods. It's surprising to learn that even sweet foods or those that do not taste salty can have high levels of sodium. One easy way to decrease sodium intake is to lessen the consumption of processed foods, limit fast food, and use herbs and spices for added flavor.

8.5.3. Reduce the Heart Disease Risk with Healthy Eating

Consuming a diverse range of foods is advantageous for our health and can decrease our chances of developing diseases, such as heart disease. Aim to include a broad selection of foods from the 5 food groups in the recommended proportions. This not only supports a balanced and enjoyable diet but also delivers crucial nutrients to the body.

The Heart Foundation recommends:

- Plenty of vegetables, fruits and wholegrain.
- A range of nutritious protein sources, including nuts, seeds, legumes (like beans and lentils), fish, and shellfish. Lean poultry and eggs in smaller quantities can also be a part of a diet that promotes heart health. Make sure the red meat you choose is lean, and only eat it once or twice a week.
- Cheese, yogurt, and milk without flavorings. People who have elevated blood cholesterol levels ought to select low-fat options.
- Healthy fat choices – nuts, seeds, olives, avocados, and the cooking oils they produce.
- Use spices and herbs to flavor food rather than salt.

People should also be aware of how much they eat and whether they are consuming a lot of unhealthy foods. People are eating more than they need to because portion sizes have increased over time, which can raise their risk of cardiovascular disease and lead to obesity. A balanced plate should ideally have portions of ½ vegetables, ¼ carbs, and ¼ protein.

Follow these steps to lower your chances of developing heart disease:

- Limit fried fast food and processed foods.
- Swap out saturated fats like butter, coconut oil, and cream with nutrient-rich unsaturated fats found in seeds and plants. Opt for healthy alternatives like extra virgin olive oil, avocado, sunflower, canola, safflower, peanut, soybean, and sesame. Make sure to incorporate foods like nuts, seeds, avocado, olives, and soy into your diet for added benefits.
- Eat a greater variety and quantity of plant-based foods, such as wholegrain cereals, fruits, and vegetables.
- Limit your consumption of refined carbs with higher glycemic indices, such as added sugar-containing foods.
- A weekly maximum of 350 g (cooked weight) of unprocessed red meat (beef, veal, mutton, lamb, pork, kangaroo, rabbit, and other game meats) should be consumed; processed meat (sausages, ham, salami, and prosciutto) should be avoided.
- Trim meat of any visible fat, and take the skin off of poultry.
- Consume legumes on a regular basis, such as tofu, soybeans, lentils, and baked beans with less salt.
- Most days of the week, have a handful of raw, unsalted nuts as a snack (especially walnuts and almonds).
- Eat oily fish at least once per week.
- Reduce salt intake – Eat less fast food and salty food, stay away from packaged and processed foods. For added taste, substitute herbs and spices for salt when cooking and at the table.
- Examine the sodium level in foods and select those with the lowest sodium content.

- Eat no more than seven eggs a week and convert to low-fat or non-fat dairy products if a person has high cholesterol.
- If you consume alcohol, limit your intake to no more than two standard drinks in a single day. Drinking too much alcohol can raise blood triglycerides and blood pressure.

8.5.4. Foods Important for Heart Health

While there isn't a single "magic" food that can reduce our risk of heart disease, there is evidence to suggest that certain foods are critical for heart health. Among them are:

- Oily fish – like salmon, tuna, sardines, mackerel, and others that are high in omega-3 fatty acids. This kind of fat has been demonstrated to raise HDL cholesterol levels, lower triglycerides, improve blood vessel elasticity, and thin the blood, reducing the risk of blood clots and blood flow obstructions.
- Some vegetable oils – omega-6 fatty acids, like those found in corn, soy, and safflower, and omega-3 fatty acids, like those found in canola and olive oil. All of these, when substituted for saturated fats like butter, can help reduce LDL cholesterol.
- Fruit and vegetables – The fiber, potassium, and various micronutrients found in fruits and vegetables protect against heart disease. Additionally, they are rich in folate, which helps reduce blood levels of homocysteine, an amino acid linked to a higher risk of heart disease.
- Wholegrain – A diet high in fiber from wholegrain cereals is associated with a lower risk of heart disease and lower levels of LDL cholesterol. Foods high in soluble fiber, such as barley, legumes, and oats, are excellent at reducing total cholesterol.
- Unrefined carbohydrate sources with a low glycemic load – Foods like wholegrain breads, legumes, specific types of rice and pasta, and a variety of fruits and vegetables can assist in decreasing blood triglycerides and glucose levels, as well as aiding in diabetes management and lowering the risk of heart disease.
- Legumes, nuts and seeds – provide healthy fats, fiber, plant proteins, and micronutrients that can help reduce the risk of cardiovascular disease.
- Tea – according to certain studies, the antioxidants in tea may help stop fatty deposits from accumulating in the arteries. They might also enhance blood vessel dilatation to permit greater blood flow and function as an anti-blood clotting agent.
- Foods containing vitamin E – according to certain research, vitamin E functions as an antioxidant and helps shield the body from LDL cholesterol.
- Sources of vitamin E include – avocados, wholegrain products, vegetable oils, and dark green vegetables. Consume vitamin E-rich foods instead of supplements, as the latter does not appear to offer the same level of protection.
- Garlic – Allicin, a substance found in fresh garlic, has been shown to lower blood levels of both LDL and

total cholesterol, which lowers the risk of heart disease.

- Foods enriched with plant sterols – In healthy individuals as well as those with high cholesterol or diabetes, a daily consumption of 2 to 3 g of phytosterols/stanols reduces LDL cholesterol levels by about 10%. This is equivalent to two or three servings of foods high in phytosterols, such as milk, yogurt, cereal for breakfast, and margarine spreads.

8.5.5. Prevention of Heart Disease Through Diet Therapy

By implementing good eating practices that lower the chance of contracting the illness, diet therapy can prevent heart disease. Reducing the consumption of trans and saturated fats is one of the most important actions. Whereas trans fats are present in processed foods like baked goods, fried foods, and snack foods, saturated fats are found in animal products like meat, butter, and cheese. Both kinds of fats have the ability to elevate cholesterol, which raises the risk of heart disease.

People should prioritize eating healthy fats like monounsaturated and polyunsaturated fats over consuming saturated and trans fats. Nuts, seeds, avocados, fatty fish, and vegetable oils are some of the foods that contain these fats. These fats lower LDL (bad) cholesterol levels, which can help lower the risk of heart disease. Increasing fruit and vegetable consumption is another important dietary practice to lower the risk of heart disease. These foods lower the risk of heart disease because they are high in nutrients like vitamins, minerals, and fiber. Additionally, they have antioxidants in them that guard the heart against harm from free radicals.

Another vital element of a diet high in heart-healthy foods is whole grains. These foods lower the risk of heart disease because they are high in fiber, vitamins, and minerals. Additionally, they lessen the chance of diabetes, which raises the risk of heart disease, and help control blood sugar levels.

8.5.6. Management of Heart Disease Through Diet Therapy

Using diet therapy to manage heart disease entails forming wholesome eating habits that can both help control the condition and stop future complications. Cutting back on salt consumption is one of the most important things to do. Sodium has the potential to induce fluid retention, which raises blood pressure and puts stress on the heart. Aim to limit sodium intake in people with heart disease to no more than 2,300 mg daily.

One of the most important aspects of managing heart disease is diet therapy. Here are some dietary suggestions for the management of heart disease.

1. **Reduce Intake of Saturated and Trans Fats:** Heart disease may be exacerbated by elevated blood levels of low-density lipoprotein (LDL) cholesterol caused by trans and saturated fats. Rather, concentrate on eating fats that are better for you, like monounsaturated and polyunsaturated fats, as they can lower cholesterol.
2. **Increase Intake of Fruits and Vegetables:** Antioxidants, fiber, and other nutrients found in fruits and vegetables can lower the risk of heart disease.
3. Choose whole grains over refined grains: Compared to refined grains,

whole grains are higher in fiber and nutrients, which can lower cholesterol and lower the risk of heart disease.

4. **Reduce Sodium Intake:** Excessive salt consumption raises blood pressure, which raises the risk of heart disease. Instead, go for low-sodium options or flavor food with herbs and spices.
5. **Limit Intake of Added Sugars:** Overindulging in added sugars can raise the risk of heart disease, diabetes, and obesity. Instead, limit sugar-filled beverages and opt for naturally sweet foods like fruits.
6. **Choose Lean Protein Sources:** Overindulging in animal-based saturated fat raises the risk of heart disease. Pick lean protein sources like legumes, chicken, and fish.
7. **Limit Alcohol Intake:** Overindulgence in alcohol can raise blood pressure and worsen heart disease. If someone decides to consume alcohol, they should only have one drink for women and two for men per day.

To effectively manage heart disease, it is imperative to collaborate with a healthcare provider, such as a registered dietitian, to create a personalized nutrition plan.

8.5.7. Plant-Based Diet for Heart Health

A diet high in plants may benefit the heart. In comparison to a diet high in meat, a person who eats mostly or exclusively fruits, vegetables, nuts, beans, whole grains, and meat substitutes like soy has a lower risk of developing heart disease, high blood pressure, high cholesterol, and type 2 diabetes.

There are numerous varieties of diets based solely on plants. The three that are most typical are:

- **Vegan:** No dairy products, meat, or eggs from animals.
- **Lacto-Vegetarian:** No eggs or meat, but dairy products are acceptable.
- **Lacto-Ovo-Vegetarian:** No meat, but dairy products and eggs are OK.

It is possible to follow a plant-based diet without becoming an atheist. Some people identify as semi-vegetarians or flexitarians, meaning they eat fish, poultry, meat, or pork on occasion. You may also hear the term "pescatarian," which refers to someone who eats both fish and plant-based foods. It is advised to start consuming more whole grains, beans, nuts, seeds, fruits, and vegetables. One can reduce or completely give up animal products, depending on how extreme one wants to go. To make sure the diet is balanced, speaking with a dietitian is advised.

To meet your protein needs without meat, choose beans, lentils, nuts, seeds, quinoa, or tofu. Follow your doctor's advice on fat, calories, sugar, and salt intake. It's important to monitor these factors, regardless of whether you consume animal products (Traverso-Yepez & Hunter, 2016; Downs et al., 2020).

8.6. DIET THERAPY FOR FOOD ALLERGIES AND INTOLERANCES

An immunological reaction is what's known as a food allergy. Mild to severe symptoms may affect the skin, gastrointestinal tract, respiratory system, or circulatory system of the body. Food intolerance is more prevalent but can be challenging to identify. Food intolerances are not the result of an immunological response. Rather, a variety of factors contribute to their occurrence, leading to uncomfortable symptoms that could persist for several hours or even days.

An allergic response to a food is a reaction of the immune system to that food. Even though they are frequently not too serious, they can occasionally be fatal. Numerous chemicals will be released if the immune system misinterprets the food and believes it to be a threat. Symptoms appear when the chemicals are released. A food allergy can cause swelling of the face, lips, or tongue, vomiting, an itchy throat, mouth, or ears, and a rash (or hives).

Nearly every food can trigger an allergic response, and each person's reaction to a given food is unique. However, some foods are known to frequently trigger allergies, particularly in young children. Fish, shellfish, eggs, peanuts, milk, and eggs are common allergens. Even though intolerances are more common than food allergies, they are difficult to diagnose. Even though they are not life-threatening, symptoms can make a person feel ill and make daily living difficult.

Many food intolerances have unknown causes because food intolerance reactions do not involve the immune system. Symptoms frequently appear hours later and occasionally persist for several days. Food intolerance is typically accompanied by gut-related symptoms like bloating, constipation, diarrhea, and skin issues like eczema. People who have a food intolerance frequently have several symptoms, not all of which are obvious. Non-specific symptoms like headaches, fogging of the brain, and lethargy can occur. It can be challenging to identify which foods are causing a chronic illness when a person has a food intolerance because it is possible to be intolerant to multiple foods.

There are several factors that can lead to food intolerance. The way a person lives and eats can impact how their body responds to certain foods. If someone believes their lifestyle or dietary choices could be playing a role, they should consider seeking guidance from a nutrition expert. A professional in this field can help identify problem foods and provide recommendations for overcoming intolerances.

8.6.1. The Role of Diet Therapy in Managing Food Allergies and Intolerances

To manage food allergies and intolerances, diet therapy is essential because it entails removing or minimizing particular foods that cause negative reactions and making sure that the body is getting enough nutrients to stay healthy overall. Strict avoidance of the allergen is part of diet therapy for people with food allergies. This can necessitate closely reading labels and preventing cross-contamination when preparing food. Emergency drugs like epinephrine auto-injectors may also be prescribed in extreme circumstances.

Diet therapy for people with food intolerances may include cutting back on or avoiding foods that contain the offending ingredient. For instance, people who are lactose intolerant may use lactase enzyme supplements or abstain from dairy products altogether.

Ensuring appropriate nutrient intake is a key component of diet therapy for food allergies and intolerances. It's crucial to replace foods that may cause nutrient deficiencies with other sources of nutrition when eliminating them from your diet. For instance, people who are allergic to milk might need to get their calcium from other sources, like supplements or fortified plant-based milk.

In addition, diet therapy includes educating and supporting patients as well as their families. A registered dietitian can assist in creating a customized meal plan, offer advice on what foods to choose and what to substitute, and support when making dietary adjustments. All things considered, diet therapy plays a vital role in the management of food allergies and intolerances, assisting people in avoiding trigger foods, ensuring appropriate nutrient intake, and enhancing their quality of life.

8.6.2. Elimination Diets for Food Allergies and Intolerances

An elimination diet is a dietary approach that excludes particular foods or food groups thought to be responsible for causing negative reactions, also known as "food intolerances." By removing these foods for a period of time and then reintroducing them later through a "challenge," individuals can determine which foods may be triggering symptoms or exacerbating them. Food reactions are not always immediate and can vary, with some not necessarily linked to the immune system.

Food intolerances can be triggered by natural compounds found in foods, like sugars or proteins, as well as common additives such as colors, preservatives, and flavor enhancers that can cause reactions through different pathways in the body. While there is ongoing debate about the specific mechanisms involved in food reactions, many tests to identify the suspected triggers may not be reliable.

Clinical experience suggests that an elimination diet is a valuable tool for pinpointing food culprits and is generally safe as long as a variety of foods are consumed to ensure essential nutrient needs are met.

To ascertain whether a person has an allergy or intolerance to particular foods, elimination diets include cutting out particular foods or food groups from the diet for a predetermined amount of time. The plan is to remove any allergens or irritants from their diet and observe any improvement in their symptoms.

There are two main types of elimination diets:

1. **Simple Elimination Diet:** One method involves excluding one or two foods from your diet for a few weeks to observe if symptoms improve. If there is an improvement, the foods can be reintroduced one at a time to determine if they are triggering the symptoms.
2. **Comprehensive Elimination Diet:** This entails removing a greater variety of foods from the diet for an extended length of time, usually four to six weeks. Following this time, progressively add each food back into your diet one at a time to determine which ones might be contributing to your symptoms.

It is crucial to remember that elimination diets should only be carried out with the advice of a medical expert, such as an allergist or registered dietitian. This is since elimination diets can be challenging to stick to and, if done incorrectly, can result in nutritional deficiencies.

8.6.3. Reintroduction Protocols for Food Allergies and Intolerances

Strategies to safely reintroduce problematic foods into an individual's diet are known as reintroduction protocols for food allergies and intolerances. To maintain a healthy, well-balanced diet, these protocols are intended to assist patients in recognizing and managing their dietary allergies and intolerances. Reintroduction schedules differ based on the particular food intolerance or allergy and the severity of the patient's symptoms.

- **Allergies to Food:** Food allergies are immune system responses that happen when a person's body interprets a protein in food as dangerous and sets off an allergic reaction. Food allergies frequently cause anaphylaxis, vomiting, diarrhea, hives, itching, and swelling. The most frequently occurring food allergies are to peanuts, tree nuts, fish, shellfish, eggs, wheat, and soy. Reintroducing a food allergen into a person's diet requires first confirming the diagnosis with allergy testing. This can be accomplished by measuring the amounts of particular antibodies in the blood through blood tests or skin prick tests.

Starting a reintroduction protocol is the next step after a diagnosis has been confirmed. According to the protocol, the allergen is progressively added to the person's diet in tiny doses over the course of a few days or weeks. Usually, this begins with a very small amount. The goal of this procedure, called "desensitization," is to progressively hone the person's immune system to tolerate the allergen. A close watch should be kept on the person during the reintroduction protocol in case of any allergic reactions. If any symptoms appear, it's critical to halt the protocol right away and, if required, seek medical assistance. Having emergency drugs on hand, like epinephrine, is also crucial in case of severe allergic reactions. If the reintroduction goes well, the person can progressively increase how much of the allergen they eat until they are able to handle a typical serving size. To maintain tolerance, it's crucial to keep ingesting the allergen on a regular basis. Food sensitivities. Food intolerances are non-immunological reactions to specific foods that arise when a person's digestive system cannot properly break down or digest particular food ingredients. Food intolerances frequently cause bloating, gas, diarrhea, and abdominal pain as symptoms. Lactose intolerance, fructose intolerance, and gluten intolerance are common food intolerances. To begin the process of reintroducing a food intolerance into a person's diet, diagnostic testing must be used to confirm the diagnosis. An elimination diet, blood test, or breath test can be used to do this. Starting a reintroduction protocol is the next step after a diagnosis has been confirmed. The protocol entails progressively adding small doses of the problematic food to the person's diet over time; usually, the dose is increased gradually over a few days or weeks, starting small.

8.6.4. Managing Food Allergies and Intolerances in Children

An allergy is an unfounded fear that the immune system of the body raises in reaction to benign substances found in our surroundings, such as food proteins or pollen. Consequently, in reaction to the foreign materials (also referred to as antigens), our bodies produce particular antibodies. The release of chemicals resulting from a reaction between the antigens and antibodies causes an allergic reaction, such as itching, skin rash, stomach pain, or dyspnea, when the body comes into contact with these substances again. Take away from the child's diet anything that makes them react negatively to allergens. Request a thorough allergy assessment from your doctor to prevent your child from needlessly removing necessary foods from their diet, like cow's milk.

Allergies are often triggered by:

- Cow's milk, hens' eggs and soya in infants
- Foods like fish, soy, cow's milk, hens' eggs, tree nuts, peanuts, and wheat cereal for young children.
- Adults should consume vegetables, fruit, nuts, spices, hen's eggs, cow's milk, and fish.

The two most prevalent food allergies are lactose intolerance (allergy to lactose in milk and milk products) and celiac disease (allergy to gluten found in wheat, oat, and barley).

When food allergies in infants and toddlers are avoided during their early years, they frequently go away by the time the child enters school. This occurs because the food has not been absorbed by the body for a considerable amount of time.

8.6.4.1. Tests and Diagnosis

Provocation (challenge) testing, blood testing, skin testing, and other diagnostic procedures are performed during a medical examination.

- **Treatment:** Avoiding the food that is thought to be the cause of the allergy is the only proven treatment for a food allergy.
- **Foods That Should be Avoided:** Carefully review the ingredient list of any product. The ingredients list will make it easier to identify any allergies in food that is packaged. In addition to the basic ingredient list, NESTLÉ has been giving customers additional information for a long time.

  Even though the product may only contain traces of these ingredients, they are listed on the packaging because they are known to frequently cause allergies or intolerances. This is because ingredients may find their way into the product due to other products being made on the same production line, for example.

- Alternatives for offending foods:
- **Lactose Intolerance:** Soy milk is a suitable substitute for those who are allergic to cow's milk. In some situations, dairy products like yogurts can also be administered. You can give foods like eggs, meats, fish, poultry, soy, tofu, fortified juices, and commercially made low-lactose milks. Under a doctor's supervision, a small amount of milk combined with other non-dairy food can be given, depending on the severity.
- **Celiac Disease:** It is necessary to replace wheat, rye, oats, and barley with rice, soybean, maize, and corn. Each food item's label should be carefully read. Wheat is also found in baked goods like cakes, breads, and biscuits. The pediatrician should be consulted when selecting food items.

Suitable alternative toppings for bread include:

- Non-dairy margarines
- Cold meat cuts
- Fruit based jams and spreads
- Vegetable pastes etc.

Get a child to eat low-lactose foods like yogurts and yogurt-based desserts to make sure they get the calcium they require even if they are lactose intolerant and cannot have any milk or dairy products. Because yogurts contain lactic acid rather than lactose, they are more palatable than milk.

8.6.5. Food Allergy and Intolerance Testing Methods and Their Limitations

Foods that trigger an allergic or intolerance reaction in a person are identified through testing procedures for food allergies and intolerances. The following are some popular testing techniques along with some of their drawbacks.

1. **Skin Prick Test:** The suspected allergen is lightly pricked onto the skin, and the response is noted. Although this test is simple to administer and yields results quickly, it has a high false positive rate and some individuals may have severe reactions when taking it.
2. **Blood Test:** To check for the presence of particular antibodies to the putative allergen, a blood sample is tested. Although this test can be costly and time-consuming, it helps identify allergens that might be causing a delayed reaction.
3. **Elimination Diet:** The suspected allergen is cut out of the diet, and any reactions are tracked by closely observing the symptoms. Although it can take weeks or months to finish, this test is dependable, and figuring out which foods are triggering the reaction can be challenging.
4. **Oral Food Challenge:** In a controlled setting, the suspected allergen is consumed, and symptoms are noted. The most accurate test, however it takes a long time and needs to be done under close medical supervision.
5. **IgG Antibody Testing:** A blood test that quantifies the amount of IgG antibodies to particular meals. IgG testing for food intolerance is controversial, though, and some allergy societies do not advise it.

All things considered, every testing technique has its limitations, and no test can be guaranteed to be 100% accurate.

Therefore, when attempting to identify food allergies or intolerances, it is crucial to consult a qualified healthcare professional.

8.6.6. Psychological Impacts of Food Allergies and Intolerances

The psychological health of an individual can be significantly impacted by food allergies and intolerances. Typical psychological effects of food allergies and intolerances include the following:

1. **Anxiety and Stress:** People who have dietary allergies or intolerances frequently suffer from anxiety and stress because they worry about inadvertently consuming a trigger food or having a negative reaction.
2. **Social Isolation:** A lower quality of life and social isolation can result from the fear of experiencing an allergic reaction in public or from not knowing where to get safe food options.
3. **Depression:** Food allergies and intolerances are chronic conditions that can cause despair, hopelessness, and hopelessness.
4. **Eating Disorders:** As a coping method for their dietary allergies or intolerances, some people may develop eating disorders like bulimia or anorexia.
5. **Reduced Quality of Life:** Food allergies and intolerances can place dietary restrictions on an individual that can make it difficult for them to travel, attend social events, and engage in other activities.
6. **Fear and Hyper Vigilance:** A diminished quality of life and hypervigilance can result from a persistent fear of a reaction.

To manage the psychological effects of their condition, people with food allergies or intolerances should learn coping mechanisms and seek out support from licensed healthcare providers (Traverso-Yepez & Hunter, 2016; Downs et al., 2020).

CASE STUDY

OPTIMAL NUTRITION FOR PERFORMANCE

Raj, a dedicated vegan, wanted to ensure he was getting all the essential nutrients to fuel his rigorous training. Holistic Nutrition Therapy assessed his dietary intake, lifestyle, and health history. A plan was developed to include plant-based protein sources, supplemented with essential vitamins and minerals. Emphasis was also placed on gut health to improve nutrient absorption.

- **Health History**: Previous iron-deficiency anemia, and occasional muscle cramps.
- **Issue:** Raj's vegan diet lacked certain essential nutrients, leading to iron-deficiency anemia and muscle cramps, which affected his training.
- **Changes Made:** Raj incorporated iron-rich plant foods like lentils, chickpeas, and spinach into his diet. He also started taking B12 and Omega-3 supplements. His meal plans were adjusted to ensure adequate protein intake from varied sources.
  - Incorporated a variety of iron-rich plant foods, including lentils, chickpeas, quinoa, and fortified cereals.
  - Added a daily B12 supplement and an algae-based Omega-3 supplement to his regimen.
  - Introduced protein-rich snacks like chia seed pudding and hemp seed smoothies post-training.
  - Ensured hydration with electrolyte-rich drinks during intensive training sessions.
- **Outcome:** Enhanced athletic performance, improved recovery times, and strengthened immune system.

CLASS ACTIVITY

SURVEY FOR COMMERCIAL ENTERAL FEEDS AVAILABLE IN THE MARKET

- **Aim**: To conduct a market survey in the area to determine which commercial enteral feeds are offered in India.
- **Survey**: Conduct a survey in your community, such as the neighborhood market, pharmacy, or hospital. To determine which commercial enteral feed varieties are offered in your area. Using the format provided, describe the unique qualities, nutritional makeup, and composition of any five enteral feeds that you have identified.

Begin the activity with an overview of the enteral feeds.

- **Introduction:** Provide a succinct overview of enteral feedings, including their types, advantages, and when to use them in the space provided here.
- **Survey Result** (Using the format provided here, present the survey results for any five commercial enteral feeds).
- Enteral Feed

Enteral Feed (Name and Type)	Composition	Nutritive Content	Indicative for Which Disease Condition	Cost

Conclusion (Comment on the cost and availability of the commercial feeds available in the market in the space provided herewith).

SUMMARY

- Food is a basic prerequisite for our survival. The quality and quantity of food determine the health and nutritional status of an individual. An individual's nutritional status depends upon the provision of sufficient nutrients through food.
- Diet therapy promotes a balanced selection of foods vital for good health. By combining foods appropriate for each individual and drinking enough water, one can help maintain the best possible health. Eating a proper diet is critical for the health of individuals, groups with special medical and dietary needs, and entire populations afflicted with malnutrition.
- A particular modified diet is prescribed specifically for each individual. Those individuals who have medical conditions or who are sensitive to certain foods need to be very compliant and cautious about what they eat.
- The goal of nutrition intervention is to supply adequate nutrients to meet the patient's nutrient requirement by the most physiologic, safety and cost-effective route. Sometimes person cannot eat any or enough food because of an illness. The stomach or bowel may not be working quite right, or a person may have had surgery to remove part or all of these organs. Under those conditions, nutrition must be supplied in a different way i.e., through enteral and parenteral route.
- A therapeutic diet is a menu plan that limits the consumption of particular foods or nutrients. It is a component of the therapy of a medical condition and is typically recommended by a doctor and organized by a dietician.
- The dietary habits characteristic of Mediterranean countries with the consumption of olive oil (monounsaturated fatty acid) as the main fat source appear to provide optimal health benefits with a low incidence of coronary heart disease. The Mediterranean diet consists of large amounts of fruit, vegetables, pulses, nuts, cereal products, and fish, while generally only small amounts of meat and dairy foods are consumed.
- Diet therapy for weight management is a highly effective approach to achieving and maintaining a healthy weight. With an increasing number of individuals around the world struggling with weight-related issues, diet therapy offers a valuable solution to the problem. Through this approach, individuals can create a calorie deficit, promoting weight loss and improving overall health outcomes.
- The goal of diet therapy for diabetes management is to achieve and maintain healthy blood sugar levels while also meeting nutritional needs. This is typically achieved through a balanced diet that includes a variety of foods from all food groups in appropriate portions. A registered dietitian can help develop an individualized meal plan based on an individual's needs and preferences.
- Heart disease is the leading cause of death globally, and diet therapy is an essential component of prevention and management. Eating a variety of foods is beneficial to our health and can help reduce our risk of disease (including heart disease).

- A food allergy is a reaction caused by the immune system. Symptoms can range from mild to severe and may affect the gut, skin, breathing ability or the body's circulation system. Food intolerance is more common, though can be difficult to diagnose. Food intolerances are not caused by an immune system reaction. Instead, they are caused by a number of factors, resulting in unpleasant symptoms which may last for several hours or even days.
- Diet therapy plays a crucial role in managing food allergies and intolerances as it involves eliminating or reducing specific foods that trigger adverse reactions and ensuring adequate nutrient intake to maintain overall health.

REVIEW QUESTIONS

1. What is the main objective of diet therapy?
2. What does diet therapy consist of?
3. How does a high-fiber diet benefit the body?
4. Can you reduce your risk of heart disease through healthy eating? If so, how?
5. How can diet therapy help in the management of heart disease?
6. What is the difference between food allergies and intolerances, and why is it important to understand them?
7. What are reintroduction protocols for food allergies and intolerances, and why are they important?

REFERENCES

1. Bergman, C., Tian, Y., Moreo, A., & Raab, C. (2021). Menu engineering and dietary behavior impact on young adults' kilocalorie choice. *Nutrients, 13*(7), 2329. https://doi.org/10.3390/nu13072329.
2. Castro-Barquero, S., Ruiz-León, A. M., Sierra-Pérez, M., Estruch, R., & Casas, R. (2020). Dietary strategies for metabolic syndrome: A comprehensive review. *Nutrients, 12*(10), 2983. https://doi.org/10.3390/nu12102983.
3. Downs, S. M., Ahmed, S., Fanzo, J., & Herforth, A. (2020). Food environment typology: Advancing an expanded definition, framework, and methodological approach for improved characterization of wild, cultivated, and built food environments toward sustainable diets. *Foods, 9*(4), 532. https://doi.org/10.3390/foods9040532.
4. Dworatzek, P. D., Arcudi, K., Gougeon, R., Husein, N., Sievenpiper, J. L., & Williams, S. L. (2013). Nutrition therapy. *Canadian Journal of Diabetes, 37*(S45–S55). https://doi.org/10.1016/j.jcjd.2013.01.019.
5. Fulton, S. L., McKinley, M. C., Young, I. S., Cardwell, C. R., & Woodside, J. V. (2016). The effect of increasing fruit and vegetable consumption on overall diet: A systematic review and meta-analysis. *Critical Reviews in Food Science and Nutrition, 56*(5), 802–816. https://doi.org/10.1080/10408398.2012.727917.
6. Gittelsohn, J., & Lee, K. (2013). Integrating educational, environmental, and behavioral

economic strategies may improve the effectiveness of obesity interventions. *Applied Economic Perspectives and Policy, 35*(1), 52–68. https://doi.org/10.1093/aepp/pps044.

7. Guthrie, J., Mancino, L., & Lin, C. T. J. (2015). Nudging consumers toward better food choices: Policy approaches to changing food consumption behaviors. *Psychology and Marketing, 32*(5), 501–511. https://doi.org/10.1002/mar.20795.
8. Hawkes, C., Smith, T., Jewell, J., Wardle, J., & Hammond, R. (2015). Smart food policies for obesity prevention. *The Lancet, 385*(9985), 2410–2421. https://doi.org/10.1016/S0140-6736(14)61745-1.
9. Hemler, E. C., & Hu, F. B. (2019). Plant-based diets for cardiovascular disease prevention: All plant foods are not created equal. *Current Atherosclerosis Reports, 21*(18). https://doi.org/10.1007/s11883-019-0779-5.
10. Hu, F. B. (2003). Plant-based foods and prevention of cardiovascular disease: An overview. *The American Journal of Clinical Nutrition, 78*(3), 544S–551S. https://doi.org/10.1093/ajcn/78.3.544S.
11. Mackenzie, J., & Murray, E. (2021). Socially constructing healthy eating: A Foucauldian discourse analysis of healthy eating information and advice. *Qualitative Health Research, 31*(12), 2135–2146. https://doi.org/10.1177/10497323211023436.
12. Mozaffarian, D., Angell, S. Y., Lang, T., & Rivera, J. A. (2018). Role of government policy in nutrition—Barriers to and opportunities for healthier eating. *BMJ, 361*, k2426. https://doi.org/10.1136/bmj.k2426.
13. Popkin, B. M., Corvalan, C., & Grummer-Strawn, L. M. (2020). Dynamics of the double burden of malnutrition and the changing nutrition reality. *The Lancet, 395*(10217), 65–74. https://doi.org/10.1016/S0140-6736(19)32497-3.
14. Seferidi, P., Hone, T., Duran, A. C., Bernabe-Ortiz, A., & Millett, C. (2022). Global inequalities in the double burden of malnutrition and associations with globalization: A multilevel analysis of Demographic and Health Surveys from 55 low-income and middle-income countries, 1992–2018. *The Lancet Global Health, 10*(4), e482–e490. https://doi.org/10.1016/S2214-109X(21)00594-5.
15. Swinburn, B. A., Kraak, V. I., Allender, S., Atkins, V. J., Baker, P. I., Bogard, J. R., Brinsden, H., Calvillo, A., De Schutter, O., & Devarajan, R. (2019). The global syndemic of obesity, undernutrition, and climate change: The Lancet Commission report. *The Lancet, 393*(10173), 791–846. https://doi.org/10.1016/S0140-6736(18)32822-8.

CHAPTER 9

Healthy Eating Initiatives

LEARNING OBJECTIVES

After studying this chapter, you will be able to:

- Discuss healthy eating and healthy foods
- Understand the role of local governments in promoting healthy eating
- Focus on FDA's nutrition initiatives.

KEY TERMS FROM THIS CHAPTER

Antioxidants
Anxiety
Bowel
Calcium
Cardiovascular disease
Depression
Fiber
Fruits
Heart disease
Menu labeling
Moderation
Pancreatic cancer
Policymakers
Potassium
Stress
Temptation
Toxins
Type 2 diabetes
Vegetables
Whole grains

INTRODUCTION

A large number of people worldwide do not follow a healthy diet. The goal of Healthy People 2030 is to assist individuals in consuming the appropriate amounts of wholesome foods, such as fruits, vegetables, and whole grains, in order to lower their risk of developing chronic illnesses and enhance their general health. Promoting healthy eating aims to assist individuals in consuming the appropriate levels of important nutrients, such as potassium and calcium. Overindulgence in unhealthy foods, such as foods heavy in added sugars and saturated fat, increases the risk of obesity, heart disease, type 2 diabetes, and other health issues. People's overall health can be improved and their risk of chronic diseases can be decreased by using strategies and interventions to assist them in making healthy food choices.

Some people lack the knowledge necessary to make healthy food choices. Some people lack access to wholesome foods or lack the funds to purchase a sufficient supply of food. Reducing food insecurity and hunger while promoting health is largely dependent on public health interventions that prioritize assisting everyone in accessing nutritious foods.

9.1. HEALTHY EATING AND HEALTHY FOODS

Maintaining good health and lowering the risk of chronic diseases like cardiovascular disease, type 2 diabetes, and some cancers like bowel, breast, and pancreatic cancer require eating a nutritious, balanced diet. A healthy diet throughout life can help prevent various non-communicable diseases (NCDs) and conditions, including malnutrition in all its forms, according to the World Health Organization (WHO).

According to research, the majority of Britons do not follow government dietary guidelines. The population's intakes of sugar, salt, and saturated fat are above government recommendations, while intakes of fiber, fruits and vegetables, and oily fish are below them, according to the most recent UK National Diet and Nutrition Survey (NDNS) data. Adults are thought to consume 200–300 more calories per day on average than is advised, and children who are obese or already overweight may consume up to 500 more calories per day than is necessary for a healthy body weight.

The National Health Service (NHS) in the United Kingdom is estimated to have spent £6.5 billion on diseases associated with obesity and overweight in the fiscal year 2020–2021. This represents about 4.7% and a half of the NHS budget. Increasing healthy life expectancy requires lowering obesity and making dietary improvements. Based on recommendations from the Scientific Advisory Committee on Nutrition (SACN), government dietary recommendations are supported by the most recent scientific evidence. The mission of SACN is to evaluate the whole body of evidence and develop dietary recommendations for the UK population, including vulnerable and diverse groups, based on the assessment of the risks and benefits of nutrients, dietary patterns, food, or food components to health. Many people rely on health and care providers to give accurate, fact-based advice. We are all interested in what we eat and drink, which is why medical personnel who may consult with patients on this subject must be knowledgeable about it (Chadwick et al., 2003; Himmelgreen et al., 2011; White, 2018).

9.1.1. The Fundamentals of Healthy Eating

We all need a balance of protein, fat, carbs, fiber, vitamins, and minerals in our diets to maintain a healthy body, despite what some extreme diets may suggest. Instead of cutting out specific food groups from your diet, choose the healthiest options within each category.

Protein supports mood and cognitive function and provides you with the energy to get up and go—and keep going. People with kidney disease may be adversely affected by an excess of protein, but new research indicates that many of us require higher-quality protein, particularly as we age. This does not imply that you must consume more animal products; instead, you can make sure your body gets all the necessary protein by consuming a range of plant-based protein sources every day.

- **Fruit:** Fruits in cans, freezer packs, or fresh are excellent options. Try fruits like mango, pineapple, or kiwi fruit, which are alternatives to apples and bananas. Try frozen, canned, or dried fruit if fresh isn't available. Fruit that has been canned or dried may have syrups or additional sugars added. Select fruit varieties that are canned and packed in either their own juice or water.
- **Vegetables:** Use an herb like rosemary to give grilled or steamed veggies some variation. Vegetables can also be pan-fried or sautéed in a nonstick pan using a tiny bit of cooking spray. For a quick side dish, try frozen or canned vegetables; just microwave and serve. Select canned veggies that don't have any extra butter, salt, or cream sauces added. Try a different vegetable every week to add variety.
- **Calcium-Rich Foods:** Think about low-fat and fat-free yogurts without added sugars in addition to fat-free and low-fat milk. They can be a great dessert alternative and come in a variety of flavors.
- **Meats:** If your go-to recipe calls for breaded chicken or fried fish, consider baking or grilling as healthier alternatives. Consider substituting dry beans for meats as well. Look for calorie-conscious recipes in magazines, online, and from friends. You might be surprised to discover a new favorite meal!
- **Comfort Foods:** Even if a food is heavy in calories, fat, or added sugars, you can still enjoy it. The secret is to only occasionally eat them.

Some general tips for comfort foods:

- **Eat Them Less Often:** Reduce your daily consumption of these foods to once a week or once a month.
- **Eat Smaller Amounts:** Try a half-bar or a smaller chocolate bar if that's your favorite higher-calorie snack.
- **Try a Lower-Calorie Version:** Make food with fewer calories or in a different way. Try redoing your macaroni and cheese recipe using non-fat milk, reduced butter, low-fat cheese, fresh spinach, and tomatoes, for instance, if the original calls for whole milk, butter, and full-fat cheese. Just be careful not to eat more food than you need.

9.1.2. Making the Switch to a Healthy Diet

Making the switch to a healthier diet doesn't have to be a binary decision. It's not necessary to be flawless, to give up foods you enjoy entirely, or to make all the changes at once; doing so will almost always result in cheating or giving up on your new eating regimen.

A better strategy would be to implement a few minor adjustments at once. You can accomplish more in the long run without experiencing deprivation or overwhelm by a significant diet change if you set modest goals.

To plan a healthy diet, consider taking several small, doable steps, such as including a salad in your diet once a day. As you get used to your minor adjustments, you can add more healthful options.

9.1.3. Setting Yourself Up for Success

Attempt to keep things simple in order to position yourself for success. A healthier diet doesn't have to be difficult to follow. Consider your diet in terms of color, variety, and freshness, for instance, rather than obsessing over calorie counting. Make it a point to stay away from packaged and processed foods and, whenever you can, use more fresh ingredients.

Prepare more of your own meals. You can better control what you eat and keep an eye on exactly what goes into your food by cooking more meals at home. You'll consume fewer calories and stay away from packaged and takeaway foods' chemical additives, added sugar, and unhealthy fats, which can make you feel bloated, agitated, and worsen symptoms of anxiety, stress, and depression.

Make the right changes. It's crucial to swap out unhealthy foods in your diet for healthier ones when reducing your intake of them. Your health will improve if you replace harmful trans fats with healthy fats (for example, grilled salmon instead of fried chicken). However, substituting refined carbohydrates for animal fats (like having a donut instead of bacon for breakfast) won't make you feel better or reduce your chance of heart disease. Read the labels. It's critical to know what's in your food because, even for products that purport to be healthy, manufacturers frequently conceal high levels of sugar or harmful fats in packaged goods. Focus on how you feel after eating. This will promote new, healthful tastes and habits. You'll feel better after a meal if you eat healthier food. The likelihood that you will experience discomfort, nausea, or low energy increases with the amount of junk food you consume.

Drink plenty of water. Although drinking water helps rid our bodies of toxins and waste products, many people experience chronic dehydration, which can lead to headaches, fatigue, and low energy. Because it's common to confuse thirst with hunger, drinking plenty of water will also help you choose better foods.

9.1.4. Moderation: Important to Any Healthy Diet

To put it simply, moderation is the practice of only consuming as much food as your body requires. After a meal, you should feel full but not overstuffed. Many of us define moderation as cutting back on our current eating habits. However, it doesn't entail giving up your favorite foods. If you have a healthy lunch and dinner after eating bacon for breakfast once a week, it might be considered moderation; however, if you follow it with a box of doughnuts and a sausage pizza, it won't be. Don't consider any particular foods to be "off-limits." It's normal to crave those foods more when they're off limits, and to feel guilty if you give in to temptation. Reduce the amount of unhealthy foods you eat and cut back on how often you eat them. You might notice that you have less of a craving for or view unhealthy foods as rare treats as you cut back on your consumption.

Consider serving sizes that are smaller. These days, serving sizes have skyrocketed. When eating out, order a starter rather than an entree, share a dish with a friend, and avoid ordering anything supersized. Visual cues can assist with portion sizes at home. A standard light bulb is roughly the size of half a cup of mashed potato, rice, or pasta, and your portion of meat, fish, or chicken should be the size of a deck of cards. You can fool your mind into believing that a meal is larger by serving it in bowls or on smaller plates. Add extra leafy greens or finish the meal with fruit if you're not feeling full after eating.

Give it some time. It's crucial to take a moment to appreciate food as fuel rather than as something to quickly consume between appointments or while commuting to pick up the kids. Eat mindfully and stop eating before you feel full, as it may take your brain a few minutes to signal your body that it is full.

When possible, share meals with others. Eating by yourself frequently results in mindless overeating, especially when done in front of a TV or computer.

Eat fewer snacks at home. Take caution when choosing the foods you keep on hand. Keeping unhealthy snacks and treats on hand makes it harder to eat in moderation. Instead, surround yourself with good options, and when you feel like treating yourself to something special, go out and get it. Restrict your emotional eating. Not all of the time do we eat to satiate our hunger. In addition, a lot of us use food as a coping mechanism for stressful situations or negative feelings like loneliness, melancholy, or boredom. However, you can take back control of your eating habits and emotional state by learning more constructive coping mechanisms for stress and emotions.

- **It's Not Just What You Eat, But When You Eat**: Consume breakfast and smaller meals all day long. Healthy meals throughout the day will keep you energized, and a hearty breakfast can speed up your metabolism.

Steer clear of late-night eating. Aim for an earlier dinner and observe a fast of 14–16 hours until breakfast the following day. According to studies, controlling your weight may be aided by eating only during periods of maximum activity and taking long daily breaks for your digestive system.

9.1.5. Add More Fruit and Vegetables to Your Diet

Fruits and vegetables are rich in vitamins, minerals, antioxidants, and fiber, and low in calories. This is known as their nutrient density. Eating the suggested daily intake of five or more servings of fruits and vegetables will naturally satisfy you and reduce your intake of unhealthy foods. Half a cup of raw fruit or vegetables, or a small apple or banana, for example, constitutes a serving. The majority of us should eat twice as much as we do now.

To increase your intake:

- Enhance your preferred breakfast cereal with berries, which are high in antioxidants.
- For dessert, have a mixture of sweet fruits such as oranges, mangos, pineapples, and grapes.
- Try a colorful salad instead of your typical side of pasta or rice.
- Snackle on veggies like carrots, snow peas, or cherry tomatoes with peanut butter or a spicy hummus dip instead of processed snack foods.

9.1.6. Make Vegetables Tasty

Although simple salads and steamed vegetables can easily turn boring, there are many ways to infuse your vegetable dishes with flavor. Include color. Deeper-hued vegetables not only have higher antioxidant, vitamin, and mineral concentrations, but they also add visual appeal and taste variation to meals. Use roasted red cabbage wedges, yellow squash, sun-dried or fresh tomatoes, glazed carrots or beets, or colorful, sweet peppers to add some color. Invigorate leafy greens. Don't stop at lettuce. Chinese cabbage, broccoli, mustard greens, kale, arugula, and spinach are all rich in nutrients. Try dressing your salad greens with a spicy dressing, drizzling them with olive oil, or tossing them with almond slices, chickpeas, a little bacon, parmesan, or goat cheese. Fulfill your desire for sweets. Vegetables that are naturally sweet, like bell peppers, carrots, beets, sweet potatoes, yams, onions, and squash, enhance meal sweetness and lessen your desire for added sugar. Add them for a gratifying sweet kick to stews, pasta sauces, and soups.

Try different ways to prepare broccoli, Brussels sprouts, asparagus, and green beans. Try grilling, roasting, or pan-frying these nutritious sides with garlic, shallots, chili flakes, mushrooms, or onion as an alternative to boiling or steaming them. Alternatively, marinate in zesty lime or lemon before cooking (Chadwick et al., 2003; Himmelgreen et al., 2011; White, 2018).

9.1.7. Healthy Eating Learning Opportunities and Nutrition Education

Opportunities for children to learn about healthy eating include nutrition education and other activities incorporated into the school day that can help them make informed food and drink choices. A thorough health education program must include nutrition education since it gives kids the knowledge and abilities to choose foods and beverages that are healthy.

- **Making Opportunities for Nutrition Education in Schools is Important: In** the US, students are only receiving less than 8 hours of mandatory nutrition education each school year, which falls far short of the recommended 40 to 50 hours needed to actually change behavior. Moreover, the percentage of schools that are providing mandatory instruction on nutrition and dietary habits has decreased from 84.6% in 2000 to 74.1% in 2014.

The ideal scenario for schools would be to give students more hours of nutrition education instruction and involve parents and teachers in nutrition education activities, given the significant role that diet plays in preventing chronic diseases and supporting good health. According to research, nutrition education can help students understand how eating habits can be influenced by emotions and how a healthy diet affects emotional well-being. Nonetheless, given the workload that schools have, staff members may think of ways to incorporate nutrition education into the current curriculum.

- **Schools can Include Nutrition Education Throughout the School Day:** Schools have the flexibility to integrate nutrition education into different parts of the school day and in various areas of the school. This allows schools to use strategies that align with their specific settings, schedules, and resources.

9.1.7.1. In the Classroom

In the classroom, nutrition instruction can be integrated into other subjects like: or taught as a stand-alone health education course.

- Counting with pictures of fruits and vegetables.
- Learning fractions by measuring ingredients for a recipe.
- Examining how plants grow.

- Learning about cultural food traditions.

The National Health Education Standards should be followed, and nutrition education should include elements of a strong health education program.

9.1.7.2. Farm to School

Each school or district's farm-to-school initiatives are unique, but they frequently incorporate one or more of the following tactics:

- Purchasing and providing food for the school meal programs that is produced locally or regionally.
- Teaching pupils about nutrition, health, and agriculture.
- Involving students in practical learning experiences such as cooking classes, field trips to farms, or gardening.

Pupils who take part in farm-to-school programs eat more fruits and vegetables, are more open to trying new foods, and have greater knowledge about agriculture and nutrition.

9.1.7.3. School Gardens

School garden initiatives can improve students' understanding of nutrition, their inclination to try new foods, and their perceptions of fruits and vegetables. School gardens differ in terms of size and use. Schools may have raised beds, greenhouses, window sill gardens, or planted fields. In addition to planting seeds, harvesting fruits and vegetables, and tasting the produce, students can also prepare the garden's soil. School lunches or taste tests may include produce from school gardens. The school garden can be used by teachers to impart math, science, history, and language arts lessons.

9.1.7.4. In the Cafeteria

In these learning laboratories, cafeterias serve as a place where students can sample new foods through the school meal program, observe what balanced meals look like, and get encouragement to try new foods through taste tests or verbal cues from nutrition staff. Additionally, posters encouraging healthy eating and artwork created by students can be used to decorate cafeterias.

9.1.7.5. Other Opportunities

The following are some areas where schools can incorporate lessons about nutrition and eating well.

- Morning announcements.
- School assemblies.
- Materials sent home to parents and guardians.
- Staff meetings.
- Parent-teacher group meetings.

By using these tactics, educators can make sure that students receive and see consistent messages about healthy eating both at home and on the school campus (Chadwick et al., 2003; Himmelgreen et al., 2011; White, 2018).

9.2. THE ROLE OF LOCAL GOVERNMENTS IN PROMOTING HEALTHY EATING

Local governments play a crucial role in promoting health and well-being by implementing initiatives to improve food environments in communities. They can lead by example by offering healthier food options in government facilities, changing zoning laws to enhance access to healthy foods in lower-income areas, mandating menu labeling in restaurants, and facilitating community-wide changes. By developing the necessary infrastructure for obesity prevention programs and policies, local governments can effectively communicate the importance of healthy eating and obesity prevention to community members. It is essential for local governments to create opportunities for individuals to make healthier food choices by making the healthy option the easiest choice available.

Local governments should focus on both increasing access to healthy foods and reducing access to unhealthy foods. Currently, there are more strategies in place to enhance access to healthy foods than to limit access to unhealthy options. However, efforts should also be made to restrict access to unhealthy foods, as this could be more effective in combating obesity than simply increasing access to healthy options. For instance, implementing restrictions on sugary beverages in after-school programs may be more impactful in reducing calorie intake and preventing weight gain in young people than initiatives like farmers' markets or community gardens. Additionally, while some initiatives may not directly impact healthy eating habits, they can help build relationships among community stakeholders, which can be beneficial for advocating for more impactful initiatives in the future.

Three objectives guide the committee's recommendations for healthy eating practices and the next steps that local governments should take into account.

1. Improve access to and consumption of healthy, safe, and affordable foods.
2. Reduce access to and consumption of calorie-dense, nutrient-poor foods.
3. Raise awareness about the importance of healthy eating to prevent childhood obesity.

As mentioned in the introduction, some action steps are bolded as the most promising. However, this list of goals and the strategies and steps discussed under each are not ranked in order of importance. The best people to determine which plans of action and strategies will work best for their community's needs and circumstances are local leaders, members, and policymakers. The choice will mostly depend on elements including leadership, priorities, resources, and demographics. The objectives, plans of action, and strategies pertaining to healthy eating are listed below.

9.2.1. Goal 1: Improve Access to and Consumption of Healthy, Safe, and Affordable Food

9.2.1.1. Strategy 1: Retail Outlets

Expand the availability of healthful foods in the community by utilizing supermarkets, grocery stores, and convenience/corner stores.

- **Rationale:** Healthy foods must be easily accessible, reasonably priced, and convenient for people to eat a balanced diet. According to research, people who live in neighborhoods with better access to grocery stores and supermarkets typically eat healthier and are less likely to be obese. Increased consumption of fruits and vegetables can also result from having access to supermarkets and grocery stores. Diet is also influenced by the existence of corner stores and tiny food stores. Supermarkets and other locations carrying healthful foods are typically less accessible to lower-class, minority, and rural residents. The likelihood of these people having a high body mass index (BMI) is higher. Research indicates that lowering teenagers' BMIs can be achieved by utilizing land use regulations related to urban planning to improve access to supermarkets.

The following action steps are complicated, and the committee acknowledges that some may call for more time, money, or assistance than others. It is crucial to think about the foundation that could make these actions simpler to carry out. For instance, local governments may need to establish regulations aimed at assisting companies and other stakeholders in navigating the intricate development process to draw supermarkets to underprivileged areas.

Action Steps:

Provide incentive programs (such as tax credits, grant and loan programs, small business/economic development programs, and other economic incentives) to entice supermarkets and grocery stores to underprivileged neighborhoods.

- Adjust the bus routes or introduce alternative transportation options, such as community vans or shuttles, to guarantee that residents can conveniently and affordably reach supermarkets or grocery stores using public transportation.
- Establish incentive programs to support small food store owners in underserved areas in stocking healthier and more affordable food options. This may include providing grants or loans for purchasing refrigeration equipment to store fresh produce and low-fat dairy products, offering

free publicity, implementing a city awards program, or connecting store owners with wholesale distributors.

- Use zoning laws (such as "as of right" and "conditional use permits") to allow healthy food providers to open in underprivileged areas.
- Improve the accessibility of grocery stores by implementing public safety initiatives like increased outdoor lighting and police presence.

9.2.1.2. Strategy 2: Restaurants

Identify and make healthier food options more readily available in restaurants.

- Rationale: More Americans than ever are dining out than in their homes. Foods obtained from sources outside the home have more calories and fat than foods prepared inside. Obesity is linked to eating out more frequently, particularly at fast-food establishments. Making educated decisions at restaurants is challenging if nutrition information is not readily available and easily understood at the time of ordering. Initial research from areas where menu labeling has been implemented indicates that when customers are aware of the nutritional value of the options, they may make healthier selections.

Action Steps:

- Make it mandatory for chain restaurants to label their menus with the number of calories listed on the menu boards and in-store menus.
- Encourage independent eateries to post calorie counts on their menu boards and in-store menus for patrons to see.
- Provide rewards (like recognition or endorsements) to eateries that serve healthier options (like boosting the selection of healthier foods, serving portion sizes appropriate for the age of the customer, or making healthier default options like non-fat milk in place of soda and apples or carrots in "kids' meals").

9.2.1.3. Strategy 3: Community Food Access

Encourage initiatives to offer fruits and vegetables in diverse locations, including farmers' markets, farm stands, mobile markets, community gardens, and gardens aimed at youth.

- **Rationale:** One of the most important ways to raise the standard of diet is to increase the availability of fruits and vegetables. Due to their high fiber and water content, fruit and vegetables can increase satiety and decrease total calorie intake despite having comparatively few calories. A successful obesity prevention strategy may include swapping out higher-calorie foods like those with added sugars and fat for fruits and vegetables. Research indicates that supporting farmers' markets can raise the consumption of fruits and vegetables. Additionally, there's a chance that community gardens and garden-based nutrition intervention programs will encourage young people to taste more fruits and vegetables and increase their intake of these foods.

Action Steps:

- Farmers' markets should be encouraged to accept WIC Farmers' Market Nutrition Program coupons and Special Supplemental Nutrition Program for Women, Infants, and Children (WIC) food package vouchers. Additionally, funding for electronic payment-processing equipment should be allocated to farmers' markets so that they can accept WIC Program Electronic Benefit Transfer (EBT) cards and Supplemental Nutrition Assistance Program (SNAP, formerly the Food Stamp Program).
- To encourage residents of lower-income neighborhoods, as well as WIC and SNAP recipients, to visit farmers' markets and farm stands, and increase funding for outreach, education, and transportation.
- To encourage, protect, and expand possible locations for farmers' markets and community gardens, such as abandoned city property or underutilized parking lots, land use policies and zoning regulations should be introduced or changed.
- Provide community-based group activities (like community kitchens) that connect acquiring affordable, healthful food with enhancing one's ability to shop for and prepare it.

9.2.1.4. Strategy 4: Public Programs and Worksites

Make sure that publicly managed organizations, like after-school programs, childcare centers, recreation centers, and local government workplaces, put policies and procedures in place to encourage the consumption of healthy foods and beverages and to limit or completely remove the availability of foods high in calories and low in nutrients.

- **Rationale:** National studies repeatedly demonstrate that young people's diets do not adhere to recommended daily allowances for health, are a major cause of overweight and obesity, and put them at risk for major health consequences. An atmosphere that promotes healthy eating should be present in areas where kids congregate and spend a lot of time. Ensuring that the food served in after-school programs, child care centers, and recreation centers is healthy could help improve the diets of the many children who spend time in these settings. Studies indicate that there may be insufficient activity levels and inadequate nutritional quality of meals and snacks in child care facilities. In addition, kids imitate adult behavior. It would be implied that eating healthily is a major concern if publicly owned organizations made sure that programs and facilities serving both adults and children adhered to strict nutrition standards. Local government organizations can take the lead and set an example by implementing procedures and policies that encourage the consumption of healthful foods in public areas.

Action Steps:

- Establish and enforce strict dietary requirements for all food and drink items served in government-run or regulated child care centers, parks, recreation centers, and after-school programs. This should include restricting the availability of high-calorie, low-nutrient foods.
- Make sure that wherever foods and beverages are sold or made available, local government organizations that run cafeterias and vending machines have strict nutrition guidelines in place.
- Offer financial support or subsidies to government-managed or -regulated initiatives, as well as communities, that offer wholesome food at affordable costs and restrict the consumption of high-calorie, low-nutrient foods (such as after-school programs that serve fruits and vegetables daily and do away with such foods from vending machines or program materials).

9.2.1.5. Strategy 5: Government Nutrition Programs

Expand the number of people who take advantage of nutrition assistance programs offered by the federal, state, and local governments (such as WIC, school breakfast and lunch programs, the Child and Adult Care Food Program, the Afterschool Snacks Program, the Summer Food Service Program, and SNAP).

- **Rationale:** Programs for nutrition assistance make food for a healthy diet accessible to children and those with lower incomes. They have no connection to an increase in weight in either adults or children. Indeed, the School Nutrition Dietary Assessment Study-III recently revealed that children who took part in the program had a decreased risk of being overweight or obese. Similar correlations between School Lunch, SNAP, and the WIC program have been found in other studies. Additionally, the WIC food package was recently revised to increase the amount of fruits, vegetables, and whole grains that participants receive. This makes it even easier for lower-income mothers and their kids to eat the nutritious foods that the Dietary Guidelines recommend.

Action Steps:

- Establish rules requiring government-run and -regulated organizations in charge of managing nutrition assistance programs to work together with other organizations and programs to boost program enrollment and participation (for example, WIC agencies should make sure that eligible individuals are also taking part in SNAP, etc.).
- Make sure that the organizations in charge of licensing child care and after-school programs promote the use of nutrition assistance programs and boost the number of people enrolled in them (CACFP, the

Afterschool Snack Program, and the Summer Food Service Program).

9.2.1.6. Strategy 6: Breastfeeding

Encourage nursing and work to create communities that are welcoming to breastfeeding.

- **Rationale:** There are several health advantages to breastfeeding for both moms and babies. Studies indicate that a child's likelihood of being overweight decreases with the length of time the child is breastfed. However, a lot of obstacles prevent moms from continuing to breastfeed, even despite national recommendations to raise the rates of breastfeeding initiation and duration. Local governments have the power to enact laws and initiatives that facilitate nursing mothers. Mothers who give birth in "baby-friendly hospitals," such as those that adhere to the Baby-Friendly Hospital Initiative USA, which is a program of the United Nations Children's Fund/World Health Organization (UNICEF/WHO), are more likely to breastfeed their child exclusively.

Action Steps:

– Adopt procedures in municipal and county hospitals that align with UNICEF/WHO's Baby-Friendly Hospital Initiative USA. This program offers hospitals 10 steps to successful breastfeeding that protect, promote, and assist nursing.

– End laws and regulations that prohibit or restrict breastfeeding in public areas and promote the establishment of lactation rooms in public areas. Permit breastfeeding in public areas.

– Provide incentive schemes to compel government organizations to provide lactation rooms and other conducive work environments for nursing mothers.

– Give money to WIC clinics so they can buy breast pumps to give to participants.

9.2.1.7. Strategy 7: Drinking Water Access

To promote the use of free, safe drinking water in public spaces rather than beverages with added sugar, increase the availability of this resource.

- **Rationale:** Drinking beverages with added sugar is thought to be a major cause of childhood obesity. Children and teenagers who drink water instead of sugar-sweetened beverages tend to consume less energy overall. By adding water fountains to public spaces and buildings, overweight and obesity can be prevented and reduced as well as increased water intake.

Action Steps

– Make plain water available in all outdoor spaces managed and operated by the local government, as well as in other public areas and facilities.

– Adopt building codes (such as those governing public restrooms) that mandate access to and upkeep of fresh drinking water fountains.

9.2.2. Goal 2: Reduce Access to and Consumption of Calorie-Dense, Nutrient-Poor Foods

9.2.2.1. Strategy 8: Policies and Ordinances

Put in place financial policies and local laws (such as taxes, incentives, land use and zoning regulations) to deter people from consuming foods and beverages that are high in calories but low in nutrients.

- **Rationale:** Merely boosting the availability and intake of healthful foods won't cut down on excess calorie intake or body weight. Reducing the availability and intake of foods high in calories but low in nutrients is also necessary to cut back on excess calories and help prevent childhood obesity. These convenient and affordable foods are often high in added fats, sugars, and refined grains. Today's kids have almost constant access to foods and drinks high in calories but low in nutrients. In lower-class urban neighborhoods, fast food restaurants and other eateries serving cheap, unhealthy food are frequently concentrated. Additionally, they are frequently found close to schools, which may raise the obesity rate. Furthermore, fast-food restaurant advertising may have an impact on obesity rates. Taxing foods high in calories but low in nutrients is one way to reduce consumption. The laws governing zoning and land use that control fast food establishments may also have an impact on consumption. It is noteworthy to mention that any restrictions on advertising or levying of taxes on food and beverages must take into account local and state legal issues. In addition, different communities may have different authorities over these matters.

Action Steps:

- Encourage people to avoid consuming foods and drinks with low nutritional value, like sodas with added sugar, by implementing a tax strategy.
- Adopt zoning and land use regulations that limit the location of fast food restaurants next to public playgrounds and school properties.
- Enact local laws to prohibit the mobile sale of foods high in calories and low in nutrients close to playgrounds and schools.
- Put in place zoning regulations intended to restrict the number of fast-food restaurants in residential areas.
- Remove all promotion and advertising of high-calorie, low-nutrient foods and drinks close to school premises and other public areas that young people frequent.
- Provide reward and recognition schemes to entice supermarkets and convenience stores to cut back on the point-of-sale promotion of nutrient-poor, high-calorie items (e.g., "candy-free" checkout aisles).

9.2.3. Goal 3: Raise Awareness About the Importance of Healthy Eating to Prevent Childhood Obesity

9.2.3.1. Strategy 9: Media and Social Marketing

Encourage social media and media campaigns that promote healthy eating and the prevention of childhood obesity.

- **Rationale:** In addition to encouraging healthy eating and portion control, media can also help people become more aware of weight as a health concern and encourage them to eat fewer high-calorie, low-nutrient foods. Signage and high-frequency radio and television advertising may help to improve public perceptions of eating a healthy diet. A community's eating habits may even change as a result of a media strategy. Both the creation of local campaigns and the implementation of national message campaigns may be beneficial, depending on the available resources and the goal of the campaign. Respecting the report's emphasis on the modifications that local governments can implement to enhance children's food and physical activity environments, it is crucial to emphasize that social media campaigns and the media themselves can enhance these local environments by emphasizing the benefits of enhancing children's food and physical activity environments and encouraging the public to take advantage of new resources available to them, such as farmers' markets, new grocery stores, healthier options at local establishments, etc.

Action Steps:

- Create media campaigns that use a variety of platforms, including print, radio, the Internet, television, social media, and other promotional materials, to spread consistent messages about healthy eating and active living.
- Create a media campaign that reframes obesity as the result of environmental injustices rather than just bad personal decisions and positions community access to healthful foods as a matter of health equity.

Create media strategies for counter-advertising unhealthy products to youth, emulating those employed in the fields of alcohol and tobacco prevention (Chadwick et al., 2003; Himmelgreen et al., 2011; White, 2018).

9.3. FDA: ITS NUTRITION INITIATIVES

To guarantee that Americans have easier access to healthier foods and nutrition information that helps us all make better decisions, the FDA is giving priority to its nutrition initiatives. A greater selection of healthier foods could lead to better eating habits, which would enhance everyone's health and wellbeing. The White House unveiled a National Strategy to end hunger and promote healthy eating and physical activity by 2030 at its Conference on Hunger, Nutrition, and Health on September 28, 2022. The goal is to reduce the number of people who suffer from diet-related diseases like diabetes, obesity, and hypertension. Many FDA initiatives are part of the strategy to help speed up efforts to provide consumers with information and create a healthier food supply. These initiatives include developing a voluntary symbol and updated definition for the "healthy" nutrient content claim, developing Dietary Guidance Statements for food labels and front-of-package labeling, and gathering feedback for nutrition labeling for groceries sold online. By establishing goals to lower sodium content in food and giving the industry more latitude when it comes to salt usage in standardized foods, the strategy also addresses the need to promote the production of healthier foods. In addition to enforcing the Nutrition Facts label to disclose added sugars, the FDA will look into other strategies for lowering added sugar content in food. The FDA will host a virtual meeting and listening sessions on November 6–8, 2023, to discuss methods for lowering the country's consumption of added sugar.

The federal dietary guidelines are not in line with U.S. eating habits. In addition to consuming excessive amounts of saturated fat, sodium, and added sugars, the majority of Americans do not eat enough fruits, vegetables, dairy, whole grains, and healthy oils. Many factors, such as consumers' knowledge, preferences, and culture, as well as the availability of safe, affordable, and healthful food options, all have an impact on eating healthily. The FDA aims to influence eating habits by incentivizing the food industry to produce healthier foods and by disseminating information to enable consumers to make better decisions. Chronic diet-related diseases like obesity, diabetes, and cardiovascular disease are on the rise in the United States. Chronic but preventable diseases are the leading cause of death and disability in the United States. People who belong to racial and ethnic minority groups, have lower socioeconomic status, and reside in rural areas are disproportionately affected by diet-related chronic diseases. Poor nutrition plays a major role in these diseases. Furthermore, since individuals with obesity, diabetes, and other chronic illnesses are more likely to experience severe symptoms from COVID-19 and even die from the disease, the pandemic has made it abundantly evident that we need to improve nutrition. The COVID-19 pandemic caused a significant increase in Body Mass Index rates, according to a 2021 CDC report; people who were overweight or obese and parents of younger school-aged children saw the biggest increases.

9.3.1. The Role of the FDA

By assisting in the improvement of dietary practices in the U.S., the FDA plays a significant part in a larger, whole-of-government strategy to help lower the burden of chronic diseases and advance health equity. Food labeling, for instance, can be a potent tool for change. By providing them with knowledge, consumers are better equipped to choose healthier foods. Food labeling may promote a healthier food supply in addition to assisting consumers in making better food choices, provided that some manufacturers reformulate their products to make them healthier.

To assist consumers in making healthier decisions, the FDA has taken many important actions. To meet nutrient needs while staying within calorie limits, for instance, consuming too much added sugar can make it difficult. As an example, the FDA updated the Nutrition Facts label with updated information and a refreshed design. In order to provide customers with more nutrition information when they eat out, calories must now be disclosed on some menus and menu boards. As part of an ongoing effort to reduce sodium in the food supply, the FDA most recently issued voluntary sodium reduction targets for the industry in a wide variety of processed, packaged, and prepared foods. The FDA is aware that more has to be done. We'll keep putting our efforts into making the food supply healthier, giving consumers more power, and assisting in giving our youngest populations healthy beginnings. By implementing a comprehensive government strategy in collaboration with our federal partners, we can significantly lower the likelihood of chronic diseases linked to diet, promote health equity, and enhance the well-being of coming generations.

9.3.2. Key Elements

9.3.2.1. Sodium Reduction

The average American consumes 50% more sodium than what is advised for people 14 years of age and older. Overs oiling with salt can lead to hypertension, or elevated blood pressure, which raises the risk of heart attacks and strokes. Over 40% of adult Americans have high blood pressure, and for non-Hispanic Black adults, the percentage rises to roughly 60%. It is difficult for consumers to cut back on their sodium intake because packaged and restaurant foods account for about 70% of total sodium intake. The FDA released voluntary sodium reduction targets for the food industry in October 2021 for a wide range of processed, packaged and prepared foods. This was the first step in an iterative approach to reduce sodium across the food supply. The FDA intends to keep an eye on the food supply and create new, updated targets later on.

9.3.2.2. Maternal and Infant Health and Nutrition

For the first time, dietary recommendations for children under two are included in the Dietary Guidelines for Americans, 2020–2025. Additionally, recommendations for individuals who are pregnant or nursing are expanded.

The FDA is now able to carry out more work that impacts these vital populations thanks to these updates. The trajectory of eating habits and health-related behaviors throughout life can be influenced by adopting healthy dietary patterns early in life. Three main topics are the focus of our work:

- **FDA/EPA Fish Advice:** The FDA and the Environmental Protection Agency released updated guidelines in October 2021 to assist individuals who are considering becoming pregnant, are already pregnant or nursing, and parents or other caregivers who are feeding children in making decisions about safe and nutrient-dense fish. Nutrients and healthy fatty acids found in seafood can support children's development and growth. The FDA is still working on creating a range of consumer education tools, such as new infographics that will be available in October 2021, to help consumers better understand the recommendations and to promote the consumption of seafood that is lower in mercury among these populations. Furthermore, a study on the impact of seafood consumption on a child's growth and development has been initiated by the FDA and federal partners.
- **Infant Formula Premarket Review:** For many infants during a crucial stage of growth and development, infant formula serves as a major or even their only source of nutrition. The majority of infants in the U.S. receive infant formula by 6 months of age. Before introducing new infant formulas into interstate commerce, infant formula manufacturers are required to submit a detailed report to the FDA detailing the safety and nutritional adequacy of the new formula. The FDA has set certain nutritional requirements for infant formulas. The volume and complexity of these submissions are rising. In addition to making sure that these products adhere to the FDA's safety, nutrition, and quality standards, the agency is working to increase consumer access to infant formula products. The FDA is also keeping an eye on the most recent scientific research on novel ingredients found in infant formula, such as new "bioactive" ingredients that bear similarities to the components of human milk.
- **Promoting Nutritious and Safe Food Choices:** The FDA will collaborate with federal partners to create more educational programs and outreach campaigns to inform pregnant individuals, potential mothers, parents, and caregivers about the latest recommendations for healthy eating habits. Through the implementation of the Closer to Zero initiative, the FDA will provide parents and caregivers with easy access to reliable, science-backed information to assist them in making nutritious food choices for their children's well-being, while also striving to minimize exposure to harmful substances in food as much as possible.

9.3.2.3. Labeling and Claims

Food packages can use claims and symbols as quick signals to help customers choose foods that fit into healthy eating habits and comprehend nutrition information. Additional information on food labels can give consumers important insights into which foods are healthier. Food producers may be inspired to reformulate and create healthier foods if they wish to use specific labeling on their products. For instance, trans fat in the food supply decreased by nearly 80% after it became mandatory to disclose it on the Nutrition Facts label in 2006. The FDA has worked on claims and labeling since:

- Supplying directions to food companies regarding the application of Dietary Guidance Statements. On March 23, 2023, the FDA released draft guidelines for the industry on statements that could be used on food packages to help consumers understand how a food or food group can contribute to a nutritious eating pattern. This is in line with the Dietary Guidelines for Americans' focus on eating patterns and food groups. Updating the nutrient content claim "healthy." On September 28, 2022, the FDA released a proposed rule that would modernize the definition of the "healthy" claim that can be made on food packaging and bring it into compliance with dietary guidelines and current nutrition information. The voluntary "healthy" claim should be updated to make it easier for consumers to recognize foods that fit into healthy eating habits. Some manufacturers may also decide to create new products or reformulate existing ones to use the "healthy" claim.
- **Developing a "Healthy" Symbol:** The FDA is conducting consumer research to develop a "healthy" symbol that could appear on food packages to further assist consumers in easily identifying packaged foods that would meet an updated definition of "healthy."
- **Front-of-Pack Labeling:** The FDA is working on creating a front-of-package, or FOP, labeling system that will make it simple and quick for consumers to recognize foods that fit into a healthy eating pattern. The use of these programs has skyrocketed globally in recent years. Fair and equal access to nutrition information can be facilitated by FOP systems, especially for individuals with limited nutrition knowledge.
- Investigating suggestions for online grocery shopping regarding nutrition, ingredient, and allergen labeling information. Online grocery shopping is becoming more and more popular among consumers. The COVID-19 pandemic caused this trend to pick up speed. With more people doing their grocery shopping online, there is a chance to make sure that consumers have access to label information that will enable them to choose foods that are healthier and more informed. Because of this, the FDA must gain more knowledge about what food labeling information is made available to customers at the point of sale when they shop for groceries online. The FDA first addressed labeling in online grocery shopping during a session at the FDA's October 2021 New Era E-Commerce Summit.

9.3.2.4. Consumer Education

One of the main tenets of the FDA's Nutrition Initiatives is the provision of customized education on nutrition and labeling. One key component in assisting consumers in making better dietary choices is education. Giving consumers resources on how to interpret food labels gives them the knowledge they need to develop healthier eating habits. For instance, to promote awareness and use of the updated Nutrition Facts label, the FDA initiated a nationwide consumer education campaign in 2020. The FDA intends to persist in its educational endeavors concerning nutrition and labeling by creating and endorsing educational resources targeted at diverse demographics, such as parents and caregivers, as well as racial and ethnic minority groups.

9.3.2.5. Support Innovation

The capacity for innovation in the industry plays a role in producing healthier foods. Standards of identity (SOI) set restrictions on the ingredients and production processes that can be required or optional for a given food. Foods like milk, some cheeses, juices, breads, and many more have Special Orders (SOIs) associated with them. The industry can innovate and, in certain situations, produce healthier food by updating SOI. Our nutrition-focused SOI work consists of the following.

- Defining SOI guidelines to assist in deciding whether to create, amend, or do away with a food standard. Preserving the nutritional value of a standardized food item, encouraging innovation, and figuring out how food standards fit into a changing food supply will all be important components of these principles. The most recent Unified Agenda included an effort by the FDA and USDA to jointly publish a new proposed rule on principles for modernizing food standards.

Draft guidelines for the labeling of plant-based milk substitutes have been released. The FDA acknowledges that the popularity of plant-based foods, especially plant-based milk substitutes, has grown quickly. The FDA wants to guarantee that businesses can keep creating these kinds of goods while also making sure that consumers are aware of the product's characteristics and nutritional makeup (Chadwick et al., 2003; Himmelgreen et al., 2011; White, 2018) (Table 9.1).

Table 9.1. Progress

Initiative	Progress
Sodium Reduction	• Issued final guidance on voluntary, short-term targets for sodium (October 2021) • Published guidance to provide enforcement discretion for the use of "potassium salt" on product ingredient lists (December 2020)
Maternal and Infant Health and Nutrition	• Published updated fish advice for those who might become pregnant, are pregnant or breastfeeding, and children, in partnership with EPA (October 2021) • Co-hosted Bioactive Ingredients in Infant Formula Workshop with the National Institutes of Health (September 2021) • Issued draft guidance for industry on protein efficiency ratio studies and infant formula submissions
Labeling and Claims	• On November 16, 2023, the Reagan-Udall Foundation for the FDA will host a public meeting on front-of-package labeling. • Issued two procedural notices in January 2023 and June 2023 on consumer research plans to explore the development of a FOP nutrition labeling scheme. • Issued draft guidance about dietary guidance statements in food labeling (March 2023) • Issued a proposed rule to update the "healthy" claim (September 2022) • Issued two procedural notices on the preliminary quantitative consumer research on the healthy symbol; the first notice was issued in May 2021 and the second notice in March 2022 • Discussed labeling for online grocery shopping settings at the FDA's New Era E-Commerce Summit (October 2021) and issued an RFI for online grocery shopping (April 2023).
Consumer Education	• Co-hosted, with the National Institutes of Health, a workshop on research in risk communication, environmental contaminants in food, and the role of nutrition as they relate to child development (February 2023) • Released continuing medical education videos on the Nutrition Facts label (October 2020) • Launched Nutrition Facts label Education Campaign, which includes paid advertising, stakeholder and media outreach and a variety of new education materials, as well as a social media toolkit (March 2020) • Published menu labeling consumer education materials and social media toolkit on the website and promoted via social media (March-May 2019)

Support Innovation (Standards of Identity and Plant-based Labeling)	• Issued proposed rule to permit the use of salt substitutes in foods with standards of identity for which salt is a required or optional ingredient (April 2023) • Issued draft guidance on the labeling of plant-based milk alternatives (February 2023) • A joint FDA and USDA effort to publish a new proposed rule on principles for food standards modernization was included in the most recent Unified Agenda • Discussed horizontal approaches to SOIs with stakeholders at a public meeting (September 2019) and reviewed comments submitted to docket • Issued request for information to solicit feedback on the labeling of plant-based products using dairy terms in labeling (September 2018) • Amended SOI for yogurt (June 2021)

Source: https://www.fda.gov/food/food-labeling-nutrition/fdas-nutrition-initiatives.

CASE STUDY

LIVE LIFE WELL @ SCHOOL

Woronora River Public School is working across all aspects of school life – known as a whole-of-school approach – to implement the Live Life Well @ School program.

Students, parents and teachers are all encouraged to have a say in school decision-making, which has helped to enhance the program and foster a sense of community within the school. The school recognizes that it is important to engage students, staff and parents because children's health is strongly influenced by what happens at home, in the family, and in the wider community. One of the first steps taken by Woronora River Public School to implement the Live Life Well @ School program was to survey parents and students about their attitudes to nutrition and their eating habits. The questions focused on breakfast eating, lunchbox contents, fruit and vegetable intake and snack foods and drinks. The results of the survey contributed to the development of a statement of priorities to improve knowledge and understanding of healthy nutrition and fundamental movement skills among students, parents and staff.

Follow up surveys will help assess changes in knowledge and eating habits over time.

To help bring about positive change, Woronora River Public School has:

- established a Nutrition Committee involving student, parent and teacher representatives to improve the school canteen's nutrition standards and develop a Healthy Eating at School Policy
- developed a physical activity program for K-6 incorporating fundamental movement skills using the Get Skilled, Get Active resource.
- held a fruit and vegetable promotion during 'Fruit and Veg' week in September 2008, which included fruit and vegetable taste testing sessions with students, a fruit and vegetable promotion by the school canteen and rewards for students who had fruit and vegetables in their lunchbox
- implemented Crunch&Sip®, a set classroom break for students to eat fruit or vegetables and drink water
- run staff meetings and informal discussions with parents on Woronora River Public School's Physical Activity Policy
- conducted lunchbox and water audits to support Crunch&Sip®
- introduced fortnightly theme days at the school canteen, such as a mango and apple frappe day

Woronora River Public School has also developed links with the South Eastern Sydney and Illawarra Health Service to help support the implementation of the Live Life Well @ School program.

CLASS ACTIVITY

BASIC FACTS ABOUT NUTRIENTS

- *Project Skill*: Knowing the roles played by a few key nutrients in the body as well as the foods that contain them.

➢ **Life Skill:**

Acquiring knowledge.

Preparation

Locate at least two real-world instances or images of a nutritious food source for each nutrient specified on the form. Provide samples of foods that are grown nearby or that can be found in the veld or bush, if at all possible. Arrange the food items and/or photographs on a surface or table for the group to see. Make copies of the Basic Facts Handout.:- Before presenting this activity, make sure you have read and comprehended this handout. If there were not enough copies of the Basic Facts Handout for each participant, the information should be read aloud, point by point, and after the activity, everyone should be allowed to record the information in their note books.

Introduction

1. Display your Work pad Poster from Activity 1.1 and summarize the Food Functions points.

Inform the people in your group that the purpose of this activity is to teach them about the sources of certain essential nutrients and how the body uses them.

2. Inform your group members that they will begin this activity with a warm-up game that focuses on the roles that nutrients play in the body's defense.
3. Ask everyone to form into 3 groups:

- **Group A: Germs Make People Sick:-** This group should plan ways to make a lot of noise and "attack" Group B during the game.
- **Group B: Healthy People:-** When group A tries to attack this group during the game, ask them to come up with ways to demonstrate that they are content, healthy, and at ease.
- **Group C: Protective Nutrients:-** This group must consider how they will defend Group B from Group A's attack as vital nutrients.

4. After giving the groups time to rehearse, let them play for five to 10 minutes.
5. Request that everyone in the group talk about and clarify what they took away from the game regarding the role that nutrients play in preventing disease and maintaining the health of our bodies.

6. Remind everyone that while protecting the body is one of the functions of nutrients, it is not their only one. Describe how our bodies need to receive the proper nutrients in order to ensure that ALL of their bodily functions are met.

Process

1. Gather everyone in a big group and distribute copies of the handout, "Basic Facts about Nutrients."
2. Go through the information on each nutrient in the handout with great care. Asking questions will help your group members consider and discuss the material as you work through the handout, such as:
   - Does anyone know why protein is important to the body?
   - What foods are good sources of vitamin A?
   - Do you know of wild veld foods that might be good sources of iron?
3. Display the photographs or examples of food sources you have gathered for each nutrient as you discuss its sources. Encourage group members to continue sharing their knowledge of the roles played by nutrients in food sources.
4. The focus on vitamin A, niacin, iodine, and iron is because many Namibians do not get enough of these nutrients, which leads to malnutrition and even deficiency diseases (a "deficiency disease" is not a result of a bacterial infection; rather, it occurs when the body does not receive enough of a particular nutrient).

Apply

1. Conclude this activity by engaging in a discussion about what your group members have gained from it. Guide the conversation by prompting them with the following questions.
   - Did you know most of these facts before we began this activity?
   - Did any of these facts surprise you?
   - What did you learn from playing the game?
   - How do you think that playing such games helps with learning? Why?
   - Can you think of better ideas for learning games?
   - Why is it important to learn these basic facts about nutrients?
   - What could happen if people don't know these basic facts?
   - How will you use your new knowledge about the nutrients needed by our bodies?
   - What happens to the body if we do not eat enough vitamins and minerals?
   - What do you think is the general function of Vitamins in our body?
   - What is the function of Iron in the body?
   - What local foods are good sources of iron?
   - What happens to your body if you do not eat enough Iron?

SUMMARY

- Eating a healthy, balanced diet is an important part of maintaining good health and reducing the risk of chronic diseases such as cardiovascular disease, type 2 diabetes, and some cancers, such as bowel, breast and pancreatic cancer.
- While some extreme diets may suggest otherwise, we all need a balance of protein, fat, carbohydrates, fiber, vitamins, and minerals in our diets to sustain a healthy body. You don't need to eliminate certain categories of food from your diet, but rather select the healthiest options from each category.
- Switching to a healthy diet doesn't have to be an all or nothing proposition. You don't have to be perfect, you don't have to completely eliminate foods you enjoy, and you don't have to change everything all at once—that usually only leads to cheating or giving up on your new eating plan.
- Eating a healthier diet doesn't have to be complicated. Instead of being overly concerned with counting calories, for example, think of your diet in terms of color, variety, and freshness. Focus on avoiding packaged and processed foods and opting for more fresh ingredients whenever possible.
- Moderation means eating only as much food as your body needs. You should feel satisfied at the end of a meal, but not stuffed. For many of us, moderation means eating less than we do now.
- Fruit and vegetables are low in calories and nutrient dense, which means they are packed with vitamins, minerals, antioxidants, and fiber. Focus on eating the recommended daily amount of at least five servings of fruit and vegetables and it will naturally fill you up and help you cut back on unhealthy foods.
- Healthy eating learning opportunities includes nutrition education and other activities integrated into the school day that can give children knowledge and skills to help choose and consume healthy foods and beverages.
- Local governments can have a strong and direct impact on people's health and well-being and are well positioned to make positive changes in food environments in communities. Local governments can provide strategic leadership, such as providing improved access to healthy foods in lower-income areas, using zoning laws to change local food environments, requiring menu labeling in restaurants, serving as a catalyst for community change by offering healthier foods at government facilities, developing the infrastructure necessary for obesity prevention policies and programs to be implemented and evaluated at the local level, and communicating the importance of healthy eating and obesity prevention to community members.
- The FDA is prioritizing its nutrition initiatives to ensure people in the United States have greater access to healthier foods and nutrition information we can all use to identify healthier choices more easily. Increasing the availability of healthier foods could improve eating patterns and, as a result, improve everyone's health and wellness.

- The FDA plays a key role within a broader, whole-of-government approach to help reduce the burden of chronic diseases and advance health equity by helping to improve dietary patterns in the U.S.

REVIEW QUESTIONS

1. What is the initiative that promotes healthy eating?
2. How can we promote healthy eating?
3. What are examples of healthy eating foods?
4. How can local governments promote healthy foods?
5. What are the FDA recommendations for diet?

REFERENCES

1. Barth, M. M., Bell, R. A., & Grimmer, K. (2020). *Public Health Nutrition*. Springer Publishing Company. https://doi.org/10.1891/9780826146854.
2. Chadwick, R., Henson, S., Moseley, B., Koenen, G., Liakopoulos, M., Midden, C., Palou, A., Rechkemmer, G., Schröder, D., & Wright, A. (2003). Nutrition and health. In *Functional foods* (pp. 39–60). Springer Berlin Heidelberg. https://doi.org/10.1007/978-3-662-05115-3_4.
3. Finn, S., Culligan, E. P., Snelling, W. J., & Sleator, R. D. (2018). Early life nutrition. *Science Progress, 101*(4), 332–359. https://doi.org/10.3184/003685018x15360040523721.
4. Himmelgreen, D. A., Romero Daza, N., & Noble, C. A. (2011). Nutrition and health. In *A Companion to Medical Anthropology* (pp. 305–321). Wiley-Blackwell. https://doi.org/10.1002/9781444395303.ch15.
5. Jackson, A. A. (2017). Nutrition and liver health. *Digestive Diseases, 35*(4), 411–417. https://doi.org/10.1159/000456596.
6. Prusinski, J. (2017). White lupin (Lupinus albus L.)–Nutritional and health values in human nutrition–A review. *Czech Journal of Food Sciences, 35*(2), 95–105. https://doi.org/10.17221/114/2016-cjfs.
7. Kilgour, O. F. G. (1985). Diet and health. In *Mastering Nutrition* (pp. 214–242). Macmillan Education UK. https://doi.org/10.1007/978-1-349-17814-8_10.
8. Mulyani, S. (2019). The differences of body image, energy intake level, and nutritional status between health and non-health female student of Jenderal Soedirman University. *Jurnal Gizi dan Pangan Soedirman, 2*(2), 78. https://doi.org/10.20884/1.jgps.2018.2.2.1382.
9. Potestio, K. (2020). Nutrition. In *Men's Health* (pp. 158–173). Routledge. https://doi.org/10.4324/9781351022620-17.
10. Raine, K. D. (2014). Improving nutritional health of the public through social change: Finding our roles in collective action. *Canadian Journal of Dietetic Practice and Research, 75*(3), 160–164. https://doi.org/10.3148/cjdpr-2014-017.

11. Richmonds, R. (2004). Nutrition. In *Encyclopedia of Women's Health* (pp. 910–912). Springer US. https://doi.org/10.1007/978-0-306-48113-0_304.
12. Saxena, K. B., Kumar, R. V., & Sultana, R. (2010). Quality nutrition through pigeon pea—A review. *Health, 2*(11), 1335–1344. https://doi.org/10.4236/health.2010.211199.
13. Sharkey, J. R., Locher, J., Sahyoun, N., & Wilcox, S. (2012). Nutrition and aging: Nutritional health inequity. *Journal of Aging Research, 2012*, 1. https://doi.org/10.1155/2012/164106.
14. White, J. (2018). Nutrition. In *Health and Safety Management* (pp. 83–91). CRC Press. https://doi.org/10.1201/b22392-13.
15. Yeh, M. C., & Bruning, N. (2012). Nutrition. In *Encyclopedia of Immigrant Health* (pp. 1137–1140). Springer New York. https://doi.org/10.1007/978-1-4419-5659-0_550.
16. Zirakzadeh, A. (2012). Hypertension and nutrition. In *Nutritional Health* (pp. 247–270). Humana Press. https://doi.org/10.1007/978-1-61779-894-8_11.

CHAPTER 10

Food Packaging Regulations

LEARNING OBJECTIVES

After studying this chapter, you will be able to:

- Understand the purpose of packaging
- Identify materials used in food packaging
- Know the regulation of food packaging
- Focus on EPA guidelines for the management of MSW.

KEY TERMS FROM THIS CHAPTER

Abrasions	Aluminum
Assessment	Cardboard
Containment	Deterioration
Glass	Hazards
Metals	Microorganisms
Packaging	Plastic
Polyethylene	Polyvinyl chloride
Shelf-life	Steel
Transportation	

INTRODUCTION

The most basic human need—food—has turned into a profitable industry as well. Packaged food is becoming more and more popular as people's hectic lifestyles leave them with little time to prepare meals. Ensuring that food products are delivered to consumers safely and healthily is dependent on proper packaging. Humans have used packaging in one form or another for thousands of years. Packaging has existed since the beginning of human habitation. Food was originally transported using bark, leaves, and skins. Neolithic humans discovered pottery and used metal containers, while Mesolithic humans used baskets. Rodent-proof sealed pottery jars were used four millennia ago, and glass production was a significant Egyptian industry in 1550 BC. Iron could be tin-plated starting in AD 1200, and after AD 1600, when steel supplanted iron, this technique started to be useful. Aluminum was first extracted by Oersted in 1825. Plastics were developed more recently; in particular, the first commercial plastics were developed in the United States between 1935 and 1942. In the last 30 years, packaging has become one of the most important aspects of food production, both in terms of volume and importance.

10.1. PURPOSE OF PACKAGING

Food packaging plays a crucial role in preserving food products from external factors and potential harm, as well as containing the food and offering consumers details about ingredients and nutrition. Secondary functions such as traceability, convenience, and tamper-proof features are becoming more essential. The primary objective of food packaging is to efficiently contain food while meeting industry standards and consumer preferences, ensuring food safety, and reducing environmental impact.

10.1.1. Protection/Preservation

Food packaging can prolong shelf life, prevent product deterioration, preserve the positive effects of processing, and improve or preserve food quality and safety. Packaging shields against three main categories of outside influences: chemical, biological, and physical. Chemical protection reduces the number of compositional changes brought on by exposure to gases (usually oxygen), moisture (gain or loss), and light (visible, infrared, ultrasound).

An array of packaging materials is capable of offering a chemical barrier. Although metal and glass offer almost complete protection against chemicals and other environmental agents, most packages aren't made entirely of these materials because closure mechanisms are often included to make filling and emptying easier. Closure devices can be made of materials that permit very little permeability. For instance, plastic caps and the gasket materials used in metal can lids to enable sealing after filling both have some permeability to gases and vapors. While plastic packaging has many barrier qualities, it is typically more permeable than metal or glass. Biological protection acts as a defense against microorganisms, insects, rodents, and other animals to prevent disease and spoilage. It also helps control the ripening and aging of products. These barriers work in various ways, such as blocking access to products, preventing odors from spreading, and maintaining the package's internal environment. On the other hand, physical protection shields food from mechanical damage by providing cushioning against shocks and vibrations during distribution.

Typically made from paperboard and corrugated materials, these barriers resist impacts, abrasions, and crushing damage, making them ideal for shipping containers and packaging fragile foods like eggs and fresh fruits. Proper physical packaging also ensures consumer safety by preventing access to harmful products, like child-resistant closures for dangerous items. Additionally, the use of plastic packaging instead of glass containers for products like shampoo and soda bottles has reduced the risk of injuries from broken glass.

10.1.2. Containment and Food Waste Reduction

The benefits of reducing food waste along the entire supply chain must be taken into account in any evaluation of the environmental effects of food packaging. Many countries have reported significant food waste, with percentages ranging from 25% for food grains to 50% for fruits and vegetables. Food waste has been attributed to inadequate storage, transportation, and preservation/protection. By keeping food fresher for longer and increasing its shelf life, packaging helps cut down on waste overall. The per capita waste generated in Mexico City contained less packaging, more food waste, and one-third more total waste than generated in comparable U.S. cities. Additionally, the total waste from packaged foods is 2.5%, whereas fresh foods account for 50% of waste. This is partly because agricultural by-products collected at the processing plant are put to different uses, whereas those generated at home are usually thrown away. Packaging may therefore help to lower the overall amount of solid waste.

10.1.3. Marketing and Information

A product's packaging serves as its public face and is frequently the only one that buyers see before making a purchase. Thus, in a competitive market, unique or creative packaging can increase sales. The purpose of the package design could be to improve the product's reputation or set it apart from competitors. For recipes, for instance, larger labels might be utilized. Additionally, packaging informs the customer. For instance, product identification, nutritional value, ingredient disclosure, net weight, and manufacturer information are all legally required to be included on packaging labels. Important details about the product are also communicated on the packaging, including price, brand identification, and cooking instructions. The disposal of waste may be impacted by each of these improvements.

10.1.4. Traceability

"The ability to follow the movement of a food through specified stage(s) of production, processing, and distribution" is how the Codex Alimentarius Commission defines traceability. Enhancing supply management, facilitating trace-back for food safety and quality purposes, and differentiating and marketing foods with imperceptible or subtle quality attributes are the three goals of traceability. Food manufacturing companies can track their products throughout the distribution process by adding unique codes to the product labels. Codes can be read manually or automatically and come in a variety of formats (such as printed barcodes or electronic radio frequency identification [RFID]).

10.1.5. Convenience

The convenience features of packaging, such as easy access, handling, and disposal, as well as product visibility, reseal ability, and microwavability, are key factors driving package innovation. Packaging plays a crucial role in reducing the time and effort needed to prepare and serve food. Innovations like oven-safe trays, boil-in bags, and microwavable packaging allow consumers to cook meals with minimal preparation. New closure designs make opening and resealing packages easier, while special dispensing features enhance convenience. For example, a cookie manufacturer has introduced a flexible bag with a scored section for easy access to the cookies. A peelable seal covers the opening, allowing for reclosure after use. These advancements in food packaging have led to the development of modern retail formats that offer one-stop shopping and access to internationally sourced products. While these convenience features add value and competitive advantages to products, they may also impact the amount and type of packaging waste that needs to be disposed of.

10.1.6. Tamper Indication

Special packaging features intended to lessen or eliminate the risk of tampering and adulteration have been developed as a result of deliberate tampering with food and pharmaceutical products. Tamper-evident features are difficult to replace, although any package can be compromised. Banding, unique membranes, breakaway closures, and unique printing—such as text or graphics—on bottle liners or composite cans that change irrevocably when opened are examples of tamper-evident features. Additional features of special printing include holograms that are difficult to copy. Although tamper-evident packaging typically necessitates extra packing materials, which makes disposal problems worse, the advantages usually exceed any disadvantages. One instance of a tamper-evident characteristic that doesn't need extra packaging materials is the heat seal found on medical packaging that is designed to change color when opened chemically.

10.1.7. Other Functions

Packaging can also be used for other purposes, such as holding promotional items (like a gift, extra product, or coupon) or as storage containers for the home. The ability to reuse packaging helps reduce the amount of waste being thrown away (Paine, 1993; Porta et al., 2020).

10.2. MATERIALS USED IN FOOD PACKAGING

In addition to keeping food safe to eat and fresh, many materials used in food packaging can also be used to market and sell goods. Food packaging comes in a variety of forms, each with pros and cons. Plastic wrap is one of the most widely used varieties. Plastic wrap works well to keep food fresh and is inexpensive. It is not recyclable, though, and it is not very strong. Glass jars are another popular kind of food packaging. Glass jars can be recycled and are more robust than plastic wrap. They are less practical to use and cost more, though. An additional option are tin cans. They can be recycled and are incredibly robust. They are not as practical to use and cost more than plastic wrap (Figure 10.1).

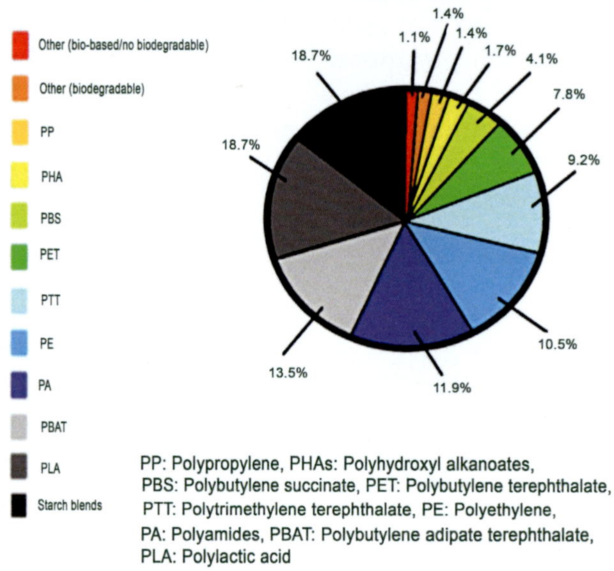

Figure 10.1. *Food packaging materials.*

Source: Panou A, Karabagias IK. Biodegradable packaging materials for foods preservation: Sources, advantages, limitations, and future perspectives. Coatings. 2023; 13(7):1176. https://doi.org/10.3390/coatings13071176.

Cardboard boxes are a flexible material choice. They can be printed with logos or other marketing messages and are available in a variety of sizes and shapes. Additionally recyclable are cardboard boxes. They are susceptible to moisture damage, though, and they are not as strong as some alternative solutions. A more recent development in food packaging that has gained popularity recently is pouches. Pouches can be resealed after opening and are composed of plastic or foil. Pouches are the best option for travelers or those with limited storage space because they are lightweight and require less space than other forms of packaging. Nevertheless, pouches can be challenging to open and are not always as sturdy as some other options. Food packaging materials come in a huge range of varieties. Every material has advantages as well as disadvantages. These are the materials that are commonly used for food packaging.

10.2.1. Plastics

One of the most popular forms of packaging in the food industry is plastic, which is frequently used to package both wet and dry goods. One inexpensive packing material is plastic. Businesses that produce food employ a wide variety of plastic packaging materials.

10.2.1.1. Polyethylene

Polymers with low and high densities are the most commonly used materials for packaging in the world. Polythene's chemical and physical characteristics allow it to be used with nearly any food. Polyethylene packaging comes in a wide variety of subtypes. These consist of LDPE, PET, and HDPE. The most widely used of these is PET (polyethylene terephthalate), which is found in milk and soft drink bottles among other things. The primary impetus behind the substitution of PET bottles for glass in beverage packaging was their introduction. Glass bottles and jars were fragile and heavy, making shipment hazardous and challenging. Glass bottled water solutions, juices, or syrups run the risk of breaking and causing food waste or lost goods. Because plastic bottles allay these worries, food manufacturing companies have largely switched from glass bottles to PET containers.

10.2.2. Polyvinyl Chloride

Another material that is frequently used for plastic packaging is polyvinyl chloride or PVC. You may have encountered it in medication packaging, which includes adhesive tape, water bottle sleeves, cling film, and blister packs for tablets.

1. **Advantages:**
   - Low-cost food packaging materials
   - Can handle both dry and wet food products
   - Resistant to factors such as moisture
   - Plastics provide the option of direct labeling
   - Flexible packaging material
2. **Disadvantages:**
   - Harmful to the environment
   - Most plastic packaging reacts badly with hot food items
   - Not easily recyclable material

10.2.3. Metals

Without a doubt, some of the best materials for food packaging are metals. They offer many food products the best protection and are durable and inflexible. Common forms of metal packaging include cans, boxes, and containers made of metal. Food packaging uses a wide variety of metals, just like it does plastics.

- **Tin:** As early as 1810, tin was the first metal to be used for food packaging. The main drawback of it is corrosion. When pure tin comes into direct contact with food, it corrodes. Tin is therefore covered in layers of protection when it comes to packaging.
- **Aluminum:** Tin was the pioneer of metal packaging, but in the modern world, aluminum is what really makes it happen. Households

frequently utilize aluminum foil to preserve food. Restaurant containers and beverage cans made of laminated aluminum are also frequently utilized. Cans made of aluminum have the best recycling potential of any material.

- **Steel:** Steel still has a wide range of uses, despite not being as widely used as packaging materials like tin or aluminum. For example, metal cans, caps, and containers are made from steel sheets.

1. **Advantages:**

- Rigid packaging material
- Protects during transportation as well as on the shelf
- Pest resistant
- Moisture resistant
- Prevents air contamination
- Suitable for wet or dry food items
- One of the most recyclable materials.

2. **Disadvantages:**

- Expensive packaging method
- Prone to corrosion without protection layers
- Non-flexible packaging material
- Reacts to certain food items (acidic foods)

10.2.4. Glass

Among the earliest materials for packaging is glass. Glass bottles have even been discovered by archaeologists close to ancient Egyptian tombs. Glass can be used to pack anything, whether it's dry, wet, or even acidic. It's also one of the materials with the least amount of food contact toxicity. Glass is widely used to create glass bottles and containers that are used as the main packaging for liquids such as juices, oils, and syrups.

1. **Advantages:**

- Highly corrosion-resistant and food-safe material
- Glass containers are esthetically pleasing
- Transparent containers reveal the contents to consumers
- Non-corrosive and inert to acids
- Ideal for preserving food long term
- Hygienic and the safest packing material for human health
- Rigid packaging material

2. **Disadvantages:**

- Highly brittle – glass containers and bottles can easily break
- Heavy and expensive compared to other alternatives

10.2.5. Wood Derivatives

Although many wood derivatives are used in the food industry, wooden packaging is not. Cardboard and paper are two of the most widely used wood-derived products in packaging.

- **Cardboard:** Corrugated cardboard boxes and folding cartons are frequently used in tertiary packaging, regardless of the primary or secondary packaging method. Retail stores remove this packaging before putting food items on the shelf, so customers do not see it or interact with it.
- **Paper and Paper Board:** Food manufacturers often use paper and paperboard as primary and secondary packaging materials. While paper is commonly used for dry food products, it is not ideal for wet foods due to its weak nature. Paperboard, on the other hand, is

a stronger alternative with added barrier properties. This material can be used to create boxes, trays, and cartons for packaging various food items. There are different types of paperboard available, each suitable for different types of food products.

1. **Advantages:**

- Low-cost food packaging materials
- Very lightweight
- When it comes to tertiary packaging materials, cardboard is unmatched due to its affordability, lightweight, and rigidity.
- Easy to print on paper and cardboard
- Paper-based products and corrugated boxes can be recycled. Packing materials frequently contain recycled paper.

2. **Disadvantages:**

- Limited to the packaging of dry food items
- Paper packaging is weak and easily tears

10.2.6. Ceramics

Ceramics aid in the removal of hazardous substances from food packaging. Clay and quartz, two minerals that are present in soil, are used to make ceramics. Ceramic packaging, an all-natural material, is less expensive than glass and metal packaging.

1. **Advantages:**

- Non-corrosive and nontoxic food packaging material
- Suitable for packaging wet and dry food products
- Preserves food quality due to being non-corrosive
- Prevents food contamination
- Aesthetically pleasing packages
- The raw material is biodegradable

2. **Disadvantages:**

- Not entirely food-safe
- Brittle and breaks easily, like glass
- More expensive than plastic packaging and wood derivatives
- Limited package sizes

10.2.7. Cellulose

A common misconception is that cellulose is a kind of plastic packaging.

As cellulose is a product of plants, which makes it one of the rare organic materials found in packaging. Cellulose is used to make shrink wrap and cellophane sheets. Despite having a plastic-like appearance, this material is entirely biodegradable.

1. **Advantages:**

- Plant-based products that are sustainable and environmentally friendly.
- Biodegradable in a few months without the need for additional care.
- Moisture and heat-resistant food contact material

2. **Disadvantages:**

- Weak and offers no protection against impact

Unsuitable for food items with a long shelf life (Paine, 1993; Porta et al., 2020)

10.3. THE REGULATION OF FOOD PACKAGING

Although the legal community is generally aware of food laws, few people are aware of the laws and rules pertaining to food packaging, despite the fact that these laws and regulations may have a big influence on a food company's ability to successfully market new products. This section describes American regulations pertaining to food packaging and how the U.S. regulatory framework contrasts with those of other countries around the globe. We also explain the history of many of the present food packaging laws and regulations, including some recent revisions. Beyond simply keeping food and beverages contained, food packaging fulfills several vital purposes. These include information transmission, convenience, and physical protection as well as delaying the deterioration of products. Susceptors, which help foods cooked in microwave ovens brown, features that react chemically or biologically with their contents to prolong shelf life, and the capacity to track the food's condition are examples of more sophisticated food packaging. Regulations controlling the type and application of food packaging have evolved to take these developments into account.

10.3.1. The Basics

Let's begin by discussing the fundamentals. Food packaging regulations typically set out safety standards that aim to safeguard the quality and sanitation of food products. These standards usually prevent harmful substances from leaching into food from the packaging, which could pose risks to human health or alter the taste, composition, or odor of the food. Although the majority of international food packaging regulations are founded on comparable scientific ideas, various jurisdictions may take quite different approaches. These various strategies may also affect the information needed to approve a material for use in food packaging. For example, in the U.S., substances that may not reasonably be expected to become components of food and are not likely to present any public health problem can be cleared based on analytical chemistry data and extrapolations that show such components present no cause for toxicological concern due to minimal dietary exposure. On the other hand, regardless of exposure, nearly all food packaging materials in the European Union (EU) must be expressly cleared based on a toxicological evaluation and published regulations.

10.3.2. The Beginning of Food Packaging Regulation in the U.S.

Official guidelines for food packaging in the U.S. started when the Food Additives Amendment was passed in 1958. Among other things, this amendment amended the Federal Food, Drug, and Cosmetic Act (FFDCA) by adding a new section that defined "food additive" as any substance whose intended use causes or may reasonably be expected to cause it to become a component or otherwise affect any food's characteristics unless that substance is covered by one of the Act's several exceptions or exclusions, or is generally recognized as safe (GRAS).

This definition is in charge of one unique aspect of the U.S. food contact substances that may reasonably be expected to migrate to food are regulated as food additives, similar to the way that substances added directly to food, like sugar substitutes or emulsifiers, are regulated. This food packaging regulatory scheme is unique to other jurisdictions. Crucially, food additives that are not governed by regulations or otherwise approved by the FDA for their intended use are deemed to adulterate food under Section 409 of the Act. In these situations, there is no need to demonstrate the possibility of harm in order to prove adulteration.

Under the Act, there are three significant exceptions. The first is known as the exception for "prior sanctioned" substances. Before 1958, some businesses that produced materials for food packaging would contact the FDA or the U.S. USDA Department of Agriculture (USDA) if it would object to a specific material being used to package food. This was frequently done to reassure consumers that the product was safe to use around food. The agency that received the request would issue a "no objection" letter if neither the FDA nor the USDA objected to the use of the substance in question. The contents of these letters were released from the need for additional FDA clearance or authorization to be used as food additives after the Food Additives Amendment was passed in 1958, making them known as "prior sanctioned" substances. It was a grandfathering clause of sorts.

Additives that are classified as generally recognized as safe (GRAS) are likewise excluded from this definition. A GRAS substance is one that has been sufficiently demonstrated through scientific procedures (or, in the case of a substance used in food before January 1, 1958, through either scientific procedures or experience based on common use in food) to be safe under the conditions of its intended use, among experts qualified by scientific training and experience to evaluate its safety. FDA approval or authorization is not required for these substances before they can be sold for use. FDA has, nevertheless, always permitted manufacturers or users to submit the evidence supporting a GRAS determination for FDA review. In this regard, the FDA published a final rule in August 2016 regarding its GRAS notification procedure. The kinds of information that can be used to bolster a GRAS conclusion are described in the final rule.

The third class of substances excluded from this definition of importance for food additives are those that are in contact with food or may be used in food packaging, but it is not reasonable to expect them to become parts of food. These substances can be used without prior FDA authorization or clearance because they are not by definition food additives. Further discussion of the background to this so-called no migration exemption follows below. The functional barrier doctrine is a subset of the no migration exclusion policy. Food additives do not include substances that come into contact with food but are kept apart from it by a functional barrier that stops the substances from migrating. It is required to prove that the functional barrier genuinely stops migration to apply this doctrine. When making this kind of determination, it's crucial to consider the material's physical and chemical properties, composition, and thickness as well as how they separate the substances from the food. The kind of food and the length of time it was exposed to are additional variables to take into account.

10.3.3. The Food Additive Petition Process

A procedure to petition the FDA to permit the use of an additive in food through the issuance of a "Food Additive Regulation" was included in the Food Additives Amendments of 1958. Through a notice and comment rulemaking process, food additives (including materials used in food packaging that fit the definition of a food additive) are cleared through the food additives petition process. In addition to information on the food additive's intended use, including food types, temperature conditions, and duration of contact with food, a food additive petition must include comprehensive details about the identity, composition, and manufacturing process of the substance in question, as well as chemistry and toxicology data confirming the food additive's safety for the intended use. After a food additive passes this inspection, the FDA issues a rule. As long as all requirements and restrictions outlined in the regulation are fulfilled, anyone can rely on these general clearances.

After the establishment of the food additive petition process, the FDA declared in 1961 that they would no longer issue "no objection" letters based on migration data. Following this announcement, there was debate over whether a company could determine on their own that a substance was not a food additive due to "no detectable migration," and therefore bypass the requirement of submitting a food additive petition before using the substance in food packaging.

Before then, the FDA had not released clear guidelines on how to decide whether a material could reasonably be anticipated to find its way into food. However, there were two primary sources of advice on this matter. First, the FDA proposed in 1969 that if a material migrates to food in amounts of no more than 50 parts per billion (ppb), it would not be necessary to regulate it as a food additive. Although the FDA never formally adopted this proposal, it did take off and became known as the "Ramsey Proposal."

Furthermore, in Monsanto v. Kennedy, which the FDA issued a regulatory ruling that forbade the sale of drink bottles constructed of a specific kind of polymer, the U.S. the Court of Appeals for the District of Columbia determined that Congress had to have intended for the FDA "to determine with a fair degree of confidence that a substance migrates into food in more than insignificant amounts" in order for the agency to classify a material as a food additive. The FDA will now review the case in accordance with this standard, the court remanded it to them.

Eventually, the FDA passed a rule that partially resolved this problem. To formally establish a procedure for determining when the migration of a substance that comes into contact with food is "so trivial as not to require regulation of the substance as a food additive," the agency amended the food additives regulations in 1995. Based on exposure and toxicology, the FDA may exempt a food contact material from regulation under the Threshold of Regulation (TOR) Rule. The material must have a dietary exposure of less than 0.5 ppb or, if it is already approved for use as a directive additive, less than 1% of the Accepted Daily Intake (ADI) must be exposed through food contact use. According to toxicology data, the material that comes into contact with food cannot cause cancer or contain any contaminants that have the potential to cause cancer. Upon receipt of a request to exempt a food contact substance from regulation as a food additive due to insignificant migration to food, the FDA will examine the available data and issue a TOR exemption letter if the criteria are met.

These letters are available to everyone and are not proprietary. Initially, the FDA concentrated on preventing polymer migration into food when creating regulations for the safe use of polymers in contact with food. This resulted in the "basic polymer doctrine," which is the regulation of polymers as a whole as opposed to the intricate variations in the polymerization process. This doctrine states that materials used in good manufacturing practices, such as catalysts, chain transfer agents, and initiators, are implicitly cleared with the polymer and are part of the basic polymer. These materials must be essential to the polymerization process, utilized sparingly, wash out or integrate into the polymer backbone, and have no effect on the appropriate level of polymer purity.One final feature of food packaging regulation in the U.S. that needs to be discussed is the alleged "housewares exemption."

This exemption, which developed from the legislative background of the 1958 Act and was repeatedly acknowledged by FDA officials in the years that followed, exempts materials used in houseware products that come into contact with food from the requirement for FDA premarket clearance. Generally speaking, housewares are defined as items that are sold empty and are used to store, prepare, or serve food by customers at home or in a business setting. Substances found in housewares are nevertheless subject to the FFDCA's general safety regulations.

10.3.4. The Food Contact Notification Program

The implementation of the Food Contact Notification (FCN) program in January 2000 marked a significant advancement in the regulation of food packaging in the United States. It has emerged as the go-to technique for removing materials that come into contact with food since its creation. The expediency of the FCN program over the food additive petition process is one of its main advantages. A Food Certificate Notice (FCN) takes effect 120 days after the FDA accepts the notification unless the agency objects in writing. However, clearing the use of a food additive through the petition process can take two to four years or longer. FCNs and food additive regulations differ significantly in that FCNs belong exclusively to the substance's manufacturer and its clients.

While there are some distinctions, the data requirements for an FCN and a food additive petition are comparable. Both require the notifier to estimate how much of a substance in food contact is anticipated to be consumed by humans, i.e., the Dietary Intake Estimate (EDI). This can be computed by multiplying the predicted migration amount of the substance by the approximate fraction of the daily diet predicted to come into contact with materials containing the substance. One distinction between the data needed for a petition and an FCN is that the latter must contain the cumulative estimated dietary intake (CEDI) from all relevant sources, not just the food contact substance under consideration. The nature of the substance and its CEDI will determine the kind and quantity of toxicology data needed to approve the food contact material for its intended use.In order for the FDA to issue a finding of no significant impact or, if necessary, complete an environmental impact statement, FCNs, TOR requests, and food additive petitions must additionally include information about the potential environmental impact of their use. The National Environmental Policy Act (NEPA), which mandates that federal agencies evaluate the possible environmental impact of "major" actions, is the source of this requirement.

10.3.5. Food Packaging Regulation in the EU and Other Jurisdiction

In the modern global economy, trade between nations may be impacted by the various food packaging regulations in place in various jurisdictions. To better align with other jurisdictions, particularly the EU, some countries have revised or are currently revising their laws and regulations governing food packaging.

To comprehend the distinction between the U.S. and European food regulatory frameworks, let's begin with a quick review of EU legislation on food packaging. All materials that come into contact with food in the EU are governed by the Framework Regulation, including housewares, food processing equipment, and containers for food transportation. Additionally, it sets safety guidelines and gives the European Commission (EC) the authority to enact regulations for particular substances or materials that come into contact with food.

The Plastics Regulation is the most comprehensive piece of EU legislation on a particular class of materials that come into contact with food to date. This regulation includes certain polymer production aids, additives (apart from colorants) and a positive list (also called the Union list) of monomers and other starting materials that can be used in the manufacturing of food contact plastics. Additionally, unless there is an exception for their use, the Plastics Regulation sets an overall migration limit (OML) of 10 mg/dm^2 for all plastic materials in contact with food. For specific substances, there are also specified residual quantities in material (QM) limits and specific migration limits (SMLs).

The Plastics Regulation is "harmonized," meaning that, with very few exceptions, Member States of the EU are prohibited from imposing additional or different requirements on plastics meant to be used with food. One of the few categories of food-contact materials that Europe has yet to harmonize is plastics. However, member states continue to have regulatory control over other kinds of materials (e.g., coatings, metal, paper, etc.).

Positive lists are increasingly being used to regulate the materials used in food packaging. Regardless of exposure, all substances under a positive list system must be expressly cleared based on a toxicological evaluation and published in regulations; using substances that are not on required positive lists is typically forbidden. These lists typically relate to particular classes of materials that come into contact with food, like additives or monomers.

The following are some of the authorities whose food packaging laws currently include positive lists, or are in the process of doing so.

- MERCOSUR (Mercado Común Del Sur or the "Common Market of the South") is made up of Brazil, Argentina, Paraguay, Uruguay, and Venezuela. It is the biggest trading bloc in South America. This body proposes GMC resolutions, which need to be ratified by all member nations to become law. The EU's

Plastic Regulation clearances serve as the basis for MERCOSUR's resolutions on the Positive List of Additives for Plastic Materials Intended for the Manufacture of Packages and Equipment in Contact with Foods and the Positive List of Monomers, Other Starting Substances, and Polymers Authorized for the Manufacturing of Plastic Packaging and Equipment that Come Into Contact with Food.

- The GCC (Gulf Cooperation Council) Food contact materials are among the topics covered by the standards and technical regulations that the Standardization Organization (GSO) creates for Yemen and the GCC Member States (the Sultanate of Oman, the State of Qatar, the Kingdom of Saudi Arabia, the United Arab Emirates, and the Kingdom of Bahrain). In the GCC Member States, adherence to these standards is optional. A list of approved monomers, additional starting materials, macromolecules produced through microbial fermentation, additives, and polymer production aids (PPA) is established by GSO 1863/2013 and can be utilized in the production of plastic packaging.
- China's Food Safety Law mandates that all products related to food, including materials used in food packaging, must be approved. Hygienic Standards for Uses of Additives in Food Containers and Packaging Materials, one of China's standards, covers polymers, coatings, and adhesives in addition to providing a positive list of additives that are allowed to be used in food-packaging materials.
- Intending to implement a positive list system for food contact materials by April 2020, Japan is currently in the process of making this change. Based on the current positive lists created by the Japan Hygienic Olefin and Styrene Plastics Association and the Japan Hygienic PVC Association, a proposed positive list of 32 resins has been developed.

10.3.6. Food Packaging in the Spotlight

Over the past few years, food packaging has received more attention, some of which has come from heightened scrutiny regarding the safety of materials that come into contact with food. The concern is partly caused by technological advancements and the FDA's need to revisit some of its earlier decisions, but it is also a result of attacks on the safety of food packaging materials by environmental organizations and other self-described "public interest" groups. Chemicals that are perfluroinated are one example.

In March 2015, the FDA received a petition to delist three perfluoroalkyl ethyl compounds that were used in grease and water-repellent coatings for paper and paperboard that were used to package fatty and aqueous foods. The FDA was unable to ascertain the dietary exposure from the substances' use in food contact during the petition review process. Furthermore, no toxicological research was conducted to demonstrate the toxicity of the three substances for development or reproduction; however, more recent information has demonstrated the toxicity of substances with comparable structures. The FDA revised the food additive regulations in January 2016 to prohibit the use of the three substances after concluding that there was no longer a reasonable certainty of no harm from their use in food contact. It's interesting to note that none of the three drugs had been used in the U.S. Since 2011, the organizations who filed the petition have expressed concern over the importation of food packaging containing these materials from other nations.Certain states or local governments may also outlaw the use of substances that come into contact with food. A law prohibiting the use of perfluoroalkyl and polyfluoroalkyl compounds in food packaging was signed into law by the governor of Washington State in March 2018. Unless a safer alternative can be found by the state Department of Ecology by then, the ban will go into effect in 2022. 2019 will see the implementation of a ban in San Francisco on the sale of food service wares containing fluorinated chemicals. It goes without saying that Proposition 65 (Safe Drinking Water and Toxic Enforcement Act of 1986) in California imposes warning requirements on materials used in food packaging.

The food and food packaging industries will need to stay up to date with the latest scientific developments and how they align with the legal and regulatory frameworks intended to ensure the safety of materials used for food packaging, particularly as food packaging continues to garner increased attention regarding the safety of individual components (Paine, 1993; Porta et al., 2020)

10.4. EPA GUIDELINES FOR MANAGEMENT OF MSW

To safeguard the environment, human health, and natural resources, proper waste management is essential. The Environmental Protection Agency (EPA) uses nonregulatory strategies like WasteWise and pay-as-you-throw to encourage behavioral change in solid waste management. Residents are charged for MSW services based on how much trash they discard in pay-as-you-throw systems. This incentivizes the production of less garbage and increases material recovery via composting and recycling. Pay-as-you-throw programs typically result in waste reductions of between 14–27% annually for participating communities. EPA and the U.S. voluntary partnership WasteWise was established in 1994. Businesses, establishments, charitable groups, and governmental bodies to encourage recycling, avoid waste, and purchase goods with recycled content. In 2005, over 1800 organizations took part in the WasteWise program. Additionally, when compared to other products serving the same purpose, the EPA Environmentally Preferable Purchasing program assists federal agencies and other organizations in purchasing goods with less or no impact on the environment and human health. The main goal is to prevent pollution, which encompasses more environmental issues than just waste reduction.

The utilization of a hierarchical, integrated management approach—which includes source reduction, recycling, composting, combustion, and landfilling—is emphasized in the EPA's solid waste management regulations. Subtitle D of the Resource Conservation and Recovery Act governs landfilling and combustion as waste disposal techniques (40 CFR Parts 239–259).

10.4.1. Source Reduction

Source reduction, also known as waste prevention, involves reducing the amount and toxicity of waste produced by making changes to the design, manufacture, purchase, or use of materials and products.

The Environmental Protection Agency (EPA) views source reduction as the most effective method for reducing the environmental impact of solid waste, as it eliminates waste generation altogether. Source reduction strategies include using less packaging, creating longer-lasting products, and reusing materials. Examples of source reduction practices include using lighter packaging materials, buying durable goods, opting for larger sizes or refillable containers, and choosing non-toxic products. Overall, source reduction offers numerous environmental benefits such as resource conservation, environmental protection, and reduction of greenhouse gas emissions.

10.4.1.1. Lightweighting

Lightweighting, or using thinner gauges of packaging materials—either by lowering the amount used or by using alternative materials—is one method of achieving source reduction.

Between 1992 and 2002, the average weight of glass containers dropped by almost 50%. Similarly, compared to 1975, aluminum cans were 26% lighter in 2005, with one pound of aluminum used to make 34 cans as opposed to 27 in 1975. Anaheuser-Busch Companies Inc. claims to the EPA. Lightened their 24-ounce aluminum cans in 2003, resulting in a 5.1 million pounds decrease in the amount of aluminum used. Additionally, less aluminum is now utilized in foil laminates. In addition, steel cans have been made lighter; they are now at least 40% lighter than they were in 1970. Tin content per ton of tinplate steel has decreased significantly from pre-World War II levels of 50 pounds to the average of 6 pounds today. Plastic containers, though a relatively new material for packaging, are also lighter. Since 1977, the weight of 2-liter PETE soft drink bottles has dropped by 25% (from 68 to 51 g), saving over 206 million pounds of plastic packaging annually. Similarly, during the past 20 years, the weight of the 1-gallon plastic milk jug has decreased by 30%. In the paperboard industry, lightweighting has been accomplished by utilizing materials with thinner gauge. For instance, Anheuser-Busch reduced the thickness of its 12-pack bottle packaging, saving 7.5 million pounds of paperboard.

10.4.1.2. Reusable and Refillable Containers

Another method of achieving source reduction is by reusing items. For instance, glass containers like bottles can be washed with strong detergents and reused multiple times. Refillable plastic containers are typically made from materials like PETE, PEN, or high-density polyethylene, although the use of polycarbonate is declining. This is due in part to the logistical challenges of collecting, transporting, and cleaning these containers, leading manufacturers to prefer single-use containers.

Additionally, producers have reduced their sources by providing refill products, especially for non-food items like cleaning supplies for the home. Due to transportation costs and cleaning requirements, thinner one-way glass or plastic containers have largely replaced refillable glass beverage containers. Reusable glass containers are still common in other nations, though. In an attempt to prevent contamination from post-consumer waste streams, PETE containers have been depolymerized and repolymerized; however, this process has not proven to be economically viable.

10.4.2. Recycling

Materials from the waste stream are diverted to material recovery through recycling. Reusing a returned product in its original form is called reuse; recycling, on the other hand, entails processing the material to create new products. Materials and products that have been recycled are typically collected, sorted, processed, manufactured, and sold as part of a recycling program. There needs to be a market for recycled goods and materials for recycling to be profitable.

Procurement of products manufactured from recycled materials is encouraged by the EPA's Comprehensive Procurement Guidelines (CPG). The EPA identifies goods that can be manufactured using recovered materials and suggests purchasing procedures for these goods. Procuring agencies are mandated to buy products with the maximum level of recovered material content after the EPA designates a particular product. Under the CPG program, EPA has chosen more than 60 products with recycled content and suggested a number of other products.Most packaging materials such as glass, metal, thermoplastics, paper, and paperboards can be recycled. The economic assessment of recycling involves various factors such as collection costs, separation, cleaning, reprocessing, and transportation. In addition, there must be a market for recycled products and competing materials should be taken into consideration. Metals and glass reclaimed through recycling are deemed safe for food contact containers due to the heat used during the melting process, which kills microorganisms and pyrolyzes organic contaminants. However, plastics undergo reprocessing using heat to destroy microorganisms, but this process is insufficient to pyrolyze all organic contaminants, making post-consumer recycled plastics unsuitable for food contact applications. Overall, recycling rates are increasing as evidenced by the 30 million tons of containers and packaging recycled in 2005, which accounted for 40% of the amount generated. Glass recycling has particularly seen growth thanks to improved collection methods and a higher demand for recycled glass. Approximately 90% of recycled crushed glass is used to create new containers. Additionally, aluminum can recycling has risen to 52% in 2005, up from 50% in 2003. Plastic recycling rates, specifically for PETE and high-density polyethylene bottles, have also significantly increased since the 1990s.

10.4.3. Composting

According to the EPA, composting is a type of recycling. The process of controlled aerobic or biological degradation of organic materials, including food scraps and yard waste, is called composting.

As a result, it entails stacking organic materials and providing enough moisture to allow microorganisms to break down aerobically. To prevent anaerobic conditions, turning the piles on a regular basis encourages aeration.

The resultant humus, which resembles soil, is utilized as a natural fertilizer to lessen the need for synthetic fertilizers. Because organic materials still make up a sizable portion of MSW overall (about 25% for food scraps and yard trimmings; see Figure 10.2), composting is a highly advantageous waste disposal option.

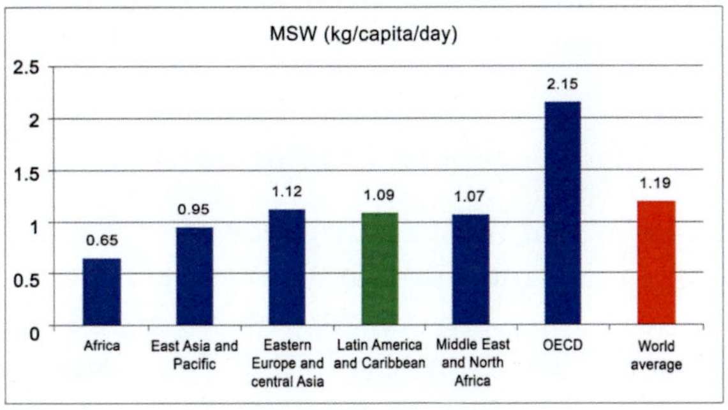

Figure 10.2. *Materials MSW generation (kg/capita/day) by various regions in the world.*

Source: *Hettiarachchi H, Ryu S, Caucci S, Silva R. Municipal solid waste management in latin america and the caribbean: issues and potential solutions from the governance perspective. Recycling. 2018; 3(2):19. https://doi.org/10.3390/recycling3020019 Hettiarachchi H, Ryu S, Caucci S, Silva R. Municipal solid waste management in Latin America and the Caribbean: Issues and potential solutions from the governance perspective. Recycling. 2018; 3(2):19. https://doi.org/10.3390/recycling3020019.*

10.4.4. Combustion/Incineration

Waste that cannot be recycled or composted is finding an increasingly appealing alternative in combustion, which is the controlled burning of waste in a designated facility. Combustion incinerators can be designed to produce steam (also known as waste-to-energy combustors or WTE facilities) that can produce heat or electricity while reducing the volume of MSW by 70–90%.

In actuality, plastics are made from petroleum feedstocks and have a high heat content, which makes them useful for burning waste to produce energy. With 89 WTE facilities out of 94 combustion facilities in the US in 2004, these facilities could process roughly 95,000 tons of MSW per day, or 13% of the total. Mass-burn incinerators, refuse-derived fuel incinerators, and modular combustors are the three varieties of incinerators, commonly referred to as municipal waste combustors (MWCs).

10.4.4.1. Mass-burn Incinerators

All forms of MSW are accepted as-is by mass-burn incinerators, except materials that are too big to fit through the feed system. To facilitate full combustion, integrated waste is positioned on a grate that travels through the combustor while air is driven into the system above and below the grate.

Mass-burn incinerators are different from other MWCs in that they are usually built on-site and burn the waste in a single stationary chamber. Boilers are installed in the majority of mass-burn facilities in order to recover combustion heat for energy production. In the United States, mass-burn technology wa+s used in 65 (77%) of the 89 WTE facilities in 2004 to process about 22 million tons of MSW.

10.4.4.2. Refuse-derived Fuel Incinerators

Refuse that has been preprocessed to remove recyclables and non-combustibles is used in refuse-derived fuel (RDF) incinerators. Shredded combustibles produce a homogenous fuel with a higher heating value. An RDF facility can be set up for combustion, processing, or both. 10 RDF facilities in the United States were split equally between processing and combustion in 2004; the other half handled both processing and combustion. Eight million tons of MSW could be burned in RDF incinerators in 2004.

10.4.4.3. Modular Combustors

Modular combustors, which are usually smaller than mass burn incinerators, accept all waste without preprocessing, just like mass-burn incinerators. They can be swiftly assembled wherever they are needed because they are typically prefabricated off-site. Modular combustors accounted for about 10% (9 out of the total 89) of the total U.S. MWC units in 2004.

10.4.5. Landfilling

Any leftover MSW as well as the byproducts of recycling and combustion processes can be disposed of in an environmentally responsible manner at landfills. Federal and state regulations control the location and operation of landfills. Modern landfills are carefully constructed buildings that keep waste separate from the surrounding environment and groundwater. Methane and other landfill gases are collected and managed by a well-designed MSW landfill to potentially be used as an energy source. Leachate, which is waste that has passed through or emerged from a landfill, contains materials that are soluble, suspended, or miscible. The EPA is looking into a bioreactor, a change to landfill architecture that can improve leachate and organic waste's aerobic and/or anaerobic degradation.

The development of advanced landfill technology, environmental regulations for landfills, and biodegradable packaging materials has been spurred by the growing awareness of environmental issues, including the increased use of synthetic packaging materials coupled with slow degradation in landfills. Contemporary landfills are carefully designed to avoid contaminating the environment and are monitored to guarantee adherence to federal (40 CFR Part 258) or comparable state regulations. The EPA has developed a landfill reclamation strategy that prevents the purchase of land for new landfills and permits the expansion of the capacity of the current MSW landfill. The voluntary landfill methane outreach program, which encourages the use of landfill gas as a renewable energy source, is also managed by the EPA.

The EPA has created bioreactor landfills, which are intended to quickly degrade organic waste by adding liquid or air to speed up microbial processes after biodegradation was shown to be a minor benefit in landfills. Aerobic, anaerobic, and hybrid bioreactors are the three different varieties. The EPA is in the process of coming up with guidelines for bioreactor standards or operating parameters.

10.4.5.1. Other Disposal Methods

10.4.5.1.1. Anaerobic Degradation

Anaerobic digestion, or degradation, is the primary process that takes place in landfills. In the absence of oxygen, microorganisms gradually break down solid waste, mostly organic materials like wood and paper, into carbon dioxide, methane, and ammonia during anaerobic degradation or digestion. This process can be accelerated by gathering and pumping leachate through the compacted solid waste, which inoculates the mass and acts as a moisture source to encourage additional degradation. Leachate should be contained in a system, typically a mix of liners and storage systems, to avoid contaminating groundwater. In the end, a treatment facility processes leachate to create a stable residue that can be disposed of securely. Organic waste pollutants and biosolids (sewage sludge) are the two main applications for anaerobic degradation. To fully utilize anaerobic degradation in the management of solid waste, more research is required.

10.4.5.1.2. Biodegradable Polymers

Renewable supplied agricultural feedstocks, animal sources, waste products from the marine food processing sector, or microbial sources are the sources of biodegradable polymers. Biodegradable materials decompose to produce environmentally friendly products like carbon dioxide, water, and high-quality compost, in addition to renewable raw ingredients.

There have been biodegradable polymers derived from cellulose and starches for many years; the first demonstration of a cellulose-based polymer, which launched the plastics industry, took place in 1862. The most popular biopolymer derived from cellulose is cellophane. Amylose, hydroxylpropylated starch, and dextrin are examples of starch-based polymers that expand and change when they come into contact with moisture. Polylactide, polyhydroxyalkanoate (PHA), polyhydroxybuterate (PHB), and a copolymer of PHB and valeric acid (PHB/V) are additional starch-based polymers. Polylactide, which is made of lactic acid produced by microbiological fermentation of starch derivatives, is resistant to moisture degradation. Bacteria acting on starches also form PHA, PHB, and PHB/V. Additionally, chitosan, which is made from the chitin found in the exoskeletons of insects and crustaceans, can be used to create biodegradable films. Chitin is a biopolymer that shares a cellulose-like chemical structure.

Another type of biodegradable polymer is edible film, which is a thin layer of edible materials placed on or between food components or applied to food as a coating. They improve the food's mechanical integrity or handling qualities in addition to preventing the migration of moisture, gases, and aromas. Collagen, a component of skin, tendons, and connective tissue, zein (corn protein), whey (milk protein), and gelatin (a byproduct of partial hydrolysis of collagen) are some of the plant and animal sources from which edible films are made.

Additionally, biopolymers, bioactive compounds, and biodegradable ingredients like starches can be added to synthetic polymers to partially degrade them. To disintegrate the polymer into smaller parts, the biocomponents are broken down. Bioactive substances function in several ways. For instance, they might be combined with swelling agents, which, when exposed to moisture, cause the plastic's molecular structure to enlarge, facilitating the breakdown of the plastic by the bioactive substances.

The development of biodegradable polymers is backed by many reasons, including the need to address litter and solid waste disposal issues as well as the replacement of non-renewable raw materials like oil, coal, and natural gas with renewable ones derived from plants. The use of biodegradable materials is not a panacea for all issues related to solid waste management, despite certain benefits. The presence of blended or modified polymers in the recycling stream could complicate recycling, but switching from synthetic to biopolymers won't have much of an impact on source reduction or incineration. In a landfill, biodegradable plastics don't really help because the oxygen and moisture needed for biodegradation are usually not present. It is debatable if there will be enough plant material to produce enough packaging polymers if biopolymers are used extensively and if crop optimization for biopolymers will affect food production. Since bioplastics are currently more expensive than the majority of petroleum-based polymers, substitution is probably going to result in higher packaging costs.

The benefits of biodegradable packaging are substantial for some applications, even though it is not generally feasible. To the extent that biodegradable plastics tend to decompose and become less noticeable after being littered, the litter argument in favor of biodegradable plastics is valid. In the marine environment, where litter presents risks to marine life, biodegradability is crucial. Additionally, biodegradability can be helpful in military settings where conventional disposal methods are insufficient. Biodegradable polymers have specific but minor roles in preventing lipid migration between food components, moisture, and smell.

The process of commercializing bioplastics is underway. NatureWorks, LLC (a stand-alone company wholly owned by Cargill Inc.) manufactures polylactide from natural products (corn sugar). The polymer can be hydrolyzed to recover lactic acid after the initial use, getting closer to the cradle-to-cradle goal of having no negative effects on future generations. Furthermore, Wal-Mart Inc. is packaging freshly cut produce with polylactide and biopolymers.

Although it is theoretically necessary to sort all plastics, in practice recycling is frequently limited to systems and polymers that are easily recognized, such as PETE and high-density polyethylene milk bottles. Sorting by specific polymer is lessened when other polymers are mixed into thermoplastic resins used to make park benches and playground equipment. To fully capitalize on this advantage, polylactide needs its own code and sorting mechanisms because it is intended for commercial composting.

10.4.5.1.3. Litter

The incorrect disposal of solid waste is known as littering. 11 states have implemented bottle bills to guarantee a high rate of recycling or reuse and to lessen litter because drink containers are a noticeable part of it. A minimum refundable deposit on beer, soft drinks, and other beverage containers is required by these bottle bills or container deposit laws, which create an economic incentive to guarantee the return of used bottles. The most notable recycling successes have been beverage containers made of glass, metal, and plastic because they are easily recognizable and composed of recyclable single materials. As an alternative, biodegradable packaging could gradually assist in eliminating litter's unsightliness and the risks it poses to wildlife and aquatic life. The availability of biodegradable containers, however, might encourage carelessness with disposal, which would impede recycling initiatives (Paine, 1993; Porta et al., 2020).

CASE STUDY

CUSTOM TRAINING BEARS FRUIT FOR COMMODITY GROUPS

The Client

Creating Nutrition Facts labels for fresh agricultural products requires a keen understanding of FDA requirements – perhaps unfamiliar territory for commodity boards and growers. This was the case for a veteran industry consultant whose role includes educating growers of these agricultural products about how to use the label to inform consumers about the nutritional benefits of their products. Recognizing that food labeling is not her area of expertise and wanting to provide her clients with accurate and the most up-to-date information on FDA regulations, the consultant turned to Food Consulting Company to augment her knowledge and create a custom training program to help educate her client.

The Challenge

Declaring nutrients on fresh agricultural products is voluntary, but not a trivial exercise. Unlike manufactured or branded foods that have a consistent nutrient profile, there is inherent variation in agricultural products. Commodity food labelers often use USDA nutrient database values derived from continuous product sampling. Ensuring products are within those bounds and knowing what nutrients and product attributes are allowed in product promotion is challenging indeed. Adding to the challenge is knowing how or when to use the FDA values for fresh fruits and vegetables if not using values from the USDA nutrient database.

The Solution

An expert from Food Consulting Company worked side-by-side with the consultant to develop a strategy to help educate the growers and provide a resource guide for them to refer to over time.

The Food Consulting Company expert compiled the resource guide with an overview of what labeling involves and the intricacies of the regulations, customized specifically for the commodity grown. It included compliant Nutrition Facts labels using the information in the USDA nutrient database and how to apply all FDA rules related to rounding and percent daily values. In a follow-up webinar for growers, packers, commodity board leaders and their marketing team, the Food Consulting Company expert led a training that stepped through the resource guide. She provided an overview of the food labeling process, how nutrients are to be listed, what graphical requirements are needed and other pertinent FDA requirements. Putting it all together, she provided example food labels to illustrate compliant Nutrition Facts for use in the marketplace.

The Results

Education about the benefits of Nutrition Facts labeling was of great value to the commodity growers and leadership team who wanted to brand and promote their fresh produce.

The custom-developed resource guide and webinar proved very useful for the audience to understand how FDA regulations coincide with the USDA nutrient database, what is needed for compliant Nutrition Facts labels and clear examples of what the labels look like – a helpful reference for the future given that many commodity boards do not have in-house expertise in this area.

CLASS ACTIVITY

CRUISIN' FOR A BRUISIN' FOOD PACKAGING SPECIALIST

Purpose

During this project, students will create a safe package for perishable fruit shipment, demonstrating how product packaging balances function, food safety, and the economy. The best design will be shipped to test the product's durability after each package is evaluated by the class.

Estimated Time

Six 50-minute sessions

Materials Needed

For the teacher:

- Packaging Grading Rubric
- Examples of fruit packaging materials
- Packaging materials for students
- Samples of one type of fruit for students to measure and weigh

For each group:

- Packaging materials for fruit package design
- Internet access
- Scale

For each student:

- Packaging rubric
- Cruisin' for a Bruisin' lab worksheet
- Packaging materials

Engage

1. Students should be asked to name a few of their favorite fruits. Write a list on the whiteboard. Mention that the neighborhood grocery store is probably where you can find these fruits. But were they grown close by? In certain cases, maybe. However, we frequently eat fruit (and other foods) that were grown in different parts of our nation or even abroad due to different climates and growing seasons.
2. Fruit is a perishable item. Inquire with students about the process of how fruit travels from the farmer to the grocery store and ultimately to the consumer.
3. In this activity, students will:

- determine the components required and create the packaging for a novel food item;
- discover the qualities of successful packaging; and.
- discover that math, science, engineering, and technology are all involved in product packaging.

Explore and Explain

- **Day 1**

1. Fruit is used specifically as an example in this activity, which focuses on the science of food packaging. Give the students several fruit packaging samples to look at. Straw clam shells, sacks of oranges, and cardboard trays with slits to accommodate individual pears or apples are a few examples. Show the examples from the class's online resources if any are not available. Find out from the students why they believe fruit packaging matters. Write down some ideas on the board.
2. Utilize the background data to inform the creation of the board's list of people involved in food packaging, its significance, and potential career paths. Farmers value efficient food packaging because they want their products to arrive to customers looking good and tasting fresh. But farmers also have to think about how much the packaging will cost.

   Costly packaging may lower the amount of money farmers must make from the sale of their goods. Customers want to buy fruit that tastes and smells good, is clean, hasn't been bruised or damaged, and has been packaged with safe materials and food handling techniques. For this reason, product packaging is important.

3. Inform students that they will be competing to create the best package possible for a single piece of fruit, and that they will be playing the role of food packaging experts. The packaging should not only preserve the fruit but also be economical and ecologically sustainable. One idea would be for a group to put a piece of fruit inside a very big box that is filled to the brim with bubble wrap. Although the fruit might not be harmed by this box, its size and additional materials needed for assembly and shipping would make it expensive and wasteful.
4. Give out the packaging rubric and the lab worksheet for "Bruisin for a Bruisin." Utilizing the rubric, describe the procedure and assessment.
5. Divide the class into groups. Tell them that each group member will create and test a package prototype on their own. Based on ratings on package durability and design provided by a rubric, the best package in the group will be chosen. After that, the group will have a chance to collaborate to refine the package that best represents their group's design and present it to the class. After that, the class will vote on which package should be sent out with a piece of fruit inside.

- **Day 2**

1. Give groups instructions to look into materials and designs for their fruit package as the first step in the design process. One piece of fruit, like an orange, pear, or apple, should fit inside a package. Choose as a class what kind of fruit the packaging will be made for. This will maintain uniformity in the cost of package materials and shipping.

- The fruit should be shipped in packaging that is appropriate for the U.S. Postal Service.
- Monte Packaging Company is an example of a website that includes packaging materials.
- Inform the students that they should refrain from buying items from the websites, but instead use them as a resource to explore different materials suitable for packaging fruit.

2. Students should evaluate the viability of various materials and designs that they may use for their fruit packages after conducting material research. On their lab sheet, each student should doodle their own design concepts along with a list of components and measurements.
3. Every group should come up with concepts for their company name and create an attractive label to be placed on the fruit. This needs to be noted on the laboratory sheet.
4. Each student will gather the supplies they need for packaging as homework. Encourage students to check their school or home recycling bins. If necessary, provide some basic supplies like cardboard or tape. Cardboard, wood shavings, corn packing peanuts, newspaper, tissue paper, cellophane, poster board, foam board, lint, and wool are a few examples of packing materials.

- **Day 3**

1. In their respective groups, students will collaborate on designing and constructing their own prototype packages. While students are allowed to use sample fruit pieces to determine dimensions for the package design, evaluation and testing of the packages will be conducted without any fruit inside.

- **Day 4**

1. After each student presents their prototype package to the class, the group will use the packaging rubric to assess the prototype's craftsmanship, esthetics, and material usage.
2. After creating their package prototype, each student will test its durability by subjecting it to tearing and crushing. They will then share their results with their group and use the rubric to assess the durability of their packaging.
3. The package design that receives the highest rating on the rubric will be selected for additional work.

- **Day 5**

1. Based on enhancing the design of the package that obtained the highest score on the rubric, the group will create a single final package for their fruit. The team needs to monitor the cost of materials, package size, shipping charges, and logistics.

- **Day 6**

1. The class will be presented with package designs, material costs, and shipping costs by each group.
2. Students should be asked to consider themselves as fruit-growing farmers. This farmer is looking for a design that will last long enough to deliver their fruit to clients undamaged. The farmer also wants to select a package that is reasonably priced.
3. Give the students a vote to determine which fruit package would be most effective to send.
4. Once the package has been selected, it should be shipped to the class address and dropped off at the post office. The class will assess the fruit and package condition when the package gets to school and come up with suggestions for improvements.
5. Use the Packaging Grading Rubric for grading.

- **Variations**
  - Assign each group to mail a separate fruit package.
  - As groups, compare the costs offered by various shipping companies.
- **ELL Adaptations**
  - There are practical exercises in this activity. For English language learners, kinesthetic learning events offer a great learning environment.
  - Before starting the lab, walk the class through each step of the procedure.
  - Create a word wall with new vocabulary and match pictures to the words.

Elaborate

- Display the Red Blossom California Strawberries website to your students. Visit www.rbtrace.com to view the video clip on strawberry packing. Go over the trace back example with customers to see how they can find out what farm grew their strawberries, the variety, and the date the strawberries were picked by entering the code on the bottom of their Red Blossom clamshell container.
- Visit a fruit packaging facility in person or take a virtual tour of a fruit packaging operation on YouTube.

Evaluate

Review and enumerate the following important ideas after completing these exercises.

- Experts in food packaging contribute to the secure delivery of food from the farm to the customer.
- Because they can be shipped over great distances, a greater variety of foods can be found in our diets.

SUMMARY

- The principal roles of food packaging are to protect food products from outside influences and damage, to contain the food, and to provide consumers with ingredient and nutritional information.
- Food packaging can retard product deterioration, retain the beneficial effects of processing, extend shelf-life, and maintain or increase the quality and safety of food. In doing so, packaging protects 3 major classes of external influences: chemical, biological, and physical.
- Chemical protection minimizes compositional changes triggered by environmental influences such as exposure to gasses (typically oxygen), moisture (gain or loss), or light (visible, infrared, or ultraviolet). Many different packaging materials can provide a chemical barrier.
- Biological protection provides a barrier to microorganisms (pathogens and spoiling agents), insects, rodents, and other animals, thereby preventing disease and spoilage. In addition, biological barriers maintain conditions to control senescence (ripening and aging). Such barriers function via a multiplicity of mechanisms, including preventing access to the product, preventing odor transmission, and maintaining the internal environment of the package.
- Physical protection shields food from mechanical damage and includes cushioning against the shock and vibration encountered during distribution. Typically developed from paperboard and corrugated materials, physical barriers resist impacts, abrasions, and crushing damage, so they are widely used as shipping containers and as packaging for delicate foods such as eggs and fresh fruits.
- A package is the face of a product and often is the only product exposure consumers experience before purchase. Consequently, distinctive or innovative packaging can boost sales in a competitive environment.
- Many materials used in food packaging help keep food fresh and safe to eat and can also be used to promote and sell products. There are many types of food packaging, such as plastic wrap, glass jars, tin cans, cardboard boxes, pouches, etc.
- Most food packaging regulatory systems include general safety requirements that are intended to preserve the physical, chemical, and sanitary integrity of the contents of food packages. These requirements generally prohibit the transfer of substances from the packaging to the food that may be deleterious to human health or bring about an unacceptable change in composition, taste, or odor of food.
- In today's global economy, different systems for regulating food packaging in different jurisdictions can impact trade between countries.
- Proper waste management is important to protect human health and the environment and to preserve natural resources. EPA strives to motivate behavioral change in solid waste management through nonregulatory approaches, including pay-as-you-throw and Waste Wise.

REVIEW QUESTIONS

1. What are the types of food packaging?
2. What materials are suitable for food packaging?
3. What are packaging laws and regulations?
4. What are the basic requirements of food packaging?
5. What does the EPA stand for in waste management?

REFERENCES

1. Bässler, H. J., & Lehmann, F. (2013). *Containment Technology: Progress in the Pharmaceutical and Food Processing Industry*. Springer. https://doi.org/10.1007/978-3642392917.
2. Cooksey, K. (2001). Antimicrobial food packaging materials. *Additives for Polymers, 2001*(8), 6–10. https://doi.org/10.1016/s0306-3747(01)80187-1.
3. Ma, L. X., & Wang, K. Q. (2013). Development trend of food packaging materials. *Applied Mechanics and Materials, 477-478*, 1533–1536. https://doi.org/10.4028/www.scientific.net/amm.477-478.1533.
4. Paine, F. (1993). Packaging food with flexible materials. *Nutrition & Food Science, 93*(2), 20–22. https://doi.org/10.1108/eum0000000000984.
5. Porta, R., Sabbah, M., & Di Pierro, P. (2020). Biopolymers as food packaging materials. *International Journal of Molecular Sciences, 21*(14), 4942. https://doi.org/10.3390/ijms21144942.
6. Saha, N. C., Ghosh, A. K., Garg, M., & Dey Sadhu, S. (2022). *Food Packaging*. Springer Nature Singapore. https://doi.org/10.1007/978-981-16-4233-3.
7. Votavová, L., Dobiáš, J., Vold⊠ich, M., & ⊠ížková, H. (2009). Migration of nonylphenols from polymer packaging materials into food simulants. *Czech Journal of Food Sciences, 27*(4), 293–299. https://doi.org/10.17221/152/2008-cjfs.
8. Yap, R. C. C., Amegadze, P. S. K., He, J., & Li, X. (2016). Functions of nano-materials in food packaging. *Journal of Molecular and Engineering Materials, 4*(4), 1640015. https://doi.org/10.1142/s2251237316400153.
9. Zabashta, A., Shalimova, T., & Basov, V. (2020). *Egg Processing Technology*. INFRA-M Academic Publishing LLC. https://doi.org/10.12737/1085371.

Index

Kolibri 360 GmbH
Ferdinand-Jühlke-Straße 7
99095 Erfurt, Deutschland
produktsicherheit@kolibri360.de